# Interstellar and Intergalactic Medium

This concise textbook, the first volume in the Ohio State Astrophysics Series, covers all aspects of the interstellar and intergalactic medium for graduate students and advanced undergraduates. This series aims to impart the essential knowledge on a topic that every astrophysics graduate student should know, without going into encyclopedic depth. This text includes a full discussion of the circumgalactic medium, which bridges the space between the interstellar and intergalactic gas, and the hot intracluster gas that fills clusters of galaxies. Its breadth of coverage is innovative, as most current textbooks treat the interstellar medium in isolation. The authors emphasize an order-of-magnitude understanding of the physical processes that heat and cool the low-density gas in the universe, as well as the processes of ionization, recombination, and molecule formation. Problems at the end of each chapter are supplemented by online projects, data sets, and other resources.

BARBARA RYDEN received her Ph.D. in Astrophysical Sciences from Princeton University. After postdocs at the Harvard-Smithsonian Center for Astrophysics and the Canadian Institute for Theoretical Astrophysics, she joined the astronomy faculty at The Ohio State University, where she is now a full professor. She has more than 25 years of experience in teaching, at levels ranging from introductory undergraduate courses to advanced graduate seminars. She won the Chambliss Astronomical Writing Award for her textbook *Introduction to Cosmology*. She is co-author, with Bradley M. Peterson, of one of the market leading astrophysics texts, *Foundations of Astrophysics*.

RICHARD W. POGGE received his Ph.D. in Astronomy & Astrophysics from the University of California Santa Cruz. Following postdocs at the University of Texas at Austin and The Ohio State University, he joined the astronomy faculty at OSU where he is a full professor and vice-chair for instrumentation. His research includes observational spectrophotometry of the astrophysics of gaseous nebulae from interstellar gas to active galactic nuclei using ground- and space-based observatories from radio to X-ray. He has over 40 years of experience teaching both undergraduate and graduate courses.

# Interstellar and Intergalactic Medium

**Barbara Ryden**

*The Ohio State University*

**Richard W. Pogge**

*The Ohio State University*

 CAMBRIDGE
UNIVERSITY PRESS

# CAMBRIDGE
## UNIVERSITY PRESS

University Printing House, Cambridge CB2 8BS, United Kingdom

One Liberty Plaza, 20th Floor, New York, NY 10006, USA

477 Williamstown Road, Port Melbourne, VIC 3207, Australia

314–321, 3rd Floor, Plot 3, Splendor Forum, Jasola District Centre, New Delhi – 110025, India

79 Anson Road, #06–04/06, Singapore 079906

Cambridge University Press is part of the University of Cambridge.

It furthers the University's mission by disseminating knowledge in the pursuit of education, learning, and research at the highest international levels of excellence.

www.cambridge.org
Information on this title: www.cambridge.org/9781108478977
DOI: 10.1017/9781108781596

First published 2021

Printed in the United Kingdom by TJ Books Limited, Padstow Cornwall

*A catalogue record for this publication is available from the British Library.*

ISBN 978-1-108-47897-7 Hardback
ISBN 978-1-108-74877-3 Paperback

Additional resources for this publication at www.cambridge.org/osas_isigm

*For Jill Knapp and Bruce Draine, as a belated thank you.* BR

*For Donald Edward Osterbrock (1924–2007), who taught me how to be a scientist.* RWP

# Contents

# Preface

The idea of writing this textbook was triggered in the 2012/2013 academic year, when The Ohio State University switched from a quarter-based calendar to a semester-based calendar. In the revised curriculum, first-year graduate students take a five credit-hour course "Observed Properties of Astronomical Systems." This is followed by six courses, each of two or three credit-hours: "Atomic and Radiative Processes in Astrophysics," "Stellar Structure and Evolution," "Dynamics," "Cosmology," "Numerical and Statistical Methods in Astrophysics," and "The Interstellar Medium and the Intergalactic Medium." The philosophy of the OSU graduate program, however, is best encapsulated in the two credit-hour course "Order of Magnitude Astrophysics," which is offered every year to first- and second-year students. In this course, students work together to solve a wide range of astrophysical problems, using basic physical principles to find back-of-envelope solutions.

The Ohio State Astrophysics Series (OSAS), of which this is the first volume, is a projected series of books based on lecture notes for the six core courses and the first-year "Observed Properties" course. These textbooks will not be exhaustive monographs but will instead adopt the back-of-envelope philosophy of the "Order of Magnitude" course to emphasize the most important physical principles in each subfield of astrophysics. The goal is to make our series a point of entry into the deeper and more detailed classic textbooks in our field. Although each volume in OSAS will stand on its own, care will be taken to unify the notation and vocabulary as much as possible across volumes.

*Interstellar and Intergalactic Medium* is based on the semester-long class "The Interstellar Medium and the Intergalactic Medium." The textbook uses the cgs (centimeter, gram, second) system of units, as is common in graduate education in astronomy. It also uses the most common astronomical distance units: the solar radius ($R_\odot$), the astronomical unit (au), and the parsec (pc). In addition, masses are given in units of the solar mass ($M_\odot$) and luminosities in units of the solar luminosity ($L_\odot$). On small scales, when we examine individual photons and other particles, the electron-volt (eV) will be a useful small unit of energy, with $1\,\mathrm{eV} = 1.602 \times 10^{-12}$ erg. Other helpful conversion factors, and the values of physical and astronomical constants, are included in tables at the back of the book.

To keep the length of this book under control, we do not provide a review of spectroscopic notation. (This subject is covered in the course "Atomic and Radiative Processes" at Ohio State University.) If spectroscopic symbols such as $^3\mathrm{P}_0$ and $^1\Sigma_g^+$ are not part of your symbolic vocabulary, our brief guide to spectroscopic notation can be downloaded from the Cambridge University Press website for this book. The website contains other ancillary materials such as additional end-of-chapter problems and links to Jupyter notebooks for recreating and modifying figures in the textbook.

A preliminary version of this text was made as an electronic book during the 2014 Book Launch program of the Office of Distance Education and eLearning (ODEE) at Ohio State University. Ashley Miller, Michael Shiflet, and the entire ODEE team were a perpetual source of assistance. The volume you are now reading is a thorough revision of the 2014 e-book. The revision was greatly aided by the graduate students of Adam Leroy (Ohio State University) and Evan Skillman (University of Minnesota). These students, when invited to critique the e-book, gave a wide array of useful (and mostly tactful) feedback.

Many of the figures and images in this book are derived from works in the published astronomical literature. We are grateful to the authors and journals who promptly granted permission to use their figures. We are especially grateful to those of our colleagues who dug out their original data for us to replot for this volume. Particular thanks are due to Adwin Boogert (University of Hawaii) for Figure 6.7, Bruce Draine (Princeton University) and Aigen Li (University of Missouri) for Figure 6.8, Joseph Weingartner (George Mason University) for Figure 6.10, Sanskriti Das and Smita Mathur (Ohio State University) for Figures 8.2, 8.3, and 10.6, Fabio Gastaldello (INAF-IASF Milan) for Figure 8.6, Bill Keel (University of Alabama) for Figure 9.1, Xiaohui Fan and Feige Wang (University of Arizona) for Figure 9.2, Stephan Frank (Ohio State University) for Figure 9.7, and Davide Martizzi (Copenhagen) for Figures 10.2, 10.3, and 10.4.

# Introduction

I

*The Interstellar Medium is anything not in stars.*

Professor Donald E. Osterbrock (1924–2007)
[from RWP's graduate ISM course notes,
January 13, 1984]

A seemingly disproportionate share of astronomy is devoted to the properties of baryonic matter and its interaction with photons. Baryonic matter, in practice, is matter that is made of protons and neutrons, with the addition of enough (non-baryonic) electrons to preserve charge neutrality on large scales. In the consensus model of cosmology, less than 5% of the density of the universe is provided by baryonic matter; the majority of the universe is made of dark energy (~69% of the total density) and dark matter (~26% of the total). This book rudely snubs the dark side of the universe; dark energy is ignored until we discuss cosmic evolution in Chapter 9, and dark matter is important only because it provides gravitational potential wells for baryonic matter to be trapped in. However, devoting an entire book to the baryonic five percent is easily justified by the rich variety of physical phenomena that arise from the interactions among baryons, electrons, and photons.

To begin, let's make an extremely rough census of the baryonic matter present in the universe today (Figure 1.1). About 7% of the baryonic matter is in the form of **stars** and other compact objects such as stellar remnants, brown dwarfs, and planets. Roughly 2% is in the diffuse gas of the **interstellar medium (ISM)** filling the volume between the stars within a galaxy. About 5% is in the gas of the **circumgalactic medium (CGM)**, bound within the dark matter halo of a galaxy but outside the main distribution of stars. Roughly 4% is in the hot gas of the **intracluster medium (ICM)** of clusters of galaxies, bound to the cluster as a whole but not to any individual galaxy. About 38% of the baryonic matter present today is in the **diffuse intergalactic medium (DIM)**, made of low-density, mostly photoionized gas, at temperatures $T < 10^5$ K. The final ~44% of the baryonic component of the universe is in the **warm-hot intergalactic medium**

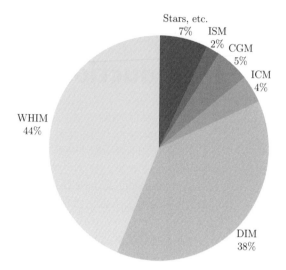

Stars, etc.
7%  ISM
2%  CGM
5%
ICM
4%

WHIM
44%

DIM
38%

**Figure 1.1** Approximate division of the baryonic mass density of the current universe. [Data from Nicastro et al. 2018, Martizzi et al. 2019, and references therein]

**(WHIM)**, made of shock-heated gas at temperatures $10^5$ K $< T < 10^7$ K. (All percentages in this paragraph are approximate, especially for the imperfectly surveyed intergalactic medium.)

This book deals with the diffuse baryonic gas, typically at a density $\rho < 10^{-18}$ g cm$^{-3}$, that hasn't curdled into stars, brown dwarfs, planets, stellar remnants, or other dense massive lumps. However, we also discuss interstellar **dust**, consisting of tiny grains of solid material. "Tiny," in this context, means radii of less than one micron. Dust grains can consist of silicates, ices, carbonaceous materials, graphite, and even diamond. Why does a book dealing mainly with low-density gas condescend to consider solid material? This is partly for historical reasons. In the twentieth century, courses on the interstellar medium dealt with the grab-bag category of "non-stellar stuff"; thus, generations of astronomers were conditioned to lump gas and dust together. However, there are also sound physical reasons for considering interstellar gas in conjunction with interstellar dust. The dust and gas strongly influence each other. Dust reprocesses sunlight, altering the radiation field passing through the gas. Dust is made of elements with a high condensation temperature, so creating dust alters the chemical abundances of the surrounding gas. Dust grains are a leading source of free electrons in the interstellar gas. The surfaces of dust grains are the matrices where gas molecules form. Thus, in the denser, dustier regions of interstellar space, it makes little sense to study interstellar gas in isolation from interstellar dust. (The transparency of intergalactic space places fairly strong constraints on the amount of *intergalactic* dust.)

Interstellar gas, though it contributes only ~2% of the baryons in the universe, is particularly interesting and will be the subject of much of this book. Because it occupies the same region as the stellar portion of a galaxy, it interacts in a variety of ways with stars. Stars are made from interstellar gas, they emit stellar winds into the ISM over the course of their lives, and, when massive stars reach the end of their lifetimes, they inject chemically enriched gas at high speeds into the surrounding interstellar gas. Stars also emit photons; the higher-energy stellar photons are capable of exciting the interstellar gas. Ionized nebulae, such as the Orion Nebula, would be dim at visible wavelengths if they weren't excited by ultraviolet light from stars. As it is, the emission lines of nebulae have strong diagnostic power, enabling us to determine the densities, temperatures, and ionization states of interstellar gas.

The interstellar gas in our galaxy (the Milky Way Galaxy) has five differ-ent phases. At the moment, we simply list and describe them. Later, we will investigate *why* the interstellar gas has five phases, and what physical processes determine their observed properties. Table 1.1 summarizes the properties of the phases of the interstellar medium. The typical temperature $T$ and number density $n_H$ of hydrogen nuclei are given for locations in the midplane of our galaxy's disk, at the Sun's distance from the galactic center, $R = 8.2\,\mathrm{kpc}$. The mass fraction is the contribution from each phase to the total mass surface density of gas, $\Sigma_{gas} \approx 10\,\mathrm{M_\odot\,pc^{-2}}$, at a distance $R = 8.2\,\mathrm{kpc}$ from the galactic center. (For comparison, the mass surface density of stars at $R = 8.2\,\mathrm{kpc}$ is $\Sigma_\star \approx 50\,\mathrm{M_\odot\,pc^{-2}} \sim 5\Sigma_{gas}$.) An approximate scale height $h_z$ is given for each phase in the direction perpendicular to the galaxy's disk; the cooler phases are seen to stay closer to the midplane.

**Molecular clouds** have $H_2$ as their dominant form of hydrogen. The densest portions of molecular clouds, where the number density can be as high as $n \sim 10^6\,\mathrm{cm^{-3}}$, are self-gravitating regions where stars can form. Although molec-ular clouds occupy a minuscule fraction of the ISM's volume, their relatively high density means they contribute about a fifth of the ISM's mass. A useful method

**Table 1.1** Phases of the ISM in the Milky Way Galaxy[a]

| Name | $T$ [K] | $n_H$ [cm$^{-3}$] | Mass fraction | $h_z$ [pc] |
|---|---|---|---|---|
| Molecular Clouds | 15 | >100 | 20% | 75 |
| Cold Neutral Medium | 80 | 40 | 30% | 100 |
| Warm Neutral Medium | 6000 | 0.4 | 35% | 300 |
| Warm Ionized Medium | 8000 | 0.2 | 12% | 900 |
| Hot Ionized Medium | $10^6$ | 0.004 | 3% | 3000 |

[a] *Data from Ferrière 2001 and Tielens 2005*

of observing molecular clouds is by looking for millimeter-wavelength emission from small molecules such as carbon monoxide.

The **cold neutral medium (CNM)** has atomic hydrogen, or H I, as its dominant form of hydrogen. The cold neutral medium is distributed in sheets and filaments occupying only $\sim$1% of the volume of the ISM but containing roughly a third of the ISM's mass. A useful method of observing the CNM is by looking for the UV and visible absorption lines it produces in the spectra of background stars and quasars.

The **warm neutral medium (WNM)** also has H I as its dominant form of hydrogen. In older papers, particularly those published before 1990, the WNM is sometimes referred to as the "warm intercloud medium." A useful method of observing the WNM is by looking for 21 cm radio emission from atomic hydrogen. The neutral phases of our galaxy's ISM, including molecular clouds, the CNM, and the WNM, provide most of the mass of the ISM despite occupying only $\sim$25% of its volume.

The **warm ionized medium (WIM)** has ionized hydrogen, or H II, as its dominant form of hydrogen. The WIM is sometimes referred to as "diffuse ionized gas." The warm ionized medium is primarily photoionized by hot stars. A useful method of observing the WIM is by looking for the Balmer emission lines of hydrogen. Although H II regions (around hot stars) and planetary nebulae (around newly unveiled white dwarfs) have temperatures similar to the warm ionized medium, they have much higher densities.

The **hot ionized medium (HIM)** also has H II as its dominant form of hydrogen. In older papers, the HIM is sometimes referred to as the "coronal gas," referring to the fact that its temperature is comparable to that of the Sun's corona, although its density is much lower. The hot ionized medium is primarily shock-heated by supernovae. A useful method of observing the HIM is by looking for far-ultraviolet absorption lines of ions such as O VI in the spectra of background stars; the hottest portions of the HIM also produce diffuse soft X-ray emission. The ionized phases of our galaxy's ISM, including both the WIM and the HIM, occupy most of the volume of the ISM despite providing only $\sim$15% of its mass. The Sun is located inside a bubble of hot ionized gas called the Local Bubble, roughly 100 parsecs across.

The total mass of the interstellar medium in our galaxy is not perfectly known. In part, this is due to the difficulty in doing a complete census of the different phases; in part, it is due to the necessity of drawing a somewhat arbitrary boundary between the interstellar medium and the circumgalactic medium. Most estimates yield $M_{\rm ism} \approx 7 \times 10^9\,M_\odot$, with about 1% of that mass being in the form of interstellar dust. For comparison, the total mass of stars in our galaxy is $M_\star \approx 6 \times 10^{10}\,M_\odot$, yielding an interstellar-to-stellar mass ratio $M_{\rm ism}/M_\star \approx 0.12$. This ratio varies greatly from one galaxy to another. For instance, the Small Magellanic Cloud, a gas-rich irregular galaxy, has $M_{\rm ism}/M_\star \approx 1.4$. By contrast,

**Table 1.2** Solar abundance of elements[a]

| Element | ppm by number | percentage by mass | atomic number | 1st ionization energy [eV] |
|---|---|---|---|---|
| hydrogen (H) | 910 630 | 71.10% | 1 | 13.60 |
| helium (He) | 88 250 | 27.36% | 2 | 24.59 |
| oxygen (O) | 550 | 0.68% | 8 | 13.62 |
| carbon (C) | 250 | 0.24% | 6 | 11.26 |
| neon (Ne) | 120 | 0.18% | 10 | 21.56 |
| nitrogen (N) | 75 | 0.08% | 7 | 14.53 |
| magnesium (Mg) | 36 | 0.07% | 12 | 7.65 |
| silicon (Si) | 35 | 0.08% | 14 | 8.15 |
| iron (Fe) | 30 | 0.13% | 26 | 7.90 |
| sulfur (S) | 15 | 0.04% | 16 | 10.36 |

[a] *Data from Lodders 2010*

the giant elliptical galaxy M87 has $M_{ism}/M_\star < 0.02$ in its central regions (within $r = 5\,\mathrm{kpc}$ of the galaxy's center).

The chemical composition of the ISM in our galaxy is not homogeneous; in addition, the mean chemical composition changes with time. However, we have a fairly good idea of the composition of one particular patch of the ISM at one particular time. By studying the composition of the Sun's atmosphere, supplemented by information from primitive meteorites, we can deduce the relative abundance of elements in the protosolar nebula from which the Sun formed 4.57 billion years ago. The resulting **solar abundance** for the 10 most abundant elements is shown in Table 1.2, along with their atomic numbers and first ionization energies. Of the atoms from which the Sun formed, about 91.1% were hydrogen and 8.8% were helium. This means that all the heavier elements in the periodic table together contributed only $\sim 0.1\%$ of the atoms. Astronomers frequently refer to elements heavier than helium as "metals," even though they are not all metallic by the standard chemist's definition; we will follow this convention. The **solar metallicity** $Z_\odot$ is the fraction of the Sun's initial mass made of "metals"; the numbers in Table 1.2 yield $Z_\odot = 0.015$. Although different regions in the ISM and IGM have different mixes of elements, we will use $Z_\odot$ as our benchmark metallicity and the relative abundances in Table 1.2 as our benchmark blend of metals.

In a gas of neutral atoms, the total number density of gas particles at solar abundance is

$$n = n_H + n_{He} + n_O + \cdots \approx 1.10 n_H \qquad \text{[atomic]}. \qquad (1.1)$$

However, in the limit that the hot gas in the solar abundance is completely ionized, the total number density is

$$n = 2n_H + 3n_{He} + 9n_O + \cdots \approx 2.30 n_H \qquad \text{[ionized]}, \qquad (1.2)$$

with free electrons providing $n_e \approx 1.20 n_H$ and naked atomic nuclei providing $n_{nuc} \approx 1.10 n_H$. In cold molecular gas, we can make the lowest-order approximation that all atoms other than the noble gases are in diatomic molecules such as $H_2$, OH, CH, CO, and so forth. With this approximation, the total number density is

$$n = \frac{1}{2} n_H + n_{He} + \frac{1}{2} n_O \cdots \approx 0.60 n_H \qquad \text{[molecular].} \qquad (1.3)$$

In approximate calculations, the total number density $n$ and the number density of hydrogen nuclei $n_H$ are sometimes used interchangeably. A more careful translation between $n_H$ and $n$ requires knowing the ionization state of hot gas or the degree of molecular formation in cold gas.

The values of $T$ and $n_H$ in Table 1.1 imply that each phase has a pressure $P \sim 4 \times 10^{-13}$ dyn cm$^{-2}$ $\sim 4 \times 10^{-19}$ atm, to within a factor of three. This is an extremely low pressure compared with the pressure of the air around us. Thus, your intuition about how gases behave, largely derived from your experience with air, might not be applicable to the tenuous gas of the interstellar medium. Even in laboratory settings, it is difficult to attain densities comparable to those of interstellar gas. Extremely high vacuum (XHV), defined as a pressure $P \leq 10^{-9}$ dyn cm$^{-2}$, is challenging to produce in the lab; at room temperature ($T \approx 300$ K), an XHV pressure $P = 10^{-9}$ dyn cm$^{-2}$ corresponds to a number density $n \sim 20\,000$ cm$^{-3}$. In the interstellar medium of the Milky Way, this XHV density is exceeded only in the dense cores of molecular clouds.

Although the pressure of the interstellar gas, $P_{ism} \sim 4 \times 10^{-13}$ dyn cm$^{-2}$, is low compared with a laboratory vacuum, it is higher than the pressure of the intergalactic gas. The intergalactic medium embraces a range of pressures, all of them quite low. For example, the warm-hot intergalactic medium has a typical pressure $P_{whim} \sim 4 \times 10^{-16}$ dyn cm$^{-2}$ at $T = 10^6$ K. The diffuse intergalactic medium has a still lower pressure, $P_{dim} \sim 4 \times 10^{-19}$ dyn cm$^{-2}$ at $T = 7000$ K.

The typical pressure of interstellar gas, $P = nkT \sim 4 \times 10^{-13}$ dyn cm$^{-2}$, can be converted to a thermal energy density, $\varepsilon = (3/2)nkT \sim 0.4$ eV cm$^{-3}$. In the context of interstellar space, is this a small or large energy density? Fortunately, the other energy densities in the local ISM are fairly well known. Table 1.3 gives the approximate values for energy densities in the interstellar medium near the Sun's location. All seven of these energy densities are within an order of magnitude of each other. Even if the rest of this book fades from memory, it will be useful to remember the general rule: "All energy densities in the local ISM are half an electron-volt per cubic centimeter."

We live in a galaxy where the ISM has five different phases, and has many possible energy sources, all of them comparable in density. It looks as though we have an interesting variety of physics in store. But why are there five different phases? Why should the interstellar gas have different phases at all? How did

**Table 1.3** Energy densities in the local ISM[a]

| Type | Energy density $(\mathrm{eV\,cm^{-3}})$ |
|---|---|
| Cosmic microwave background | 0.2606 |
| Thermal energy | 0.4 |
| Turbulent kinetic energy | 0.2 |
| Far-infrared from dust | 0.3 |
| Starlight | 0.6 |
| Magnetic energy | 0.9 |
| Cosmic rays | 1.4 |

[a] *Data from Draine 2011, Table 1.5 and Table 12.1*

astronomers figure out the **multi-phase** nature of the ISM? That's a long enough story to require a separate section.

## 1.1  History of Interstellar Studies

In the sixteenth century, Thomas Digges, an early follower of Copernicus, discarded the concept of a celestial sphere to which the stars are attached. This led to the question of what lies between the stars that are scattered throughout space. In the year 1669, the natural philosopher Robert Boyle recorded an ongoing "Controversie betwixt some of the Modern Atomists and the Cartesians." The atomists believed, in Boyle's words, "that betwixt the Earth and the Stars, and betwixt these themselves, there are vast Tracts of Space that are empty." The Cartesians, by contrast, thought that there could not be a perfect vacuum. Instead, they believed "that the Intervals betwixt the Stars . . . are perfectly fill'd, but by a Matter far subtiler than our Air, which some call Celestial, and others Aether." Unfortunately, the subtle aether proposed by the Cartesians was too subtle to be detected.

The idea of visible interstellar material arose in the eighteenth century, with the study of nebulae. The word **nebula** (Latin for "cloud" or "fog") was applied to any extended luminous object in the night sky. William Herschel, with his large telescopes, was able to resolve some nebulae into stars. For decades, a debate continued over whether all nebulae could be resolved with high enough angular resolution. The situation became clearer in the 1860s, when William Huggins demonstrated that some nebulae have emission line spectra, characteristic of diffuse gas, rather than the absorption line spectra produced by populations of stars.

Figure 1.2 shows Huggins' spectrum of the Cat's Eye Nebula, a relatively bright planetary nebula at a distance $d \sim 1\,\mathrm{kpc}$ from the Sun. Huggins saw three bright lines; the leftmost line (4861 Å) he correctly identified as the Balmer $\beta$ line

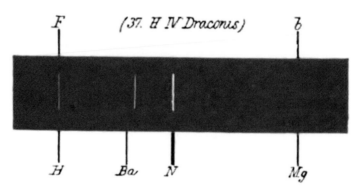

**Figure 1.2** Emission line spectrum of the Cat's Eye Nebula, seen by William Huggins on August 29, 1864. This spectrum covers the blue-green portion of the visible range, from the Fraunhofer F line (H$\beta$, at 4861 Å) to Fraunhofer b (a blend of Mg and Fe lines at 5173 Å). [Huggins 1864]

of hydrogen, the rightmost line (5007 Å) he incorrectly identified as being due to nitrogen,[1] and the central line (4959 Å) he couldn't identify at all. Eventually, higher-resolution spectra revealed that the 5007 Å line wasn't due to nitrogen, and Huggins hypothesized that the 4959 Å and 5007 Å lines were emitted by a previously unknown element. Although Huggins proposed the name "nebulum" for this element, the variant "nebulium," put forward by the astronomy writer Agnes Clerke, proved to be more popular. It wasn't until 1927 that Ira Bowen discovered that the "nebulium" lines were actually forbidden lines from the ion O III.

There's one acknowledged complication in measuring the spectra of nebulae (and other celestial objects) from the Earth's surface. At visible wavelengths, air has a refractive index $n_r \approx 1.00028$ that is slightly greater than one. Thus, wavelengths measured by Huggins (and other earthly observers) are the "air wavelength" $\lambda_{air}$ rather than the vacuum wavelength $\lambda_{vac}$ measured in the absence of a refracting medium. The two wavelengths are related by $\lambda_{air} = \lambda_{vac}/n_r$; at visible wavelengths, this means that $\lambda_{air}$ is shorter than $\lambda_{vac}$ by about one part in 3600. In this book, we adopt the common convention of using $\lambda_{air}$ rather than $\lambda_{vac}$ in the wavelength range from $\lambda_{air} = 3000$ Å to $\lambda_{air} = 1.1$ μm; this corresponds to the "optical window" at which the Earth's atmosphere is nearly transparent.[2]

Although many nebulae proved to consist of diffuse gas, which produces emission line spectra, there were a few surprises in store. For instance, in

---

[1] In his laboratory, Huggins had noted a strong emission line in the spark spectrum of nitrogen, at a wavelength $\lambda \approx 5005$ Å. He initially thought this could be identical with the nebula's 5007 Å emission line.

[2] Because of this convention, when someone says, "The 5007 Å line of O III has wavelength $\lambda = 5008.24$ Å," it doesn't mean they are inept at rounding; it means they are adept at switching between air wavelength and vacuum wavelength.

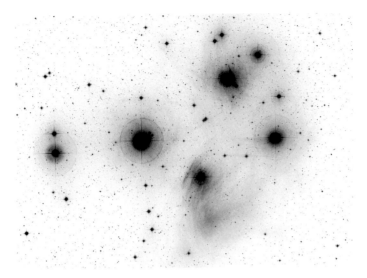

**Figure 1.3** Negative image of the Pleiades star cluster ($d \approx 136\,\text{pc}$) and its associated reflection nebula, seen at visible wavelengths. The image size is $90 \times 65\,\text{arcmin}$, corresponding to $3.6 \times 2.6\,\text{pc}$ at the distance of the Pleiades. [POSS-II]

1912, when Vesto Slipher took the spectrum of the nebulosity surrounding the Pleiades (Figure 1.3), he expected to find an emission line spectrum. Instead, he found a continuous spectrum with absorption lines superposed. Slipher correctly conjectured that he was seeing light from the stars of the Pleiades reflected from "fragmentary and disintegrated matter," or dust. The nebulosity surrounding the Pleiades is thus a **reflection nebula**.

The existence of dust had already been hinted at by the existence of **dark nebulae**, opaque objects such as Barnard 68 (Figure 1.4) that block the light of background stars. Dark nebulae were originally thought to be holes in the distribution of stars ("ein Loch im Himmel," to use William Herschel's words), but they were later recognized as being clouds of obscuring material. Dark nebulae are prominent in the brightest regions of the Milky Way.

At the beginning of the twentieth century, bright nebulae such as the Cat's Eye Nebula were thought of as isolated clouds in otherwise empty, or nearly empty, space. An additional population of interstellar gas, invisible to the eye, was revealed by the work of Johannes Hartmann. In the year 1901, Hartmann began a study of Delta Orionis, a spectroscopic binary star at a distance $d \approx 380\,\text{pc}$ from the Sun. The two stars in the binary had relatively broad absorption lines that showed the expected time-varying Doppler shifts for orbiting stars. However, Hartmann also saw a narrow calcium absorption line that did not shift back and forth in wavelength. As Hartmann wrote in 1904 (the excited italics are in his original paper): *"The calcium line at λ3934 does not share in the period displacements of the lines caused by the orbital motion of the star."*

**Figure 1.4** The dark nebula Barnard 68 ($d \sim 150\,$pc) seen at visible and near-infrared wavelengths. The image size is $4.9 \times 4.9\,$arcmin, corresponding to $\sim 0.21 \times 0.21\,$pc at the distance of Barnard 68. [FORS Team, VLT, ESO]

Hartmann concluded that the calcium absorption line was caused by a cloud of gas somewhere along the line of sight to Delta Orionis. Later, other astronomers found similar "stationary lines" along the line of sight to other bright stars. The lines were all narrow, and had strengths correlated with the distance to the background star: more distant stars showed stronger stationary lines because there is more interstellar gas along the line of sight. As higher-resolution spectrographs were used on the stationary lines, it was revealed that they had complex structures, consisting of many narrower lines with different radial velocities. This led to the realization that the ISM has a complex structure. It consists neither of smooth uniform gas nor of isolated ellipsoidal blobs drifting about in a near-vacuum.

In 1939, the astronomer Bengt Strömgren developed the idea that bright nebulae with strong emission line spectra are regions of photoionized gas, surrounding a hot star or other source of ionizing photons. The idealized **Strömgren sphere** will be discussed in Section 4.2, when we deal with the physics of ionized hydrogen. At least three types of bright ionized nebulae are recognized in the interstellar medium. First, **H II regions** are regions of interstellar gas heated and photoionized by embedded hot stars with effective surface temperature $T_{\text{eff}} > 25\,000\,$K.[3] The Orion Nebula (Figure 1.5) is an example of a nearby H II region.

---

[3] In the usual OBAFGKM classification scheme for stars, running from hot O stars to cool M stars, stars with $T_{\text{eff}} > 25\,000\,$K are of spectral type O or B.

**Figure 1.5** The Orion Nebula ($d \sim 410\,\text{pc}$) seen at visible and near-infrared wavelengths. The image size is $30 \times 30\,\text{arcmin}$, corresponding to $\sim 3.6 \times 3.6\,\text{pc}$ at the distance of the Orion Nebula. [NASA, ESA, M. Robberto & the HST Orion Treasury Project Team]

Second, **planetary nebulae** are regions of ejected stellar gas heated and photoionized by a hot remnant stellar core which is becoming a white dwarf.[4] The Ring Nebula (Figure 1.6) is among the most famous planetary nebulae; the Cat's Eye Nebula, whose spectrum was shown in Figure 1.2, is another example of a planetary nebula. The Ring Nebula has vivid colors due to line emission. The center of the nebula is blue, from the He II line at $\lambda = 4686\,\text{Å}$ (this is the Paschen $\alpha$ line of singly ionized helium). Moving away from the center, the color becomes blue-green, from the O III "nebulium" lines at $\lambda = 4959\,\text{Å}$ and $5007\,\text{Å}$. Still further out, the bright outer ring of the Ring Nebula is red, from the H$\alpha$ line at $\lambda = 6563\,\text{Å}$ and the N II lines that flank H$\alpha$ in wavelength, at $\lambda = 6548\,\text{Å}$ and $6583\,\text{Å}$.

Third, **supernova remnants** are regions of gas heated and ionized by the blastwave from a supernova explosion. The Crab Nebula, shown in Figure 1.7, is an example of a young ($t \sim 1000\,\text{yr}$) supernova remnant containing a central pulsar. Young, pulsar-containing supernova remnants like the Crab are sometimes called "plerions," after the Greek *pleres*, meaning "full." The reason is that, unlike older supernova remnants such as the Cygnus Loop (Figure 1.8), they do not resemble an empty loop or shell but are filled in with luminous gas. The Cygnus

---

[4] The name "planetary nebula" comes from William Herschel, who thought that the planetary nebulae he examined through his telescope looked somewhat like the disk of a planet.

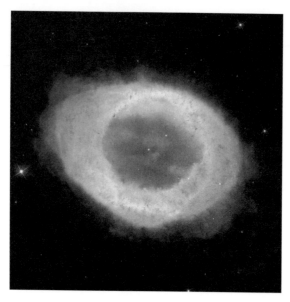

**Figure 1.6** The Ring Nebula ($d \sim 790$ pc) seen at visible and near-infrared wavelengths. The image size is $2.1 \times 2.1$ arcmin, corresponding to $\sim 0.48 \times 0.48$ pc at the distance of the Ring Nebula. [NASA, ESA, & the Hubble Heritage (STScI/AURA) – ESA/Hubble Collaboration]

**Figure 1.7** The Crab Nebula ($d \sim 2$ kpc) seen at visible wavelengths. The image size is $6 \times 6$ arcmin, corresponding to $\sim 3.5 \times 3.5$ pc at the distance of the Crab Nebula. This image shows the supernova remnant as it was $t = 946$ yr after the initial explosion. [NASA, ESA, & the Hubble Heritage Team (STScI/AURA)]

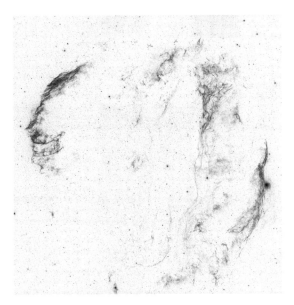

**Figure 1.8** Negative image of the Cygnus Loop ($d \approx 735\,\text{pc}$) seen at visible wave-lengths. The image size is $180 \times 180\,\text{arcmin}$, corresponding to $\sim 38 \times 38\,\text{pc}$ at the distance of the Cygnus Loop. This image shows the supernova remnant as it was $t \sim 10^4\,\text{yr}$ after the initial explosion. [T. A. Rector (University of Alaska, Anchorage) & WIYN/NOAO/AURA/NSF]

Loop is a middle-aged supernova remnant ($t \sim 10^4\,\text{yr}$); in this older remnant, most of the gas in the remnant has been moved outward by the blastwave, leaving the central parts of the supernova remnant nearly empty.

Ionized nebulae such as the Orion, Ring, and Crab Nebulae are the portions of the ISM that you are most likely to see in a coffee-table book of pretty astronomical pictures; however, they contain a minority of the ionized hydrogen in our galaxy. By mass, about 90% of the ionized hydrogen in our galaxy lies in the relatively low-density warm ionized medium (WIM). An H$\alpha$ map of the entire sky (Figure 1.9) shows that Balmer line emission from recombining hydrogen, the primary means of detecting the WIM, fills much of the sky. The Cygnus Loop ($3°$ across) appears as a tiny circle superimposed on the broader emission from the WIM; the Orion Nebula ($0.5°$ across) is seen to be just a small portion of a larger "Orion Complex," sprawling $\sim 15°$ across the constellation Orion.[5] Emission from recombining hydrogen is also seen from the Large Magellanic Cloud (LMC) and the Small Magellanic Cloud (SMC), satellite galaxies to our galaxy at a distance $d \sim 50\,\text{kpc}$.

---

[5] The Orion Complex is also known as the Orion Molecular Cloud Complex or the Orion Star Formation Complex. Its H$\alpha$ emission, seen in Figure 1.9, indicates where hot newly formed stars are exciting the surrounding gas.

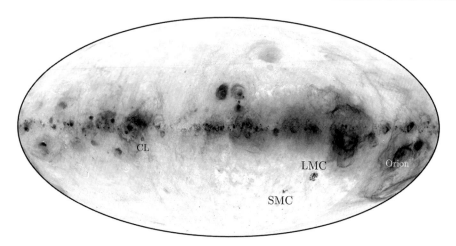

**Figure 1.9** All-sky map of H $\alpha$ ($\lambda = 6563\,\text{Å}$) emission. The photon flux is plotted on a logarithmic scale from $I = 0.5$ rayleighs (white) to $I = 200$ rayleighs (black). LMC = Large Magellanic Cloud, SMC = Small Magellanic Cloud, CL = Cygnus Loop. [Finkbeiner 2003]

## 1.2 Approaching Equilibrium

The gas of the ISM and IGM consists of individual atoms, molecules, ions, and electrons. These individual particles interact with each other. At low speeds, neutral atoms interact via elastic collisions. (At higher speeds, the situation is complicated by the fact that atoms can undergo collisional excitation and ionization.) A small atom can be approximated as a "billiard ball" with a radius $r \sim 3a_0$, where $a_0 = 5.292 \times 10^{-9}$ cm is the Bohr radius. Thus, two identical atoms will collide when their nuclei are separated by a distance $d \leq 2r$, leading to a cross section for interactions $\sigma \sim \pi(2r)^2 \sim 100a_0^2 \sim 3 \times 10^{-15}$ cm$^2$. In air at sea level, where the number density of molecules is $n \sim 2.5 \times 10^{19}$ cm$^{-3}$, the mean free path of a molecule between collisions is $\lambda_{\mathrm{mfp}} \sim 1/(n\sigma) \sim 10^{-5}$ cm $\sim 0.1\,\mu$m. In the cold neutral medium of the ISM, where the number density of atoms is $n \sim 40$ cm$^{-3}$, the mean free path is $\lambda_{\mathrm{mfp}} \sim 10^{13}$ cm $\sim 1$ au. When we are dealing with a volume of gas whose diameter is larger than the mean free path $\lambda_{\mathrm{mfp}}$, we characterize that volume by its bulk properties, such as mass density ($\rho$), pressure ($P$), number density of gas particles ($n$), and temperature ($T$).

To understand what we mean by "temperature" when discussing the ISM and IGM, we should take one step back and look at the concept of **equilibrium**. In general, the word "equilibrium" means a state of balance. We'll be talking a lot about equilibrium systems in this book, so it's important to note that the ISM and IGM are not in perfect equilibrium. (If they were, they'd be boring.) There are times, however, when the assumption of some type of equilibrium is a useful approximation. Let's start with kinetic equilibrium, then go on to excitation equilibrium, ionization equilibrium, and pressure equilibrium.

A gas of massive, non-relativistic particles is in **kinetic equilibrium** when the individual particles have a Maxwellian distribution of velocities:

$$f(\vec{v})d^3v = \left(\frac{m}{2\pi kT}\right)^{3/2} \exp\left(-\frac{mv^2}{2kT}\right) d^3v, \qquad (1.4)$$

where $m$ is the mass per gas particle and $T$ is a parameter known as the **kinetic temperature**. The mean kinetic energy per gas particle will be, integrating over Equation 1.4,

$$\langle E \rangle = \frac{1}{2}m\langle v^2 \rangle = \frac{3}{2}kT = 1.293\,\text{eV}\left(\frac{T}{10^4\,\text{K}}\right), \qquad (1.5)$$

regardless of the particle mass $m$. Because of the linear relation between kinetic temperature $T$ and mean particle kinetic energy $\langle E \rangle$, for a gas where $\langle E \rangle$ is known, we can assign a corresponding temperature,

$$T = \frac{2\langle E \rangle}{3k} = 7740\,\text{K}\left(\frac{\langle E \rangle}{1\,\text{eV}}\right), \qquad (1.6)$$

without having to verify that the gas is *exactly* in kinetic equilibrium.

The mean particle energy in the ISM ranges over five orders of magnitude, from $\sim 0.001$ eV in the coldest regions of molecular clouds to $\sim 100$ eV in the hot ionized medium. The root mean square speed of each particle is

$$v_{\text{rms}} = \left(\frac{2\langle E \rangle}{m}\right)^{1/2} = 13.8\,\text{km s}^{-1} \left(\frac{\langle E \rangle}{1\,\text{eV}}\right)^{1/2} \left(\frac{m}{m_p}\right)^{-1/2} \qquad (1.7)$$

$$= \left(\frac{3kT}{m}\right)^{1/2} = 15.7\,\text{km s}^{-1}\left(\frac{T}{10^4\,\text{K}}\right)^{1/2}\left(\frac{m}{m_p}\right)^{-1/2}, \qquad (1.8)$$

where $m_p$ is the mass of a proton. At a given kinetic temperature $T$, $^1$H atoms will travel twice as fast, on average, as $^4$He atoms, and four times as fast as $^{16}$O atoms.

A given volume will contain different types of particle, not all of which necessarily have the same kinetic temperature. In the air around you, the nitrogen molecules and oxygen molecules are in kinetic equilibrium with each other at a temperature $T \sim 300$ K and mean particle energy $\langle E \rangle \sim 0.04$ eV. This is much smaller than the energy $E = h\nu \sim 2$ eV of the visible photons traversing the air and much larger than the energy $E \sim 0.001$ eV of the cosmic neutrinos traversing the universe.

Particles reach kinetic equilibrium (or become "thermalized," to use an alternative term) by interacting with each other. Consider a gas made of identical particles with mean kinetic energy $\langle E \rangle$ but with a distribution of particle velocities that is initially non-Maxwellian. Given a mean free path $\lambda_{\text{mfp}}$ between collisions, and a typical velocity $v_{rms}$, the time required to come to kinetic equilibrium is known as the collisional time,

$$t_{coll} \sim \frac{\lambda_{mfp}}{v_{rms}} \sim \frac{1}{n\sigma} \left( \frac{m}{2\langle E \rangle} \right)^{1/2}. \tag{1.9}$$

For example, consider a gas of pure atomic hydrogen. A hydrogen atom travels through space; the typical time that elapses before it collides with another hydrogen atom is

$$t_{coll}(HH) \sim \frac{1}{n_H \sigma_{HH}} \left( \frac{m_H}{2\langle E \rangle} \right)^{1/2} \propto n_H^{-1} \langle E \rangle^{-1/2}. \tag{1.10}$$

Assuming a cross section $\sigma_{HH} \sim 3 \times 10^{-15} \, cm^2$, this becomes

$$t_{coll}(HH) \sim 2 \times 10^8 \, s \left( \frac{n_H}{1 \, cm^{-3}} \right)^{-1} \left( \frac{\langle E \rangle}{1 \, eV} \right)^{-1/2}. \tag{1.11}$$

For a dense planetary atmosphere, this time will be less than a nanosecond. However, in the central cores of molecular clouds, it will be more than an hour; in the warm neutral medium, it will be as long as half a century.

Reaching kinetic equilibrium in the ISM isn't all about banging atoms together. Even those portions of the ISM labeled as "molecular" or "neutral" have a significant number density of free electrons, produced by photoemission from dust grains. The number density of free electrons varies throughout the Milky Way Galaxy. Near the galactic center, it is as high as $n_e \sim 6 \, cm^{-3}$, and within the Local Bubble, it's only $n_e \sim 0.004 \, cm^{-3}$. However, in the midplane of our galaxy, at the Sun's distance from the galactic center, it isn't crazy to take an average number density $n_e \sim 0.03 \, cm^{-3}$ for the free electrons. Consider a neutral hydrogen atom traveling through interstellar space. The typical time elapsed before it collides with a free electron will be

$$t_{coll}(eH) \sim \frac{1}{n_e \sigma_{eH}} \left( \frac{m_e}{2\langle E \rangle} \right)^{1/2} \propto n_e^{-1} \langle E \rangle^{-1/2}. \tag{1.12}$$

Since the mass of a hydrogen atom is greater than that of an electron by a factor $\sim 1837$, the typical electron speed will be greater than that of a hydrogen atom by a factor $\sim (1837)^{1/2} \approx 42.9$; we have thus made the approximation that the atom is standing still while an electron slams into it. If we take the cross section for the collision to be $\sigma_{eH} \sim 10^{-15} \, cm^2$, then

$$t_{coll}(eH) \sim 6 \times 10^8 \, s \left( \frac{n_e}{0.03 \, cm^{-3}} \right)^{-1} \left( \frac{\langle E \rangle}{1 \, eV} \right)^{-1/2}. \tag{1.13}$$

Comparing Equations 1.11 and 1.13 we expect, given the typical free electron number density in the ISM, that free electrons will do the thermalization everywhere but in the highest density ($n_H > 0.3 \, cm^{-3}$) portions of the ISM.

The presence of free electrons brings about the possibility of electron–electron collisions. Two electrons *collide* when the electrostatic repulsion between them deflects their paths through an angle of $\sim 90°$ or more. Using an energy argument, we can say that a large deflection requires that the electrostatic potential energy at closest approach is comparable to the initial kinetic energy $E$ of an electron; that is,

$$\frac{e^2}{r_e} \approx E, \qquad (1.14)$$

where $r_e$ is the separation between the electrons at their closest approach. Thus, the effective cross section for electron–electron collisions is

$$\sigma_{ee} \approx \pi r_e^2 \sim \pi \frac{e^4}{\langle E \rangle^2} \sim 7 \times 10^{-14} \, \mathrm{cm}^2 \left( \frac{\langle E \rangle}{1 \, \mathrm{eV}} \right)^{-2}. \qquad (1.15)$$

The typical time that a free electron spends between collisions with another free electron is

$$t_{\mathrm{coll}}(ee) \sim \frac{1}{n_e \sigma_{ee}} \left( \frac{m_e}{2 \langle E \rangle} \right)^{1/2} \sim \frac{\langle E \rangle^2}{\pi n_e e^4} \left( \frac{m_e}{2 \langle E \rangle} \right)^{1/2} \propto n_e^{-1} \langle E \rangle^{3/2}. \qquad (1.16)$$

Numerically, this comes to

$$t_{\mathrm{coll}}(ee) \sim 3 \times 10^5 \, \mathrm{s} \left( \frac{n_e}{0.03 \, \mathrm{cm}^{-3}} \right)^{-1} \left( \frac{\langle E \rangle}{1 \, \mathrm{eV}} \right)^{3/2}. \qquad (1.17)$$

Thus, electrons will thermalize each other more rapidly than they thermalize the neutral atoms unless the mean electron energy is quite high (greater than $\langle E \rangle \sim 50 \, \mathrm{eV}$, found only in the hot ionized medium). The collisional time scales computed in Equations 1.11, 1.13, and 1.17 are sufficiently short that we can regard the ISM as being in kinetic equilibrium under nearly all circumstances. Thus, we can use the kinetic temperature $T$ interchangeably with the mean kinetic energy $\langle E \rangle$ per gas particle.

Although kinetic equilibrium is a safe assumption, there are other equilibrium states that we still need to consider. Imagine a system that has two possible energy states: an upper state with energy $E_u$ and a lower state with energy $E_\ell < E_u$. These states could be, for example, two electronic energy levels of an atom, two rotational states of a molecule, or two spin orientation states of a hydrogen atom (in which the electron and proton spins are either parallel or antiparallel). For a large population of such systems, we say the population is in **excitation equilibrium** if the numbers of systems in the upper and lower states follows a Boltzmann distribution:

$$\frac{n_u}{n_\ell} = \frac{g_u}{g_\ell} \exp \left( -\frac{E_{u\ell}}{kT} \right), \qquad (1.18)$$

where $g_u$ and $g_\ell$ are the statistical weights of the upper and lower level, $E_{u\ell} \equiv E_u - E_\ell > 0$, and $T$ is the kinetic temperature. In the limit $kT \gg E_{u\ell}$, the two levels will be populated according to their statistical weights; in the limit $kT \ll E_{u\ell}$, the upper level will be nearly empty.

Although the presence of the kinetic temperature $T$ in Equation 1.18 assumes the existence of *kinetic* equilibrium, a population of atoms and molecules in kinetic equilibrium is not necessarily in excitation equilibrium. For example, an astrophysical **maser** is an example of a population of molecules that is over-excited; it has more molecules in the upper energy state than could be attained in excitation equilibrium.

For any population of two-level systems, we can measure the ratio $n_u/n_\ell$ and define an **excitation temperature** $T_{\text{exc}}$ such that

$$\frac{n_u}{n_\ell} = \frac{g_u}{g_\ell} \exp\left(-\frac{E_{u\ell}}{kT_{\text{exc}}}\right), \tag{1.19}$$

meaning

$$kT_{\text{exc}} \equiv \frac{E_{u\ell}}{\ln[(g_u/g_\ell)(n_\ell/n_u)]}. \tag{1.20}$$

In general, $T_{\text{exc}} \neq T$. In fact, for a maser with inverted energy level populations ($n_u/n_\ell > g_u/g_\ell$), the excitation temperature will be negative. A statement like "the excitation temperature of this system is $-100\,\text{K}$" is not as nonsensical as it sounds. An excitation temperature, unlike the kinetic temperature, is not a measure of mean particle energy; instead, it is simply a convenient way to parameterize the excitation state of a population of two-level systems.

In addition to excitation equilibrium, we can also talk about the ionization equilibrium at some location in the ISM. The first ionization energy $I$ of most elements is $\sim 10\,\text{eV}$. (See Table 1.2 for the ionization energies of the most abundant elements.) For hydrogen, the ionization energy is $I_H = 13.60\,\text{eV}$. The most abundant element with an ionization energy lower than hydrogen is carbon, with $I_C = 11.26\,\text{eV}$; the most abundant element with an ionization energy lower than carbon is magnesium, with $I_{\text{Mg}} = 7.65\,\text{eV}$. Given this energy scale, the only regions where we expect the highly effective *collisional* ionization of neutral atoms are where $T > 10^5\,\text{K}$ and thus $\langle E \rangle > 10\,\text{eV}$. In the cooler regions of the ISM, most ionization is produced by ultraviolet photons from hot stars. If an atom of element X undergoes **photoionization**, the process can be written as

$$X^i + \gamma \rightarrow X^{i+1} + e^-, \tag{1.21}$$

where $\gamma$ represents a photon with energy greater than or equal to the ionization energy of element X from its $i$th ionization state to its $(i+1)$th ionization state. The opposite process to photoionization is **radiative recombination**:

$$X^{i+1} + e^- \rightarrow X^i + \gamma. \tag{1.22}$$

For element X to be in **ionization equilibrium**, we require a balance between photoionization and radiative recombination:

$$n(X^i)n_\gamma \sigma_{\text{pho}} c = n(X^{i+1})n_e \sigma_{\text{rr}} v, \tag{1.23}$$

where $n_\gamma$ is the number density of photoionizing photons, $\sigma_{\text{pho}}$ is the cross section for photoionization, $\sigma_{\text{rr}}$ is the cross section for radiative recombination, and $v$ is the speed of a free electron relative to the $X^{i+1}$ ion. In general $\sigma_{\text{pho}}$ is dependent on the photon energy and $\sigma_{\text{rr}}$ is dependent on the electron speed $v$. After doing appropriate averaging, we can write Equation 1.23 as

$$\frac{n(X^{i+1})n_e}{n(X^i)} = \frac{n_\gamma \langle \sigma_{\text{pho}} \rangle c}{\langle \sigma_{\text{rr}} v \rangle}. \tag{1.24}$$

The averaging of $\sigma_{\text{pho}}$ is done over the photon energy spectrum; the averaging of $\sigma_{\text{rr}} v$ is done over the Maxwellian distribution of velocities at temperature $T$. (Notice that we're assuming kinetic equilibrium, as usual.) In terms of the kinetic energy per particle $E = mv^2/2$, the Maxwellian distribution of Equation 1.4 can be written as

$$f(E)dE = \frac{2}{\sqrt{\pi}} \left( \frac{E}{kT} \right)^{1/2} \exp\left( -\frac{E}{kT} \right) \frac{dE}{kT}. \tag{1.25}$$

Thus, averaged over a Maxwellian distribution of free electron speeds,

$$\langle \sigma_{\text{rr}} v \rangle = \left( \frac{2}{m_e} \right)^{1/2} \langle \sigma_{\text{rr}} E^{1/2} \rangle$$

$$= \left( \frac{8kT}{\pi m_e} \right)^{1/2} \int_0^\infty \sigma_{\text{rr}} \frac{E}{kT} \exp\left( -\frac{E}{kT} \right) \frac{dE}{kT}. \tag{1.26}$$

For hydrogen, a fairly good approximation to the radiative recombination cross section[6] is $\sigma_{\text{rr}} \propto E^{-1}$. This approximation yields

$$\langle \sigma_{\text{rr}} v \rangle \approx \left( \frac{8}{\pi m_e kT} \right)^{1/2} \sigma_{\text{rr},0} I_H \propto T^{-1/2}, \tag{1.27}$$

where $\sigma_{\text{rr},0}$ is the cross section at an energy $I_H = 13.6\,\text{eV}$. To simplify the notation, $\langle \sigma_{\text{rr}} v \rangle$ is usually referred to as the *radiative recombination rate* and given the symbol $\alpha(T)$.

One more type of equilibrium that is of interest in the ISM is **pressure equilibrium**. When discussing Table 1.1, we pointed out that all five phases of the ISM have a pressure $P \sim 4 \times 10^{-19}$ atm, equivalent to a thermal energy density $\varepsilon \sim 0.4\,\text{eV}\,\text{cm}^{-3}$. Thus, it is tempting to assume that when two phases of the ISM (call them phase 1 and phase 2) are contiguous, they are in pressure equilibrium, with

$$n_1 kT_1 = n_2 kT_2 = 4 \times 10^{-19} \text{ atm.} \tag{1.28}$$

In fact, earlier views of the ISM, which pictured denser, cooler "clouds" in a more tenuous, warmer "intercloud medium," did assume pressure equilibrium. However, current studies of the ISM have had to reject this temptingly simple

---

[6] We'll look at more accurate fits to the recombination cross section in Chapter 4.

picture. The ubiquity of free electrons means that the ISM is coupled to the interstellar magnetic field, and thus the magnetic pressure has to be taken into account. The turbulent energy density is not negligibly small compared with the thermal energy density, so it also has to be taken into account. Explosions are going off in the ISM, increasing the temperature $T$. Hot young stars are pouring ionizing radiation into the ISM, splitting up atoms and increasing $n$. Although the ISM has tendencies toward pressure equilibrium, something is always happening to throw things out of equilibrium again.

## 1.3  Heating and Cooling in the ISM

The number density of molecules in your body is $n \sim 3 \times 10^{22}\, \mathrm{cm}^{-3}$ (mostly $H_2O$ molecules) and your internal temperature is $T \approx 310\,\mathrm{K}$. Your temperature is the result of a balance between heating and cooling mechanisms. If your temperature drops, your body can increase the heating rate (by shivering, for instance) or decrease the cooling rate (by trying to fluff out your sadly inadequate fur). If your temperature rises, your body can increase the cooling rate (by sweating, and thus increasing evaporative cooling) or decrease the heating rate (by stopping unnecessary activity). Similarly, but at much lower number densities, the temperature of the ISM also results from a balance between heating and cooling. Each phase of the ISM has a temperature where the balance is stable (or where instabilities grow extremely slowly). In this section, we give a brief outline of the different heating and cooling mechanisms available to interstellar gas. A more detailed discussion of heating and cooling processes is postponed to later sections, where we talk about each of the interstellar phases in turn.

We will attempt, in this book, to use consistent nomenclature and notation when discussing heating and cooling. The heating **gain** $G$ and cooling **loss** $L$, in $\mathrm{erg\,s^{-1}}$, are the rates at which a single gas particle gains or loses energy, on average. The **volumetric heating rate** $g$ and **volumetric cooling rate** $\ell$, in $\mathrm{erg\,cm^{-3}\,s^{-1}}$, are respectively the rate per unit volume at which the gas gains or loses energy. In addition, the **cooling function** $\Lambda$, in $\mathrm{erg\,cm^3\,s^{-1}}$, is a function that is useful when the cooling is achieved by two-body interactions. In that case, it is useful to define $\Lambda$ such that $\ell = nL = n^2\Lambda$, where $n$ is the total number density of gas particles. (Even if only one type of gas particle is losing energy, the relatively short thermalization times in the ISM mean that the loss is shared among all the gas particles.)

First, let's heat things up. What physical processes can increase the mean kinetic energy $\langle E \rangle$ of the particles in a region of interstellar space? In practice, the major heating mechanisms of the ISM involve sending free electrons speeding through space. Through collisions, the electrons share their kinetic energy with other particles and, through further collisions, the distribution of velocities approaches a Maxwellian distribution (Equation 1.4). The fast electrons may have been ejected from atoms by cosmic rays, from dust grains by photons or from atoms by photons or they may have been accelerated by shocks. Let's look at the variety of means for shooting off fast electrons into space.

**Electrons ejected from atoms by cosmic rays**   Cosmic rays are energetic charged particles that originate in supernova explosions and other high-energy events. About 90% of cosmic ray particles are protons, $\sim$8% are helium nuclei, $\sim$1% are heavier nuclei, and $\sim$1% are electrons. At particle energies $E > 3\,\mathrm{GeV}$, the flux of cosmic rays falls steeply with energy, roughly as $\Phi(E)dE \propto E^{-2.7}dE$. Cosmic rays with $E < 3\,\mathrm{GeV}$ are largely excluded from the Earth's location by the Sun's magnetized solar wind. However, the *Voyager 1* spacecraft, when it ventured beyond the heliopause into the local ISM, found that the flux of cosmic ray protons has a broad maximum in energy centered on $E \sim 25\,\mathrm{MeV}$. In the energy range $3\,\mathrm{MeV} < E < 200\,\mathrm{MeV}$, the mean flux of cosmic ray protons is

$$\Phi(E)dE \approx 2 \times 10^{-3}\,\mathrm{cm^{-2}\,s^{-1}\,sr^{-1}\,MeV^{-1}}dE. \tag{1.29}$$

This flux implies that the lower-energy, non-relativistic cosmic ray protons have a number density $n \sim 5 \times 10^{-10}\,\mathrm{cm^{-3}}$ and an energy density

$$\varepsilon(E < 200\,\mathrm{MeV}) \approx 0.03\,\mathrm{eV\,cm^{-3}}. \tag{1.30}$$

Although these lower-energy cosmic rays contribute only a few percent of the total energy density of cosmic rays ($\varepsilon_{\mathrm{cr}} \sim 1\,\mathrm{eV\,cm^{-3}}$), they are the most important for ionizing hydrogen and heating the ISM. This is so largely because of the steep drop in cosmic ray flux as $E$ increases, but partly because a cosmic ray proton's cross section for ionizing hydrogen decreases with $E$. At $E = 3\,\mathrm{MeV}$, the ionizing cross section is $\sigma \sim 10^{-17}\,\mathrm{cm^2}$; at $E = 3\,\mathrm{GeV}$, the cross section has dropped to $\sim 10^{-19}\,\mathrm{cm^2}$. When a cosmic ray proton ionizes a hydrogen atom,

$$p_{\mathrm{cr}} + \mathrm{H} \rightarrow p_{\mathrm{cr}} + p + e^{-}, \tag{1.31}$$

the ejected electron carries away a mean energy $\sim$35 eV, largely independently of the (much greater) energy of the ionizing cosmic ray proton. On average, about 10 eV of the free electron's kinetic energy goes to heat the atoms and ions in its vicinity. This leads to a cosmic ray heating gain

$$G_{\mathrm{cr}} \sim 1 \times 10^{-27}\,\mathrm{erg\,s^{-1}} \left( \frac{\zeta_{\mathrm{cr}}}{10^{-16}\,\mathrm{s^{-1}}} \right). \tag{1.32}$$

In Equation 1.32, the term $\zeta_{\mathrm{cr}}$ is the primary cosmic ray ionization rate, or the average rate at which a hydrogen atom in the ISM is ionized by cosmic rays. As the normalization in Equation 1.32 implies, in the Sun's neighborhood this is estimated to be $\zeta_{\mathrm{cr}} \sim 10^{-16}\,\mathrm{s^{-1}} \sim 3\,\mathrm{Gyr^{-1}}$. Heating by cosmic rays is the dominant heating mechanism in molecular clouds, where the dust opacity prevents high-energy photons from entering.

**Electrons ejected from dust grains by photons**   Sufficiently energetic photons can knock electrons free from small dust grains. This, as we mentioned earlier, is the main source of free electrons in the cooler regions of the ISM. When the electrons are ejected they carry kinetic energy, which can be effective at heating the surrounding gas. The *work function* for a small dust grain, which is analogous to the ionization energy of an atom, is $w \sim 5\,\text{eV}$. Thus, UV photons with $h\nu > 5\,\text{eV}$ can kick out electrons from dust grains by the photoelectric effect. In the solar neighborhood, photons in this energy range are emitted by fairly hot stars and have an energy density $\varepsilon \approx 0.045\,\text{eV}\,\text{cm}^{-3}$, representing $\sim 8\%$ of the local energy density of starlight,[7] $\varepsilon_\star \approx 0.6\,\text{eV}\,\text{cm}^{-3}$. The heating gain by photoelectrons ejected from dust grains is

$$G_{\text{pe}} \sim 2 \times 10^{-26}\,\text{erg}\,\text{s}^{-1}, \tag{1.33}$$

given the known properties of dust grains and of the ultraviolet interstellar radiation field in our galaxy. Heating by photoelectrons ejected from dust grains is the dominant heating mechanism in the cold neutral medium and the warm neutral medium of the ISM. The warm ionized medium is partially heated by photoelectric heating, but there are also contributions from photoionization heating (discussed below) and from dissipative turbulence.

**Electrons ejected from atoms by photons**   Photons in the energy range $11.26\,\text{eV} < h\nu < 13.60\,\text{eV}$ are likely to end up ionizing a carbon atom. (Photons with $h\nu < 11.26\,\text{eV}$ are incapable of ionizing carbon; photons with $h\nu > 13.60\,\text{eV}$ are likely to ionize the far more abundant hydrogen atoms.) In the solar neighborhood, photons in this energy range are emitted by hot stars and have an energy density $\varepsilon \approx 0.0060\,\text{eV}\,\text{cm}^{-3}$, representing $\sim 1\%$ of the local energy density of starlight. Neutral carbon atoms exposed to the local interstellar radiation field have a photoionization rate $\zeta_{\text{pho,C}} \sim 3 \times 10^{-10}\,\text{s}^{-1} \sim 1\,\text{century}^{-1}$. When a carbon atom is photoionized, a free electron is emitted:

$$C^0 + h\nu \rightarrow C^+ + e^-. \tag{1.34}$$

If the ionizing photon has an energy in the range $11.26\,\text{eV} < h\nu < 13.60\,\text{eV}$, the newly liberated electron has a kinetic energy $E = h\nu - 11.26\,\text{eV}$ that lies in the range $0\,\text{eV} < E < 2.34\,\text{eV}$. If we take $E \sim 1\,\text{eV}$ as the average energy carried away by a free electron, this leads to the following heating gain from the photoionization of carbon:

$$G_C \sim 5 \times 10^{-29}\,\text{erg}\,\text{s}^{-1} \left( \frac{f_{\text{n,C}}}{10^{-3}} \right) \left( \frac{\zeta_{\text{pho,C}}}{3 \times 10^{-10}\,\text{s}^{-1}} \right). \tag{1.35}$$

Here, $f_{\text{n,C}}$ is the neutral fraction of carbon, that is, the fraction of carbon that takes the form of neutral atoms rather than ionized C II. As the normalization in

---

[7] The shape and amplitude of the spectrum of interstellar ultraviolet light are somewhat uncertain; we are using the values of Mathis et al. 1983.

Equation 1.35 implies, the neutral fraction of carbon is expected to be small in the interstellar medium.[8] Thus, the heating rate by carbon photoionization (Equation 1.35) can't usually compete with the heating rate by electrons ejected from dust grains (Equation 1.33). The regions where photoionization becomes an important heating source are the (largely dust-free) H II regions within our galaxy and the (largely dust-free) diffuse intergalactic medium.

**Shock heating**   Shocks are propagating disturbances similar to sound waves but characterized by an abrupt, nearly discontinuous change in the temperature, pressure, and density on opposite sides of the shock. In the ISM, shocks can be created by the explosion of a supernova or by the collision between molecular clouds. On larger scales, shocks can be created by the collision of two clusters of galaxies. Shocks are effective at converting the kinetic energy of orderly bulk flow into the thermal energy associated with random particle motion. In the wake of a supernova shock wave, the temperature can rise to more than $10^7$ K. Shock heating is the dominant heating mechanism in the hot ionized medium of the ISM and in the warm-hot intergalactic medium.

Now that we've learned how to heat things up, let's start cooling things down. Decreasing the average kinetic energy of particles in the ISM is usually done by radiative cooling. That is, excess energy is loaded up into photons that then travel far away from the spot where they were created. When you see light from some portion of the ISM, that light has traveled a large distance from the spot where it originated; thus, visible parts of the ISM are necessarily parts of the ISM that are trying to cool. Let's look at ways of emitting photons and ensuring that they travel far away.

**Collisionally excited line emission**   In the cooler regions of the interstellar medium, cooling is performed by the emission of far-infrared photons from carbon and oxygen. The first ionization energy of oxygen ($I_O = 13.62\,\mathrm{eV}$) is nearly identical to the ionization energy of hydrogen. Thus, in the portion of the interstellar medium where the hydrogen is primarily neutral, the oxygen will be primarily in the form of neutral O I. However, the carbon will be nearly all in the form of singly ionized C II; as we saw earlier in this section, the ultraviolet radiation field in our galaxy is capable of keeping the neutral fraction of carbon down to $f_{n,C} \sim 10^{-3}$.

The electronic ground state of C II is split into a pair of fine-structure levels separated by an energy $E_{u\ell} = 7.86 \times 10^{-3}\,\mathrm{eV}$. Thus, in regions of the ISM where the temperature is $T \sim E_{u\ell}/k = 91.2\,\mathrm{K}$ (that is, in the cold neutral medium), the upper fine-structure level will be populated by collisions of the C II ion with other gas particles such as hydrogen atoms and free electrons. When a bound electron in the upper fine-structure level falls back to the lower level, it emits a

---

[8] In the CNM, a C II ion will typically spend $\sim 10^5$ yr before recombining with a free electron to become a neutral atom. However, its neutrality will last only $\sim 100$ yr before it is photoionized again.

photon of wavelength $\lambda = 158\,\mu\text{m}$, a far-infrared wavelength at which the cold neutral medium is transparent. If C II is excited by collisions with free electrons, the resulting cooling function is

$$\frac{\Lambda^e_{\text{CII}}}{10^{-27}\,\text{erg cm}^3\,\text{s}^{-1}} \approx 3.1 \left(\frac{x}{10^{-3}}\right) \left(\frac{T}{100\,\text{K}}\right)^{-0.5} \exp\left(-\frac{91.2\,\text{K}}{T}\right), \qquad (1.36)$$

where $x = n_e/n_{\text{H}}$ is the fractional ionization of the gas and the ratio $n_{\text{C}}/n_{\text{H}} = 2.7 \times 10^{-4}$ is assumed. When the fractional ionization is sufficiently low, the C II is excited primarily by collisions with hydrogen atoms, which yields a cooling function

$$\frac{\Lambda^H_{\text{CII}}}{10^{-27}\,\text{erg cm}^3\,\text{s}^{-1}} \approx 5.2 \left(\frac{T}{100\,\text{K}}\right)^{0.13} \exp\left(-\frac{91.2\,\text{K}}{T}\right). \qquad (1.37)$$

For the conditions prevailing in the cold neutral medium, we expect that collisions with both atoms and free electrons will contribute significantly to the excitation of C II.

Oxygen also plays a role in cooling the cold neutral medium. The electronic ground state of O I has a fine-structure splitting of amplitude $E_{u\ell}/k = 228\,\text{K}$. The upper fine-structure level is populated primarily by collisions with hydrogen atoms. The subsequent emission of $\lambda = 63\,\mu\text{m}$ photons results in a cooling function

$$\frac{\Lambda^H_{\text{OI}}}{10^{-27}\,\text{erg cm}^3\,\text{s}^{-1}} \approx 4.1 \left(\frac{T}{100\,\text{K}}\right)^{0.42} \exp\left(-\frac{228\,\text{K}}{T}\right), \qquad (1.38)$$

assuming that the ratio $n_{\text{O}}/n_{\text{H}} = 6.0 \times 10^{-4}$. At an oxygen-to-carbon ratio $n_{\text{O}}/n_{\text{C}} = 2.2$ (equal to that of the Sun), cooling by O I doesn't surpass cooling by C II until the temperature $T$ reaches $\sim 800\,\text{K}$. Emission of infrared light from collisionally excited fine-structure levels in carbon and oxygen is an important form of cooling in molecular clouds and the cold neutral medium. Molecular clouds can also cool by emission from the vibrational and rotational transitions of molecules.

**Collisionally excited hydrogen lines**  The first excited level of atomic hydrogen is $E_{21} = 10.20\,\text{eV}$ above the ground state. This means that the first excited level will not be highly populated by collisions until the temperature reaches $T \sim E_{21}/k = 1.18 \times 10^5\,\text{K}$. However, at solar abundance, there are $\sim 1700$ H atoms for every O atom. In addition, a hydrogen Lyman $\alpha$ photon carries away $\sim 520$ times as much energy as an O I fine-structure photon. Thus, cooling by Lyman $\alpha$ photons can compete with cooling by O I fine-structure line photons at temperatures as low as $T \approx 10^4\,\text{K}$, when the number of hydrogen atoms in the first

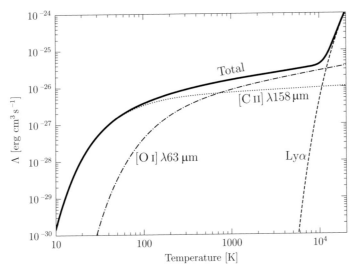

**Figure 1.10** Cooling function $\Lambda$ in the cooler interstellar medium. A fractional ionization $x = 10^{-3}$ is assumed; the carbon and oxygen abundances are those given in Table 1.2.

excited state is suppressed by a Boltzmann factor $e^{-11.8} \sim 10^{-5}$. At $T \sim 10^4$ K, the cooling function for hydrogen excited by collisions with free electrons is

$$\frac{\Lambda^e_{\mathrm{Ly}\alpha}}{10^{-27} \, \mathrm{erg \, cm^3 \, s^{-1}}} \approx 6 \times 10^5 \left(\frac{x}{10^{-3}}\right) \left(\frac{T}{10^4 \, \mathrm{K}}\right)^{-0.5} \exp\left(-\frac{1.18 \times 10^5 \, \mathrm{K}}{T}\right).$$

(1.39)

The net cooling function for gas in the temperature range $10 \, \mathrm{K} < T < 20\,000 \, \mathrm{K}$ is shown in Figure 1.10.

When the temperature of the gas is $T > 20\,000$ K, atomic hydrogen can be collisionally ionized; this is followed by radiative recombination to a high energy level and then a cascade down to the ground state. Line emission from hydrogen can be an important cooling mechanism in the warm neutral medium and warm ionized medium. These higher-energy phases of the ISM are also cooled by line emission from more highly ionized atoms such as O III, C IV, and O VI.

**Free–free emission**   In an ionized gas at $T > 10^6$ K, the free–free radiation emitted by electrons when they are accelerated by other charged particles can be a significant cooling mechanism. Free–free emission is also referred to as bremsstrahlung, a German term that literally means "braking radiation."[9] Free–free emission is important in the cooling of the hot ionized medium of the

---

[9] When Arnold Sommerfeld coined the word "Bremsstrahlung" in 1909, he emphasized the distinction between the continuous spectrum produced by bremsstrahlung and the emission line spectrum produced by "Fluoreszenzstrahlung" (or fluorescent radiation).

ISM and of the intergalactic medium. In the lowest-density regions of the diffuse intergalactic medium the gas also undergoes adiabatic cooling as the universe expands.

## 1.4 Stable and Unstable Equilibrium

Suppose that a region of gas has a fixed chemical composition and has a number density $n$ of gas particles at kinetic temperature $T$. If at these conditions the volumetric heating rate $g = nG$ is equal to the volumetric cooling rate $\ell = n^2\Lambda$, the temperature $T$ is in equilibrium, assuming no flow of heat into or out of the region. However, not every equilibrium is a stable equilibrium. If a phase of the interstellar medium is to be long-lived, its temperature equilibrium must be stable.

The landmark paper that first illuminated stable and unstable equilibria in the ISM was published by Field, Goldsmith, and Habing in 1969. The title of their paper was "Cosmic-ray heating of the interstellar gas." As the title implies, Field et al., unaware of the role played by dust in heating the ISM, assumed that collisional ionization by cosmic rays provided the bulk of the heating. Although their paper was not accurate in every detail, they did come to the key realization that a heating gain $G$ that is independent of gas temperature $T$, when combined with a cooling function $\Lambda(T)$ that is not a simple power law, can lead to multiple phases of the ISM in pressure equilibrium with each other.

Let's reproduce the argument of Field et al., using updated values of the heating gain and cooling function. At a temperature $T$, there exists an equilibrium density $n_{eq}(T)$ at which heating exactly balances cooling: $n_{eq}G = n_{eq}^2\Lambda(T)$. This equilibrium density,

$$n_{eq}(T) = \frac{G}{\Lambda(T)}, \tag{1.40}$$

is plotted in Figure 1.11, assuming photoelectric heating by dust (Equation 1.33) and cooling by C II, O I, and Lyman $\alpha$ emission (Equations 1.36 through 1.39). If every point along the equilibrium line in Figure 1.11 represented a stable equilibrium, then there could be a continuous distribution of temperatures, and thus of number densities, in the equilibrium ISM. The presence of distinct *phases* in the ISM results from the distinction between stable and unstable equilibrium.

To illustrate this distinction, Figure 1.12 shows the line of temperature equilibrium ($G = n\Lambda$), not in the $n$ versus $T$ plane but in the $P$ versus $n$ plane. If the interstellar gas is assumed to be in pressure equilibrium, then for pressures in the range $0.7 \times 10^{-13}\,\mathrm{dyn\,cm^{-2}} < P < 7 \times 10^{-13}\,\mathrm{dyn\,cm^{-2}}$, bounded by the horizontal lines in Figure 1.12, there are three possible values of $n_{eq}$ at a fixed pressure. For a pressure $P \sim 2 \times 10^{-13}\,\mathrm{dyn\,cm^{-2}}$, the three possible states of temperature equilibrium are labeled as points F, G, and H in Figure 1.12.

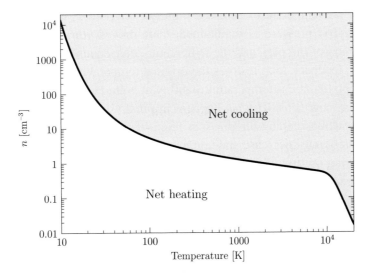

**Figure 1.11**  The heavy line gives the equilibrium density $n_{eq}(T)$ for which the temperature remains constant. Above the line, the gas will cool; below the line, the gas will heat.

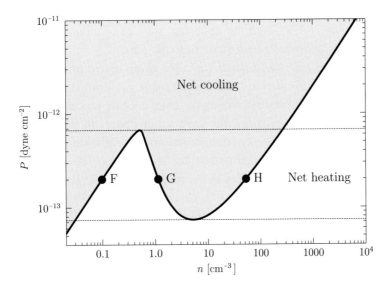

**Figure 1.12**  The heavy solid line is the line of temperature equilibrium. Above this line, the gas will cool; below the line, the gas will heat.

Consider what happens at point F of Figure 1.12 if you slightly change the temperature while keeping the pressure fixed. If the temperature increases, the number density must decrease, and you move left from point F. This moves you into the cooling portion of the plot, and the temperature consequently *decreases*. Thus, a negative feedback loop pushes you back to the original temperature.

Similarly, if the temperature decreases slightly from point F, the number density must increase if pressure is maintained. This moves you rightward into the heating portion of the plot, and the temperature consequently *increases*. Again, the negative feedback loop restores the original temperature. A similar negative feedback loop maintains temperature stability at point H.

However, now consider what happens at point G if you slightly change the temperature while keeping the pressure fixed. If the temperature increases, the number density must decrease and you move left from point G. This moves you into the heating portion of the plot, and the temperature *increases further*. Thus, a positive feedback loop pushes you to higher and higher temperatures until you reach point F. Similarly if the temperature decreases slightly from point G, the number density must increase, placing you in the cooling portion of the plot, and the temperature *decreases further*, until you reach point H. Thus, point G is in unstable equilibrium, in contrast with points F and H, which are in stable equilibrium.

As a result of their analysis, Field et al. created a **two-phase** model of the interstellar medium, consisting of cold neutral clouds, with $n \sim 10\,\text{cm}^{-3}$ and $T \sim 100\,\text{K}$, embedded within a warm intercloud medium, with $n \sim 0.1\,\text{cm}^{-3}$ and $T \sim 10^4\,\text{K}$. These stable equilibrium conditions are the equivalent of points H and F in Figure 1.12. Given the imperfect state of observations in 1969, this two-phase model provided an adequate fit to the data.

Although they are remembered for advocating a two-phase model of the interstellar medium, Field et al. added as an aside, "A third stable phase should exist above $10^6$ K, with bremsstrahlung the chief cooling process." In 1969, there was little observational evidence for such a hot phase. In the 1970s, though, detection of a diffuse soft X-ray background and of emission lines from highly ionized species such as O VI hinted at the existence of interstellar gas with $T \sim 10^6$ K. In fact, the Sun resides in the midst of a "Local Bubble" of hot gas, with $T \sim 10^6$ K and $n \sim 0.004\,\text{cm}^{-3}$. Cox and Smith, in 1974, suggested that supernova remnants could produce a bubbly hot phase, with $T \sim 10^6$ K, and that the bubbles blown by supernovae would have a porosity factor

$$q > 0.1 \left( \frac{r_{\text{SN}}}{10^{-13}\,\text{pc}^{-3}\,\text{yr}^{-1}} \right), \qquad (1.41)$$

where $r_{\text{SN}}$ is the supernova rate per unit volume and the porosity factor $q$ is essentially the fraction of the ISM volume occupied by hot bubbles. The range $0.1 < q < 0.5$ is a topologically interesting one, in which the expanding supernova remnants join to form supersized bubbles and elongated tunnels of hot gas.

McKee and Ostriker, in what is usually thought of as the "third phase" paper, made a more elaborate argument for three phases within the ISM. Their first phase is the cold neutral medium, with $T \sim 80$ K, $n_{\text{H}} \sim 40\,\text{cm}^{-3}$, and a low fractional ionization $x = n_e/n_{\text{H}} \sim 0.001$. Their second phase is the

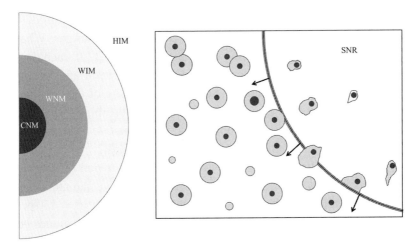

**Figure 1.13** Cartoon from the McKee and Ostriker 1977 paper on the third phase. The left panel shows a typical cold neutral cloud, surrounded by the warm medium (both neutral and ionized). The right panel shows an expanding supernova blastwave overtaking a population of cold clouds.

warm medium, containing both ionized and neutral components; in the warm medium, $T \sim 8000\,\mathrm{K}$ and $n_H \sim 0.3\,\mathrm{cm}^{-3}$, with the fractional ionization ranging from $x \sim 0.15$ in the neutral component to $x \sim 0.7$ in the ionized component. Their third phase, containing most of the volume of the ISM, is the hot ionized medium, consisting of overlapping supernova bubbles. In the hot ionized medium, $T \sim 10^6\,\mathrm{K}$ and $n_H \sim 0.002\,\mathrm{cm}^{-3}$, and ionization is nearly complete ($x \sim 1$).

Figure 1.13 shows a picture of how McKee and Ostriker envisaged a population of cold clouds being slammed by an expanding supernova blastwave; the clouds are heated and ablated, adding their mass to the Hot Ionized Medium. Although the three-phase medium of McKee and Ostriker was a closer approach to reality than the simple two-phase medium of Field et al., it still fell short of reality. The current five-phase model, described earlier, is largely empirical (not relying on assumptions about thermal pressure equilibrium) and recognizes that, in many ways, the ISM of our galaxy is a dynamic, turbulent, dusty, magnetized place.

## Exercises

1.1 The total mass of molecular clouds in our galaxy is $M \approx 1.5 \times 10^9\,M_\odot$, about 20% of the total mass of the ISM.

    (a) For simplicity, assume that every molecular cloud is a sphere of radius $r = 15\,\mathrm{pc}$ and mean density $n(H_2) = 300\,\mathrm{cm}^{-3}$. What is the mass of one such cloud? How many clouds are there in our galaxy?

(b)    Assume that the gas in a molecular cloud is mixed with dust, with the dust mass equal to 1% of the mass of molecular gas. What is the total dust mass within a single molecular cloud? If each dust grain is a sphere of graphite with radius $a = 0.1\,\mu m$ and bulk density $\rho = 2.2\,g\,cm^{-3}$, what is the number density of dust grains in the cloud? If a dust grain's cross section for absorbing light is equal to its geometric cross section, what is the mean free path of a photon in a molecular cloud before it is absorbed by dust?

(c)    Suppose that the molecular clouds described in part (a) are randomly distributed through our galaxy's disk, which we can approximate as a cylinder of radius $R = 15\,000\,pc$ and thickness $H = 150\,pc$. What is the expectation value for the number of molecular clouds between us and the galactic center? What is the probability that *zero* clouds would lie along our line of sight to the galactic center?

1.2    Figure 1.10 shows the cooling function for the low-temperature ISM as described by Equations 1.36–1.39 for fractional ionization $x = 10^{-3}$ and C and O abundances measured in the local ISM.

(a)    Recompute and replot the cooling function in Figure 1.10 for C and O abundances that are 10% of those in the local ISM (roughly the metal abundance in the Small Magellanic Cloud). Assume the same fractional ionization as for the local ISM.

(b)    Now consider the heating and cooling balance in this low-metallicity ISM, following the discussion in Section 1.4. Assume photoelectric heating by dust (Equation 1.33) and recompute and replot the curve of equilibrium density $n_{eq}(T)$ in Figure 1.11 and the F–G–H two-phase ISM equilibrium diagram in Figure 1.12.

(c)    Compare the locations of the possible temperature equilibrium points, F, G, and H, for pressure $P \approx 2 \times 10^{-13}\,dyn\,cm^{-2}$. Comment on the role of gas metallicity in determining the properties of the stable phases of the ISM in a galaxy.

# 2

# Cold Neutral Medium

*The fixed stars cannot be at less distance from the earth, than twenty-seven thousand, six hundred and sixty times the earth's distance from the sun, which is thirty-three millions of leagues: perhaps some astronomers would tell you they are further still.*

Bernard de Fontenelle (1657–1757)
"Conversations on the Plurality of Worlds"
[Elizabeth Gunning translation, 1803]

Although each of the five phases of the ISM receives its own chapter in this book, the chapters are not simply ordered from coldest to hottest (or equivalently, from highest density to lowest density). Discussing cold molecular clouds will require talking about chemistry and about the properties of dust, so we are postponing the complications of molecular clouds, and starting with something simpler: the cold neutral medium (CNM).

Cold neutral gas in the ISM gives rise to a number of narrow absorption features in the spectra of hot background stars. The most prominent absorption lines at visible wavelengths are the H and K lines of Ca II, at $\lambda = 3968\,\text{Å}$ and $3934\,\text{Å}$, plus the $D_1$ and $D_2$ doublet of Na I at $\lambda = 5896\,\text{Å}$ and $5890\,\text{Å}$. Figure 2.1, which shows the sodium $D_2$ line in the direction of three bright background stars in the Large Magellanic Cloud, illustrates the variation in absorption line profiles. The stars Sk-69° 215 and Sk-69° 224, for instance, are 5 arcminutes apart; the cold gas causing the absorption is $\sim 100\,\text{pc}$ away from Earth, meaning that 5 arcmin corresponds to a physical length scale $\sim 0.15\,\text{pc}$.

At the start of the twentieth century, it was the discovery of a narrow calcium K line that revealed to Johannes Hartmann the existence of cold interstellar gas. The sodium D absorption lines from interstellar gas were first detected in 1919 by Mary Lea Heger, looking toward Delta Orionis and Beta Scorpii. Additional interstellar absorption lines at visible wavelengths were later seen for ions such as Ti II and neutral atoms such as Ca I, K I, and Li I. An absorption line at $\lambda = 4300\,\text{Å}$ found by Theodore Dunham in 1937 was initially puzzling, since

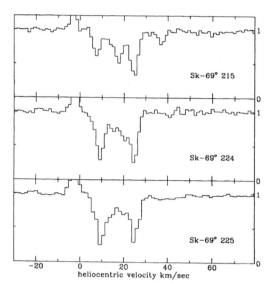

**Figure 2.1** The sodium $D_2$ interstellar absorption line seen along three lines of sight to stars within $\sim$15 arcmin of SN1987A in the Large Magellanic Cloud. [Molaro et al. 1993]

it could not be attributed to any known element. However, Swings and Rosenfeld soon pointed out that the mystery line resulted from an electronic transition of the CH molecule;[1] this constituted the discovery of interstellar molecules. Later observations found absorption features in the visible range from other diatomic molecules, such as CN, NH, $CH^+$, and $C_2$.

Although absorption lines in the visible range of the spectrum provide useful diagnostics for some atoms, many elements require an energy $E > 3\,\text{eV}$ (corresponding to a photon wavelength $\lambda < 4000\,\text{Å}$) to excite them from their electronic ground state. The study of the cold neutral medium, therefore, received a useful boost from the development and launch of orbiting ultraviolet telescopes. In particular, when hydrogen is in its ground state it can produce the Lyman series of absorption lines, corresponding to transitions upward from the $n = 1$ electronic ground state. The Lyman lines range from Lyman $\alpha$ ($n = 1 \rightarrow 2$) at $\lambda = 1216\,\text{Å}$ and $E = 10.20\,\text{eV}$, up to the Lyman limit ($n = 1 \rightarrow \infty$) at $\lambda = 912\,\text{Å}$ and $E = 13.60\,\text{eV}$. Many other atoms and ions have absorption lines in the near-ultraviolet range of the spectrum, making the range from $\sim 912\,\text{Å}$ to $\sim 4000\,\text{Å}$ a good place to study the gas of the cold neutral medium. As an example of the information carried by UV interstellar absorption lines, consider Figure 2.2, which shows absorption toward HD 93521, a hot main sequence star with effective temperature $T_{\text{eff}} \approx 31\,000\,\text{K}$. We see complex absorption lines of

---

[1] Most chemists would refer to CH as a "radical," and regard it as highly reactive and short-lived. However, at the low density of the ISM, radicals have long lifetimes, and thus astronomers don't make a verbal distinction between radicals like CH and ordinary molecules like $CH_4$.

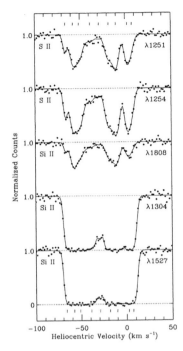
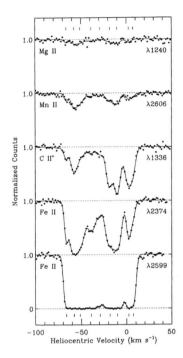

**Figure 2.2** Ultraviolet interstellar absorption lines toward the star HD 93521, seen with the High-Resolution Spectrograph on the *Hubble Space Telescope*. Notice the multiple components at different heliocentric velocities, plus the effects of line saturation on Si II and Fe II. [Spitzer and Fitzpatrick 1993]

S II, Si II, Mg II, Mn II, C II, and Fe II. The multiple components of the absorption lines are due to different structures, or "clouds," along the line of sight. The spread in velocities is due primarily to the differential rotation of our galaxy, resulting in clouds at different distances from us having different line-of-sight velocities.

## 2.1 The Equation of Radiative Transfer

The absorption and emission lines produced by interstellar gas contain information about number density, temperature, chemical abundances, ionization states, and excitation states. However, interpreting that information requires understanding the ways in which light interacts with baryonic matter. To begin, let's do a quick review of the **equation of radiative transfer**.

Light can be characterized by its **intensity** $I_\nu$, also known as the spectral radiance; its cgs units are $\mathrm{erg\,s^{-1}\,cm^{-2}\,sr^{-1}\,Hz^{-1}}$. If light travels through empty space (and if space is Euclidean and static) then its intensity remains constant. However, if light travels through intergalactic or interstellar gas, its intensity can change if the gas absorbs or emits light at the frequency $\nu$ in question. As the light travels through a small distance $ds$, the change in the intensity $I_\nu$ can be written as

$$dI_\nu = -I_\nu \kappa_\nu ds + j_\nu ds, \qquad (2.1)$$

where $\kappa_\nu$ is the **attenuation coefficient**, with cgs units $cm^{-1}$, and $j_\nu$ is the **emissivity**, with units $erg\,s^{-1}\,cm^{-3}\,sr^{-1}\,Hz^{-1}$. The attenuation coefficient $\kappa_\nu$ represents the fraction of the intensity lost per centimeter traveled; it can differ greatly for different media and at different frequencies. The emissivity $j_\nu$ represents spontaneous emission, which is independent of the intensity $I_\nu$ of the light passing through the gas. Stimulated emission, such as that produced by a maser, can be represented by a *negative* attenuation coefficient $\kappa_\nu$.

Much of astronomy consists of finding solutions to the equation of radiative transfer,

$$\frac{dI_\nu}{ds} = -\kappa_\nu I_\nu + j_\nu, \tag{2.2}$$

given appropriate values of the attenuation coefficient and emissivity. A useful quantity in talking about radiative transfer is the **optical depth** $\tau_\nu$, defined so that

$$d\tau_\nu \equiv \kappa_\nu ds. \tag{2.3}$$

The optical depth, a dimensionless number, is $\tau_\nu \gg 1$ for optically thick (or opaque) objects and $\tau_\nu \ll 1$ for optically thin (or transparent) objects. In terms of the optical depth, the equation of radiative transfer is

$$\frac{dI_\nu}{d\tau_\nu} = -I_\nu + S_\nu, \tag{2.4}$$

where $S_\nu \equiv j_\nu/\kappa_\nu$ is the **source function**.

In general, the source function $S_\nu$ is a complicated beast, since light can encounter many different types of emitters and absorbers (and scatterers) along its path, each with a different frequency dependence. However, there are some cases when we can ignore the nature of the source function. For example, suppose that $S_\nu = j_\nu = 0$; that is, the case when there is no spontaneous emission. In that case, Equation 2.4 integrates to

$$I_\nu(\tau_\nu) = I_\nu(0)e^{-\tau_\nu}. \tag{2.5}$$

In the absence of spontaneous emission, absorption causes a decrease in the intensity that is exponential in the optical depth. As another example, consider light traveling through a uniform medium for which $S_\nu$ is the same everywhere. In that case, Equation 2.4 integrates to

$$I_\nu(\tau_\nu) = S_\nu + [I_\nu(0) - S_\nu]e^{-\tau_\nu}. \tag{2.6}$$

In the limit $\tau_\nu \to \infty$, $I_\nu \to S_\nu$. That is, if you are looking at an opaque slab, it doesn't directly matter what light sources are on the other side of the slab;[2] the intensity you see is dictated by the source function of the slab itself.

---

[2] It does *indirectly* matter what light sources are present, since absorbed photons raise the temperature of the slab, and the source function $S_\nu$ is generally temperature-dependent.

The source function $S_\nu$ is particularly useful when you are dealing with a blackbody, the semi-mythical opaque object that absorbs every frequency of light that strikes it and is able to emit every frequency of light. By the end of the nineteenth century, the source function of a blackbody was fairly well known empirically, from studies using laboratory approximations of blackbodies. In 1900, Max Planck, using his radical idea of quantized energies, derived the source function for a blackbody. This source function, called the **Planck function**, is

$$B_\nu(T) = \frac{2h\nu^3/c^2}{\exp(h\nu/kT) - 1}. \tag{2.7}$$

In the limit $h\nu \ll kT$, called the Rayleigh–Jeans tail, the Planck function is proportional to $\nu^2$:

$$B_\nu(T) \approx \frac{2\nu^2}{c^2}kT. \tag{2.8}$$

In the limit $h\nu \gg kT$, called the Wien tail, the Planck function falls exponentially as $\nu \to \infty$, with

$$B_\nu(T) \approx \frac{2h\nu^3}{c^2}\exp\left(-\frac{h\nu}{kT}\right). \tag{2.9}$$

The maximum of the Planck function lies at a photon energy $h\nu_{\text{peak}} \approx 2.821kT$.

An object for which the source function $S_\nu$ is the Planck function is said to emit **thermal radiation**. An object for which the intensity $I_\nu$ is the Planck function is said to emit **blackbody radiation**. To see the difference between thermal and blackbody radiation, consider a slab of material with optical depth $\tau_\nu$ that is producing thermal radiation. If no light is falling on the far side of the slab, the intensity measured on the near side of the slab is (see Equation 2.6):

$$I_\nu = B_\nu(T)(1 - e^{-\tau_\nu}). \tag{2.10}$$

If the slab is optically thick at frequency $\nu$ ($\tau_\nu \gg 1$) then

$$I_\nu \approx B_\nu(T). \tag{2.11}$$

If the slab is optically thin ($\tau_\nu \ll 1$) then

$$I_\nu \approx \tau_\nu B_\nu(T) \ll B_\nu(T), \tag{2.12}$$

and the radiation, although thermal, will not be blackbody.

Inside a hollow object emitting blackbody radiation with temperature $T$, the energy density of photons, in $\text{erg}\,\text{cm}^{-3}\,\text{Hz}^{-1}$, is

$$\varepsilon_\nu(T) = \frac{4\pi}{c}B_\nu(T) = \frac{8\pi h\nu^3/c^3}{\exp(h\nu/kT) - 1}. \tag{2.13}$$

Integrated over frequency, the total energy density inside the blackbody is

$$\varepsilon(T) = \frac{4\pi}{c} \int B_\nu(T) d\nu = \frac{4}{c}\sigma_{sb}T^4, \tag{2.14}$$

where $\sigma_{sb}$ is the Stefan–Boltzmann constant,

$$\sigma_{sb} = \frac{2\pi^5 k^4}{15 h^3 c^2} \approx 3.539 \times 10^7 \, \text{eV cm}^{-2} \, \text{s}^{-1} \, \text{K}^{-4}. \tag{2.15}$$

For example, the energy density of the cosmic microwave background is well fitted by Equation 2.13 at a temperature $T = 2.7255$ K. This temperature implies peak emission at a photon energy $h\nu_{peak} = 0.663$ meV (in the microwave range of the spectrum) and a total energy density, from Equation 2.14, of $\varepsilon = 0.2606 \, \text{eV cm}^{-3}$.

Sometimes stars can be usefully approximated as spherical blackbodies. At a distance $r$ from a spherical blackbody with radius $R$ and temperature $T$, the energy density of blackbody radiation is diluted by the inverse square law of flux to

$$\varepsilon_\nu = \frac{\pi}{c} B_\nu(T) \left(\frac{R}{r}\right)^2. \tag{2.16}$$

The total energy density, integrated over frequency, is then

$$\varepsilon = \frac{1}{c}\sigma_{sb}T^4 \left(\frac{R}{r}\right)^2. \tag{2.17}$$

For example, the Sun can be approximated as a spherical blackbody with radius $R_\odot = 695\,700$ km and temperature $T_\odot = 5772$ K (yielding $h\nu_{peak} = 1.40$ eV, in the near-infrared). At the Earth's average distance $r = 1$ au $= 215.0\,R_\odot$ from the Sun, the energy density of sunlight is $\varepsilon = 28.34$ MeV cm$^{-3}$, nearly eight orders of magnitude greater than the average energy density of starlight in our galaxy. It is only at a distance $r \sim 7000$ au $\sim 0.034$ pc from the Sun that the energy density of sunlight no longer dominates over the galaxy's background of starlight.

## 2.2 Absorbers and Emitters

The concept of thermal emission will be useful when we investigate dust. The concept of blackbody emission will be useful when we look at the starlight that excites and ionizes interstellar gas. The spectrum of the interstellar gas itself, however, is usually either an absorption line spectrum or an emission line spectrum. Thus, to understand the light emitted by the ISM we need to use the equation of radiative transfer for lines.

For simplicity, consider as before an absorber – it could be an atom, an ion, or a molecule – that has two energy levels: an upper level with energy $E_u$ and a lower level with energy $E_\ell < E_u$. The statistical weights of the upper and lower levels are $g_u$ and $g_\ell$. If the absorber is in its lower energy state, it can be lifted to the upper energy state by absorbing a photon with energy $E \sim E_{u\ell} \equiv E_u - E_\ell$ and thus frequency $\nu \sim \nu_{u\ell} = E_{u\ell}/h$. Because of line broadening, the frequency of the photon doesn't have to exactly match $\nu_{u\ell}$. There is a number density $n_\ell$ of absorbers in the lower state, and an energy density $\varepsilon_\nu$ of photons. The rate at which the absorbers absorb photons depends linearly on both $n_\ell$ and $\varepsilon_\nu$:

$$\left(\frac{dn_u}{dt}\right)_{\ell \to u} = n_\ell B_{\ell u} \varepsilon_\nu, \qquad (2.18)$$

where $\nu = \nu_{u\ell}$, and $B_{\ell u}$ is the **Einstein B coefficient**, an intrinsic property of the absorber that is independent of density and temperature.

However, the path between levels is not a one-way upward escalator. An absorber in the upper level can transition back to the lower level via **spontaneous emission** or via **stimulated emission**, in which a photon of energy $\sim h\nu_{u\ell}$ does the stimulation. The rate at which the absorbers emit photons is

$$\left(\frac{dn_u}{dt}\right)_{u \to \ell} = -n_u(A_{u\ell} + B_{u\ell}\varepsilon_\nu), \qquad (2.19)$$

where $\nu = \nu_{u\ell}$, $A_{u\ell}$ is the **Einstein A coefficient** for spontaneous emission, and $B_{u\ell}$ is the Einstein B coefficient for downward transition due to stimulated emission.

The three Einstein coefficients, $A_{u\ell}$, $B_{u\ell}$, and $B_{\ell u}$, are all intrinsic quantum mechanical properties of the transition between the upper and lower energy levels. Having three coefficients to describe a simple two-level absorber would be discouraging were it not for the fact that the three Einstein coefficients are simply related to each other:

$$B_{u\ell} = \frac{c^3}{8\pi h \nu_{u\ell}^3} A_{u\ell} \qquad (2.20)$$

and

$$B_{\ell u} = \frac{g_u}{g_\ell} B_{u\ell} = \frac{g_u}{g_\ell} \frac{c^3}{8\pi h \nu_{u\ell}^3} A_{u\ell}. \qquad (2.21)$$

In the real universe, the Einstein A coefficient, and thus the Einstein B coefficients, can vary greatly from one type of absorber to another. In particular, for electronic transitions within an atom, ion, or molecule, the value of $A_{u\ell}$ depends on whether the transition is "permitted" or "forbidden."

**Permitted** transitions are electric dipole transitions, which have relatively high rates of spontaneous emission. Permitted transitions that produce UV or visible photons typically have $A_{u\ell} \sim 10^8 \text{ s}^{-1}$. For example, the Lyman $\alpha$ transition has $A_{u\ell} = 4.70 \times 10^8 \text{ s}^{-1}$ while the Balmer $\alpha$ transition has $A_{u\ell} = 0.44 \times 10^8 \text{ s}^{-1}$. These transitions are called "permitted" because they can occur in gases with a density comparable to that of the Earth's atmosphere. At a pressure $P \sim 1$ atm and temperature $T \sim 300$ K, an excited atom or molecule collides with other atoms and molecules at a rate $t_{coll}^{-1} \sim 3 \times 10^9 \, \text{s}^{-1}$. Although not every physical collision leads to collisional de-excitation, a spontaneous emission with $A_{u\ell} \ll t_{coll}^{-1}$ is unlikely to occur; the excited atom or molecule will be collisionally de-excited before it can undergo the spontaneous emission of a photon.

**Forbidden** transitions are magnetic dipole or electric quadrupole transitions; these have much lower rates of spontaneous emission than do electric dipole transitions. Forbidden transitions that produce visible photons typically have $A_{u\ell}$ in the range $10^{-3} \, \text{s}^{-1} \rightarrow 10 \, \text{s}^{-1}$. For example, the "nebulium" lines of O III, at $\lambda = 4959$ Å and $5007$ Å, result from forbidden transitions with $A_{u\ell} = 0.0070 \, \text{s}^{-1}$ and $0.020 \, \text{s}^{-1}$ respectively. Despite their name, forbidden transitions are not *absolutely* forbidden; they simply have such small values of $A_{u\ell}$ that they are wildly unlikely to occur under terrestrial conditions. In the cold neutral medium of the ISM, where the collision rate is a leisurely $t_{coll}^{-1} \sim 2 \times 10^{-8} \, \text{s}^{-1}$ (Equation 1.11), forbidden transitions are far more prevalent than collisional de-excitations. In Figure 2.3, the forbidden O III emission lines at $\lambda = 4959$ Å and $5007$ Å are seen to be comparable in strength to the permitted Balmer lines (H $\alpha$, H $\beta$, and so on).

Although the Einstein coefficients are a useful means of talking about absorption and emission processes, we often find it more useful to talk about cross sections. During the discussion of collisions in Section 1.2, the cross section had a natural geometric meaning. If the centers of two solid spheres are separated by a distance $d$ equal to the sum of their radii, they will bump into each other; thus, their cross section for collision is $\sigma = \pi d^2$. In the case of the absorption and emission of photons, the idea of the cross section is more abstract and doesn't involve thinking of two billiard balls bumping into each other.

If the photon energy density is $\varepsilon_\nu$ then the number density of photons per unit frequency interval is $\varepsilon_\nu/(h\nu)$. The number density of absorbers in the lower

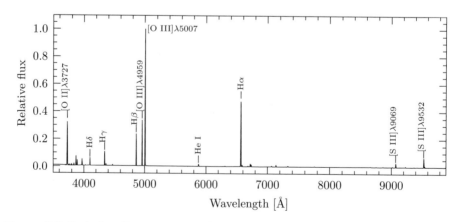

**Figure 2.3** Emission line spectrum of an H II region in the gas-rich dwarf galaxy Holmberg II ($d \sim 3$ Mpc). Atom and ion names written in square brackets, such as [O III], refer to forbidden lines. [Data provided by K. Croxall & R. Pogge]

energy level is $n_\ell$. The absorption rate, given as absorptions per unit volume per unit time, is then

$$\left(\frac{dn_u}{dt}\right)_{\ell \to u} = n_\ell \int dv\, \sigma_{\ell u}(v) c \frac{\varepsilon_v}{hv}, \tag{2.22}$$

where $\sigma_{\ell u}(v)$ is the cross section for a single absorber in the lower energy level to absorb photons of frequency $v$, taking it to the upper energy level. Since absorption lines in the cold neutral medium are narrow (that is, they cover a limited range of frequency), it is fair to write

$$\left(\frac{dn_u}{dt}\right)_{\ell \to u} \approx n_\ell c \frac{\varepsilon_v}{hv_{u\ell}} \int dv\, \sigma_{\ell u}(v), \tag{2.23}$$

where $\varepsilon_v$ is computed at the frequency $v_{u\ell}$. By combining Equations 2.18 and 2.23, we derive a relation between the absorption cross section and the Einstein B coefficient,

$$\int dv\, \sigma_{\ell u}(v) = \frac{hv_{u\ell}}{c} B_{\ell u}, \tag{2.24}$$

and thus, using Equation 2.21,

$$\int dv\, \sigma_{\ell u}(v) = \frac{g_u}{g_\ell} \frac{c^2}{8\pi v_{u\ell}^2} A_{u\ell}. \tag{2.25}$$

The absorption cross section $\sigma_{\ell u}(v)$ is generally largest at $v = v_{u\ell} = E_{u\ell}/h$. However, absorption occurs over a range of frequencies. The dependence of the absorption cross section on the frequency of the light is written as a normalized **line profile** $\Phi_v$, where

$$\sigma_{\ell u}(v) = \frac{g_u}{g_\ell} \frac{c^2}{8\pi v_{u\ell}^2} A_{u\ell} \Phi_v \tag{2.26}$$

and $\int \Phi_v dv = 1$.

The **intrinsic line width** of a line, sometimes called its natural line width, is due to the Heisenberg uncertainty principle. If an energy level of mean energy $E_u$ has a lifetime $t_u$ then the width of the energy level must be $\Delta E \sim h/t_u$, and the resulting spread in the frequency of absorbed photons is $\Delta v = \Delta E/h \sim 1/t_u$. If the only way of depopulating the upper level is through spontaneous emission to a single lower level then we expect the intrinsic line width to be $\Delta v \sim A_{u\ell}$. This means that forbidden lines, under these conditions, are intrinsically narrower than permitted lines. A quick back-of-envelope scribble reveals that the the permitted Lyman $\alpha$ line has $A_{u\ell}/v_{u\ell} \sim 3 \times 10^{-7}$, while the forbidden [O III] 5007 Å line has the tiny width $A_{u\ell}/v_{u\ell} \sim 3 \times 10^{-17}$. The intrinsic line width of [O III] 5007 Å is equivalent to the Doppler broadening you would get with $\Delta v \sim 3 \times 10^{-17} c \sim 10\,\mathrm{nm\,s^{-1}} \sim 30\,\mathrm{cm\,yr^{-1}}$.

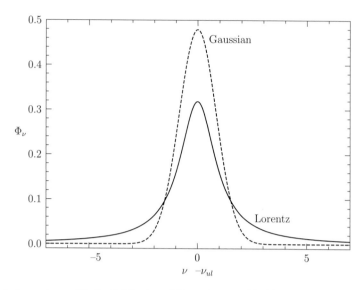

**Figure 2.4** Lorentz profile (solid line) and Gaussian profile (dashed line), normalized so that each profile has the same full width at half maximum and the same integrated area under the curve.

The intrinsic broadening produces a line profile that is well described by a **Lorentz function:**[3]

$$\Phi_\nu = \frac{4\gamma_{u\ell}}{16\pi^2(\nu - \nu_{u\ell})^2 + \gamma_{u\ell}^2}. \tag{2.27}$$

The **damping width** $\gamma_{u\ell}$ is a measure of the width of the Lorentz function. Although the standard deviation of the Lorentz function is infinite, thanks to its $\nu^{-2}$ wings, the full width at half maximum (FWHM) is

$$(\Delta\nu)_{\text{FWHM}} = \frac{\gamma_{u\ell}}{2\pi}. \tag{2.28}$$

Figure 2.4 shows for comparison a Lorentz profile and a Gaussian profile with the same FWHM and the same area under the curve; the broad wings of the Lorentz profile are obvious. For our simple two-level absorber, $\gamma_{u\ell} = A_{u\ell}$. However, for a multiple-level absorber, the upper and lower energy levels can both be broadened by transitions to other levels. If the main mechanism for depopulating levels is spontaneous decay to a lower energy level (which is often the case in low-density gas), then

$$\gamma_{u\ell} = \sum_{E_j < E_u} A_{uj} + \sum_{E_j < E_\ell} A_{\ell j}. \tag{2.29}$$

---

[3] The Lorentz function is named after Hendrik Lorentz, who in 1915 used it to describe spectral lines. The same function is also known as the Cauchy distribution (after Augustin Cauchy, who used it as a probability distribution in 1853) and as the "witch of Agnesi" (after Maria Gaetana Agnesi, who used it as an example in her 1748 calculus textbook).

This reduces to $\gamma_{u\ell} = A_{u\ell}$ when level $\ell$ is the ground state and level $u$ is the first excited state. For Lyman $\alpha$ ($n = 1 \rightarrow 2$) absorption, $\gamma_{u\ell} = A_{21} = 4.70 \times 10^8 \, \text{s}^{-1}$, leading to $\Delta v / v \approx 3 \times 10^{-8}$. For Balmer $\alpha$ ($n = 2 \rightarrow 3$) absorption, $\gamma_{u\ell} = A_{32} + A_{31} + A_{21} = 5.70 \times 10^8 \, \text{s}^{-1}$, leading to $\Delta v / v \approx 2 \times 10^{-7}$.

In the interstellar medium, the absorbers are atoms, ions, and molecules that form part of a gas with kinetic temperature $T$. In the absence of any intrinsic broadening of the line, the thermal motions would still produce Doppler broadening, resulting in a line profile

$$\Phi_v = \frac{1}{\sqrt{\pi}} \frac{1}{v_{u\ell}} \frac{c}{b} \exp\left(-\frac{v^2}{b^2}\right), \tag{2.30}$$

with the line-of-sight velocity $v$ related to the frequency $v$ by the relation $v/c = (v_{u\ell} - v)/v_{u\ell}$. The broadening parameter $b$ in Equation 2.30 is

$$b = \left(\frac{2kT}{m}\right)^{1/2} = 1.28 \, \text{km s}^{-1} \left(\frac{T}{100 \, \text{K}}\right)^{1/2} \left(\frac{m}{m_H}\right)^{-1/2}, \tag{2.31}$$

where $m$ is the mass per absorber and $m_H$ is the mass of a hydrogen atom. For the Lyman $\alpha$ line the intrinsic line width, written as a velocity equivalent, is $(\Delta v)_{\text{FWHM}} \sim 0.01 \, \text{km s}^{-1}$. Even in the relatively chilly CNM, at $T \sim 80 \, \text{K}$, the broadening parameter $b$ translates to a thermal line width $(\Delta v)_{\text{FWHM}} = 2(\ln 2)^{1/2} b \sim 2 \, \text{km s}^{-1}$.

The actual line profile for a gas of absorbers will be neither the intrinsic Lorentz profile nor the thermal Gaussian profile. Instead, it will be a convolution of the Lorentz profile with a Gaussian profile. Such a convolved line profile is called a **Voigt profile**, after the physicist and musician Woldemar Voigt, who studied its properties in 1912. The Voigt profile can be expressed as

$$\Phi_v^{\text{Voigt}} = \frac{1}{\sqrt{\pi}} \frac{1}{v_{u\ell}} \frac{c}{b} H(a, u), \tag{2.32}$$

where $H(a, u)$ is a dimensionless function that describes the shape of the Voigt profile. This function, called the Voigt–Hjerting function,[4] is given by the integral

$$H(a, u) = \frac{a}{\pi} \int_{-\infty}^{\infty} \frac{e^{-y^2} dy}{(u - y)^2 + a^2}, \tag{2.33}$$

where

$$a \equiv \frac{\gamma_{u\ell} c}{4\pi v_{u\ell} b} \tag{2.34}$$

---

[4] Frode Hjerting gets credit for tabulating this function in 1938; in the era before electronic computers, this was a non-trivial task. The Voigt–Hjerting function is better known to mathematicians as the real part of the complex Faddeeva function, named after the algebraist Vera Nikolaevna Faddeeva.

and

$$u \equiv \frac{c}{b}\left(1 - \frac{v}{v_{u\ell}}\right) = \frac{v}{b}. \tag{2.35}$$

Thus, $a$ can be thought of as the ratio of intrinsic broadening to thermal broadening and $u$ is a measure of the distance from the profile center, in units of the thermal broadening parameter $b$. The Voigt–Hjerting function doesn't generally have a simple analytic form. However, even for a permitted line such as Lyman $\alpha$, with its relatively large intrinsic broadening, and even for lines from the coolest parts of the ISM, with their relatively small thermal broadening, the parameter $a$ in the Voigt–Hjerting function is much less than one. For example, the Lyman $\alpha$ line at $T = 80\,\mathrm{K}$ has $a \approx 0.004$. Thus, we can profitably expand the Voigt–Hjerting function in a Taylor series expansion around $a = 0$:

$$H(a,u) \approx H(0,u) + a\frac{dH}{da}\Big|_{a=0}. \tag{2.36}$$

In the limit $u^2 \gg 1$, this expansion can be written as

$$H(a,u) \approx \exp(-u^2) + a\frac{1}{\sqrt{\pi}u^2}. \tag{2.37}$$

The first term in Equation 2.37 represents the Gaussian core of the Voigt–Hjerting function, provided by the thermal broadening, and the second term represents the $v^{-2}$ wings provided by the Lorentz function. To find how far we need to go from the profile center for the broad Lorentz wings to dominate, we must solve the transcendental equation

$$\exp(-u^2) = \frac{a}{\sqrt{\pi}u^2}. \tag{2.38}$$

A useful approximation is that the Lorentz wings start to dominate at a velocity

$$|u| = \frac{|v|}{b} \approx 2.8\left[1 - \ln\left(\frac{\gamma_{u\ell}/v_{u\ell}}{2 \times 10^{-7}}\right) + \ln\left(\frac{b}{1\,\mathrm{km\,s^{-1}}}\right)\right]^{1/2}, \tag{2.39}$$

where we have scaled to properties appropriate for a Lyman $\alpha$ line in the cold neutral medium. Thus, for a permitted UV or visible line in the cooler regions of the ISM, we expect the Lorentz wings to take over at roughly three times the thermal broadening. For a forbidden line, with its tiny intrinsic broadening, the Lorentz wings don't take over until more than 10 times the thermal broadening.

Of course, with computers as cheap as they are, we needn't restrict ourselves to approximations in the limit $a \ll 1$. We can crank up the old steam-powered Babbage engine (or lacking that, use a Python special function library that includes the Faddeeva function) and compute the Voigt–Hjerting function for arbitrary values of $a$. For example, the Voigt–Hjerting function, for values of $a$ ranging from $a = 0$ (a pure Gaussian, with no intrinsic broadening) to $a = 3$, is shown in Figure 2.5.

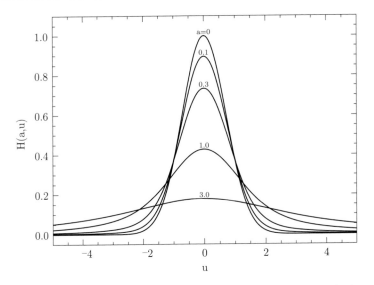

**Figure 2.5** The Voigt–Hjerting function $H(a, u)$ with the intrinsic-to-broadening ratio (see Equation 2.37) taking values $a$ in the range $0.0$–$3.0$.

## 2.3  Building Absorption Lines

When we observe a distant bright star through intervening cold neutral gas, the spectrum we see will look like Figures 2.1 or 2.2, with absorption lines on a background continuum. We can connect our theory-driven absorption cross sections ($\sigma_{\ell u}$) with the absorption lines that we actually see. If the number density of absorbers in the lower-energy state is $n_\ell$, the **attenuation coefficient** for that particular type of absorber is

$$\kappa_\nu = n_\ell \sigma_{\ell u}(\nu). \tag{2.40}$$

That is, the attenuation coefficient is the inverse of the mean free path of a photon with frequency $\nu$. Using the relation between $\sigma_{\ell u}$ and the quantum mechanical properties of the absorber (Equation 2.26), the attenuation coefficient can be written as

$$\kappa_\nu = n_\ell \frac{g_u}{g_\ell} \frac{c^2}{8\pi \nu_{u\ell}^2} A_{u\ell} \Phi_\nu, \tag{2.41}$$

where $\Phi_\nu$ is the appropriate Voigt profile for that particular absorption line at the kinetic temperature $T$ of the absorber population.[5] In the cold neutral medium, nearly all the atoms are in their electronic ground state, so spontaneous emission

---

[5] We are assuming that the contribution of stimulated emission to $\kappa_\nu$ is negligible, so whatever you do, don't apply Equation 2.41 to masers.

can be ignored and the emissivity $j_\nu$ will be zero. This leads to a simple solution to the equation of radiative transfer:

$$I_\nu(\tau_\nu) = I_\nu(0)e^{-\tau_\nu}, \tag{2.42}$$

where $I_\nu(0)$ is the intensity of the background light source and $I_\nu(\tau_\nu)$ is the intensity after the light has passed through an optical depth $\tau_\nu$ of intervening absorbers. The optical depth $\tau_\nu$, in the absence of stimulated emission, is

$$\tau_\nu = \int \kappa_\nu ds = \int n_\ell \sigma_{\ell u}(\nu)ds = \frac{g_u}{g_\ell} \frac{c^2}{8\pi \nu_{u\ell}^2} A_{u\ell} \int n_\ell \Phi_\nu ds. \tag{2.43}$$

If there is just one cloud of absorbers, at uniform temperature $T$, along the line of sight, we can assume that the line profile $\Phi_\nu$ is constant. In this case, the optical depth is proportional to the **column density** $N_\ell$ of absorbers in the lower-energy state, where

$$N_\ell = \int n_\ell ds. \tag{2.44}$$

For constant temperature $T$ along the line of sight, the optical depth of the absorption line is

$$\tau_\nu = \frac{g_u}{g_\ell} \frac{c^2}{8\pi \nu_{u\ell}^2} A_{u\ell} N_\ell \Phi_\nu, \tag{2.45}$$

or, in terms of the Voigt–Hjerting function $H(a, u)$,

$$\tau_\nu = \frac{1}{(4\pi)^{3/2}} \left[ \frac{g_u}{g_\ell} \frac{c^2}{\nu_{u\ell}^3} A_{u\ell} \right] \frac{c}{b} N_\ell H(a, u). \tag{2.46}$$

For lines whose intrinsic broadening is small compared with the thermal broadening (generally true in the ISM), we can use the $a \ll 1$ expansion of Equation 2.37 for the Voigt–Hjerting function. As long as we don't venture out to the Lorentz wings, the optical depth can be well approximated as

$$\tau_\nu \approx \tau_0 \exp\left(-\frac{v^2}{b^2}\right), \tag{2.47}$$

where the optical depth at line center is

$$\tau_0 = \frac{1}{(4\pi)^{3/2}} \left[ \frac{g_u}{g_\ell} \frac{c^2}{\nu_{u\ell}^3} A_{u\ell} \right] \frac{c}{b} N_\ell. \tag{2.48}$$

For the Lyman $\alpha$ line of hydrogen, our standby example,

$$\tau_0 \approx 0.75 \left(\frac{b}{1\,\mathrm{km\,s^{-1}}}\right)^{-1} \left(\frac{N_\ell}{10^{12}\,\mathrm{cm^{-2}}}\right), \tag{2.49}$$

where we have scaled to the thermal broadening $b$ expected for atomic hydrogen at $T \approx 80\,\mathrm{K}$. A Lyman $\alpha$ absorption line in the CNM will thus be optically thin ($\tau_0 < 1$) when $N_\ell < 10^{12}\,\mathrm{cm^{-2}}$ and optically thick ($\tau_0 > 1$) when $N_\ell > 10^{12}\,\mathrm{cm^{-2}}$.

(To give you a sense of perspective on these values, the column density of the Earth's atmosphere, looking upward from sea level, is $N \sim 2 \times 10^{25} \, \text{cm}^{-2}$.)

The shape of an absorption line depends on whether the intervening population of absorbers is optically thin or optically thick. When $\tau_\nu < 1$, Equation 2.42 reduces to

$$\frac{I_\nu(0) - I_\nu(\tau_\nu)}{I_\nu(0)} \approx \tau_\nu, \tag{2.50}$$

and the absorption line, in the case of a single cloud of uniform temperature, has the same shape as the line profile $\Phi_\nu$. The characteristic shape of a Voigt profile can be seen for the $\tau_0 = 1$ absorption line in Figure 2.6. However, when $\tau_0 \gg 1$, the absorption line saturates at its center; the intensity can't drop below zero, no matter how big $\tau$ is. Thus, the flat-bottomed absorption line no longer resembles an upside-down Voigt–Hjerting function.

We don't usually have the luxury of extremely high-resolution spectra; thus, we need to pick out the most important information available at low resolution. We can start with the median frequency ($\nu_0$) or wavelength ($\lambda_0$) of a line. If we know which particular line we are looking at, this will tell us the redshift or blueshift of the line. Next we want some measure of the strength of an absorption line; that is, a way of quantifying the difference between the observed flux $F_\nu$ within the absorption line and what the continuum flux $F_\nu(0)$ would be in the absence of absorption. Note from Figure 2.6 that the depth $F_\nu(0) - F_\nu$ of the absorption line at $\nu_0$ isn't a very useful measurement of line strength, since $F_\nu$ bottoms out at a value indistinguishable from zero at $\tau_0 > 10$. In the 1920s, as a more useful

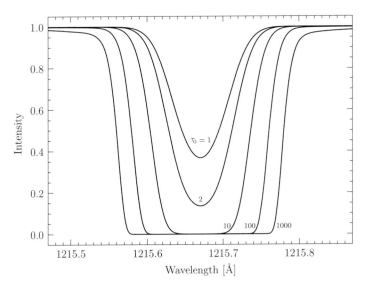

**Figure 2.6** Lyman $\alpha$ absorption lines with central optical depth $\tau_0 = 1$ through 1000 and thermal broadening $b = 10 \, \text{km s}^{-1}$.

measure of line strength, spectroscopists such as Marcel Minnaert and Theodore Dunham developed the concept of an absorption line's **equivalent width**.

In an absorption line spectrum, the measured flux $F_\nu$ at a frequency $\nu$ is related to the flux $F_\nu(0)$ in the absence of absorbers by the relation

$$F_\nu = F_\nu(0)e^{-\tau_\nu}. \tag{2.51}$$

If the original spectrum isn't too badly hacked up with absorption lines, we can make a good guess at what the underlying continuum flux $F_\nu(0)$ must be and then calculate the **dimensionless equivalent width**

$$W \equiv \int \frac{d\nu}{\nu_0}\left[1 - \frac{F_\nu}{F_\nu(0)}\right] = \int \frac{d\nu}{\nu_0}[1 - e^{-\tau_\nu}]. \tag{2.52}$$

Most equivalent widths quoted in the astronomical literature are in units of angstroms or milli-angstroms, because the usual measure of line strength is the **wavelength equivalent width**:

$$W_\lambda = \int d\lambda[1 - e^{-\tau_\lambda}] \approx \lambda_0 W. \tag{2.53}$$

Even in a low-resolution spectrum, the equivalent width $W_\lambda$ can usually be measured fairly well. Graphically speaking, the equivalent width is the width of a straight-sided, perfectly black absorption line that has the same integrated flux deficit as the actual absorption line.

As an example close at hand, consider the Fraunhofer lines in the Sun's spectrum. The calcium H and K lines have equivalent widths of 14 Å and 19 Å respectively at the center of the Sun's disk; these are the strongest absorption lines in the Sun's spectrum. The sodium $D_1$ and $D_2$ lines have equivalent widths of 0.60 Å and 0.83 Å at disk center. Although the equivalent width is fairly easy to determine, it still needs translation and interpretation. If a particular absorption line has an equivalent width of 600 mÅ, what does that tell us about the column density and temperature of the absorbers? As a first step to finding the physical properties of the absorbing gas, we need to know the numerical relation between the observed equivalent width $W_\lambda$ and the underlying optical depth $\tau_\lambda$. This relation was named by Minnaert the **curve of growth** for equivalent width.[6]

## 2.4  Curve of Growth

To study the relation between optical depth at line center, $\tau_0$, and equivalent width, $W_\lambda$, let's start with the simple case in which the absorption line is optically thin ($\tau_0 < 1$). In this case,

---

[6] Minnaert's first Ph.D. was in biology; he was thus familiar with the "growth curve" of a population or individual. Late in life, he stated that his similar phrase "curve of growth" was "a term where you are free to find a reminiscence from the biological studies of my youth."

$$W_\lambda = \int d\lambda [1 - e^{-\tau_\lambda}] \approx \int d\lambda \, \tau_\lambda \qquad [\tau_0 < 1]. \qquad (2.54)$$

For a simple interstellar absorption line, where the absorption is due to a single cloud of uniform temperature, the optical depth as a function of wavelength (or frequency) will have a Voigt profile. The optically thin line will then have equivalent width

$$W_\lambda \approx \tau_0 \int d\lambda \, H(a, u) \qquad [\tau_0 < 1]. \qquad (2.55)$$

For an optically thin line with $a \ll 1$ (intrinsic broadening much smaller than thermal Doppler broadening), the only significant contribution to the absorption will be the Gaussian core, leading to

$$W_\lambda \approx \tau_0 \int d\lambda \, e^{-u^2} \qquad [\tau_0 < 1]. \qquad (2.56)$$

In terms of the wavelength $\lambda$, as opposed to the frequency $\nu$ or line-of-sight velocity $v$, the parameter $u$ in the Voigt–Hjerting function is

$$u = \frac{c}{b}\left(\frac{\lambda}{\lambda_0} - 1\right). \qquad (2.57)$$

Thus, Equation 2.56 reduces to

$$W_\lambda = \frac{b}{c}\lambda_0\tau_0 \int e^{-u^2} du = \frac{\sqrt{\pi}b}{c}\lambda_0\tau_0 \qquad [\tau_0 < 1]. \qquad (2.58)$$

Since the equivalent width of optically thin lines grows linearly with $\tau_0$, this is called the **linear part** of the curve of growth, as seen in Figure 2.7. When we

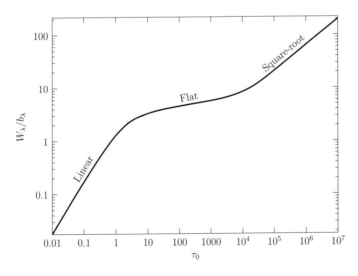

**Figure 2.7** Curve of growth for the hydrogen Lyman $\alpha$ line with $b = 10\,\mathrm{km\,s^{-1}}$, corresponding to $T \approx 6000\,\mathrm{K}$. The equivalent width is given in units of $b_\lambda \equiv b\lambda_0/c \approx 0.040\,\text{Å}$.

combine Equation 2.58 with the relation between $\tau_0$ and $N_\ell$, as given in Equation 2.48, we find

$$W_\lambda = \frac{1}{8\pi}\left[\frac{g_u}{g_\ell}\frac{c^2}{v_{u\ell}^3}A_{u\ell}\right]\lambda_0 N_\ell \qquad [\tau_0 < 1], \qquad (2.59)$$

with all dependence on the broadening parameter $b$, and hence on the temperature $T$, canceling out. The equivalent width, given our various assumptions, depends only on the properties of the absorbers and on the column density $N_\ell$ of absorbers in the lower-energy state. For Lyman $\alpha$ absorption, the relation between column density and equivalent width reduces to

$$N_\ell = 1.8 \times 10^{12}\,\text{cm}^{-2}\left(\frac{W_\lambda}{0.01\,\text{Å}}\right) \qquad [\tau_0 < 1]. \qquad (2.60)$$

This means that our requirement $N_\ell < 10^{12}\,\text{cm}^{-2}$ for optically thin Lyman $\alpha$ lines in the CNM reduces to a requirement that the equivalent width be several milliangstroms or less.

Now consider what happens when an absorption line is optically thick ($\tau_0 > 1$), but not so optically thick that the broad $v^{-2}$ wings from the intrinsic width provide a significant contribution to the absorption. The optical depth $\tau_0$ at which the $v^{-2}$ wings become important is called the **damping optical depth** $\tau_{damp}$. In the range $1 < \tau_0 < \tau_{damp}$, for a line with a Voigt profile, the equivalent width (Equation 2.53) is

$$W_\lambda = \int d\lambda\left[1 - \exp(-\tau_0 e^{-u^2})\right]$$
$$= \frac{b}{c}\lambda_0\int du\left[1 - \exp(-\tau_0 e^{-u^2})\right] \qquad [1 < \tau_0 < \tau_{damp}]. \quad (2.61)$$

(It isn't every day that we get to exponentiate an exponential.) The function in the square brackets in Equation 2.61 is significantly non-zero only when

$$\tau_0 e^{-u^2} > 1, \qquad (2.62)$$

or

$$u^2 < \ln\tau_0. \qquad (2.63)$$

With the approximation that the function in square brackets equals one when $u^2 < \ln\tau_0$ and is zero otherwise, we find

$$W_\lambda \approx \frac{2b}{c}\lambda_0\sqrt{\ln\tau_0} \qquad [1 < \tau_0 < \tau_{damp}]. \qquad (2.64)$$

In this optically thick regime, the equivalent width grows excruciatingly slowly with optical depth, going as the square root of the logarithm. Thus, this region of the curve of growth is called the **flat part**. As seen in Figure 2.7, it's pretty close to flat, but not perfectly so. In this flat part of the curve of growth, the translation from the observed equivalent width $W_\lambda$ to the central optical depth $\tau_0$ is risky, since

$$\tau_0 \approx \exp\left[\left(\frac{cW_\lambda}{2b\lambda_0}\right)^2\right] \qquad [1 < \tau_0 < \tau_{\text{damp}}], \qquad (2.65)$$

and any error in finding $W_\lambda$ (from mis-estimating the continuum flux, for instance) will propagate exponentially into an error in $\tau_0$. If we translate from optical depth to column density, using Equation 2.48, we find

$$N_\ell \approx (4\pi)^{3/2} \left[\frac{g_\ell}{g_u} \frac{v_{u\ell}^3}{c^2} \frac{1}{A_{u\ell}}\right] \frac{b}{c} \exp\left[\left(\frac{cW_\lambda}{2b\lambda_0}\right)^2\right] \qquad [1 < \tau_0 < \tau_{\text{damp}}].$$

$$(2.66)$$

On the flat part of the curve of growth, the column density at a given equivalent width *does* depend on the temperature, and thus on the thermal broadening parameter $b$. For the Lyman $\alpha$ line, Equation 2.66 becomes

$$N_\ell \approx 3.2 \times 10^{12}\,\text{cm}^{-2}\left(\frac{b}{1\,\text{km s}^{-1}}\right)\exp\left[1.5\left(\frac{1\,\text{km s}^{-1}}{b}\right)^2\left(\frac{W_\lambda}{0.01\,\text{Å}}\right)^2\right].$$

$$(2.67)$$

However, we advise you not to use the above equation unless you have a very good idea of what the equivalent width $W_\lambda$ and the thermal broadening $b$ are for the Lyman $\alpha$ line in question.

The flat part of the curve of growth does not last to arbitrarily high optical depth. At the damping optical depth, $\tau_{\text{damp}}$, the broad wings of the Voigt profile start to contribute significantly to the equivalent width. Let's suppose that the optical depth $\tau_\lambda$ doesn't fall below one until you are in the damping wings, where

$$\tau_\lambda \approx \tau_0 \frac{a}{\sqrt{\pi}u^2} \qquad [\tau_0 > \tau_{\text{damp}}]. \qquad (2.68)$$

Using this approximation for the optical depth $\tau_\lambda$, the equivalent width of a strongly damped line can be written as (Equation 2.53)

$$W_\lambda \approx \int d\lambda\left[1 - \exp(-\tau_\lambda)\right]$$

$$\approx \frac{b}{c}\lambda_0 \int du\left[1 - \exp\left(-\tau_0\frac{a}{\sqrt{\pi}u^2}\right)\right] \qquad [\tau_0 > \tau_{\text{damp}}]. \quad (2.69)$$

The optical depth approximation from Equation 2.68 admittedly overestimates $\tau_\lambda$ in the limit of small $u$. However, we don't care; as long as $\tau_\lambda \gg 1$, the function in square brackets in Equation 2.69 can be set equal to one, thus hiding the crudeness of our approximation from scrutiny. With the approximation that the function in square brackets equals one when $u^2 < \tau_0 a/\sqrt{\pi}$ and is zero otherwise, we find

$$W_\lambda \sim \frac{2b}{c} \lambda_0 \left(\frac{a\tau_0}{\sqrt{\pi}}\right)^{1/2} \qquad [\tau_0 > \tau_{\text{damp}}]. \qquad (2.70)$$

Thus, when the central optical depth $\tau_0$ is greater than a critical damping value $\tau_{\text{damp}}$, the equivalent width of the absorption line grows in proportion to the square root of $\tau_0$. Therefore this region of the curve of growth is called the **square-root part**, as displayed on the right in Figure 2.7. Being a little less cavalier about the integration, and substituting the definition of the parameter $a$, Equation 2.70 becomes

$$W_\lambda = \lambda_0 \left[\frac{b}{c} \frac{\gamma_{u\ell}}{\nu_{u\ell}} \frac{\tau_0}{\sqrt{\pi}}\right]^{1/2} \qquad [\tau_0 > \tau_{\text{damp}}]. \qquad (2.71)$$

This means that

$$\tau_0 = \sqrt{\pi} \frac{c}{b} \frac{\nu_{u\ell}}{\gamma_{u\ell}} \left(\frac{W_\lambda}{\lambda_0}\right)^2 \qquad [\tau_0 > \tau_{\text{damp}}]. \qquad (2.72)$$

In the square-root part of the curve of growth, the computed optical depth depends on the square of the measured equivalent width. The column density then becomes, for our usual Lyman $\alpha$ example,

$$N_\ell = 1.9 \times 10^{18}\,\text{cm}^{-2} \left(\frac{W_\lambda}{1\,\text{Å}}\right)^2 \qquad [\tau_0 > \tau_{\text{damp}}], \qquad (2.73)$$

which is independent of the thermal broadening parameter $b$. The damping optical depth at which the flat part of the curve of growth gives way to the square-root portion depends on the broadening $b$ as well as on the properties of the absorber. For the Lyman $\alpha$ line,

$$\tau_{\text{damp}} \approx 455 \left(\frac{b}{1\,\text{km s}^{-1}}\right) \left[1 + 0.20 \ln \left(\frac{b}{1\,\text{km s}^{-1}}\right)\right], \qquad (2.74)$$

which in the range 80 K to $10^5$ K goes from $\tau_{\text{damp}} \sim 500$ to $\tau_{\text{damp}} \sim 30\,000$.

Observing Lyman $\alpha$ lines from neutral hydrogen is useful for studying the cooler parts of the interstellar medium in our galaxy. Since Lyman $\alpha$ represents a transition from the ground state of the most abundant element in the ISM, Lyman $\alpha$ absorption lines tend to be optically thick. Figure 2.8 shows as an example the Lyman $\alpha$ lines seen in the directions of Alpha Centauri A and Alpha Centauri B, our near neighbors among the stars. Alpha Cen A and B have broad Lyman $\alpha$ emission lines from their hot chromospheres. Superimposed on their emission lines, however, are optically thick absorption lines from the atomic hydrogen of the ISM along the line of sight to the stars. The Lyman $\alpha$ emission lines of main sequence stars are non-trivial to model, so the main source of error in determining the equivalent width $W_\lambda$ of the absorption lines is modeling the

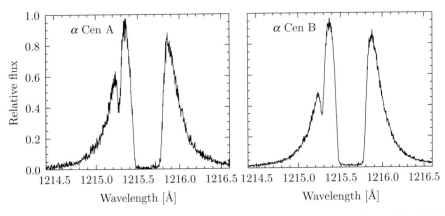

**Figure 2.8** Lyman $\alpha$ lines toward Alpha Centauri A (left) and Alpha Centauri B (right). Observations were made with the High-Resolution Spectrograph on the *Hubble Space Telescope*. [Following Linsky & Wood 1996]

stellar emission lines. For both Alpha Cen A and Alpha Cen B, the equivalent width of the Lyman $\alpha$ line is $W_\lambda \approx 0.3$ Å.

By fitting the absorption lines shown in Figure 2.8, Linsky and Wood found that the best fit, assuming a single value of the broadening parameter $b$, corresponded to $\tau_0 = 68\,000$, $b = 11.8\,\mathrm{km\,s^{-1}}$ (and thus $T = 8300\,\mathrm{K}$), and $N_\ell = 1.1 \times 10^{18}\,\mathrm{cm^{-2}}$; this represents the regime where the flat part of the curve of growth gives way to the square-root part. The calculated temperature is much higher than the $T \sim 80\,\mathrm{K}$ of the cold neutral medium; within our Local Bubble, there isn't much cold gas. If the hydrogen column density were due to gas of uniform density along the entire $d = 1.34\,\mathrm{pc}$ distance to Alpha Centauri, it would provide a number density $n_\ell \sim 0.25\,\mathrm{cm^{-3}}$. For stars outside the Local Bubble at $d \sim 100\,\mathrm{pc}$, the Lyman $\alpha$ absorption lines are well into the square-root part of the curve of growth, with $W_\lambda \sim 10$ Å, $N_\ell \sim 2 \times 10^{20}\,\mathrm{cm^{-2}}$, and $n_\ell \sim 0.6\,\mathrm{cm^{-3}}$.

On the left-hand slope of each Lyman $\alpha$ emission line in Figure 2.8, there is an optically thin absorption line. This is the absorption line of *deuterium* Lyman $\alpha$. The deuterium Lyman lines are slightly blueshifted relative to their ordinary hydrogen counterparts because of the isotopic shift due to the neutron in the deuterium nucleus. Where H Lyman $\alpha$ is at $\lambda_0 = 1215.67$ Å, deuterium Lyman $\alpha$ is at $\lambda_0 = 1215.34$ Å. The deuterium Lyman $\alpha$ lines in the spectra of Alpha Cen A and B are optically thin, with $\tau_0 = 0.68$, and thus easier to interpret than the optically thick Lyman $\alpha$ lines from ordinary hydrogen. The column density of deuterium toward Alpha Centauri, as deduced by Linsky and Wood, is $N_\ell = 6.1 \times 10^{12}\,\mathrm{cm^{-2}}$, giving a deuterium to hydrogen ratio $n_\mathrm{D}/n_\mathrm{H} \approx 6 \times 10^{-6}$ along this line of sight. This is lower than the usual $n_\mathrm{D}/n_\mathrm{H} \approx 1.6 \times 10^{-5}$ in the Local Bubble, and is much less than the primordial value of $n_\mathrm{D}/n_\mathrm{H} \approx 2.5 \times 10^{-5}$.

So, visible and UV absorption lines tell us interesting information about the cooler regions of the ISM. How much of each element and isotope is present? How hot is the gas? What are the integrated densities along the line of sight? However, we have more ranges of the electromagnetic spectrum that are of use to us. In particular, as the next chapter will show, radio waves are useful for tracing the warm neutral medium.

## Exercises

2.1 The energy density of starlight in our location in the Milky Way Galaxy is $\varepsilon_\star = 0.6\,\mathrm{eV\,cm^{-3}}$. (This excludes the contribution of sunlight.) The apparently brightest star in our night sky is Sirius, at a distance $d = 2.64$ pc. Sirius is actually a binary system: Sirius A is a main sequence star with luminosity $L_A = 25.4\,L_\odot$ while Sirius B is a white dwarf with luminosity $L_B = 0.056\,L_\odot$.

   (a) What fraction of $\varepsilon_\star$ at the solar system's location comes from Sirius A? What fraction of $\varepsilon_\star$ comes from Sirius B?

   (b) At our location in the Milky Way Galaxy, the energy density of photons with $11.26\,\mathrm{eV} < h\nu < 13.60\,\mathrm{eV}$, which are thus capable of ionizing carbon, is $\varepsilon_C = 0.006\,\mathrm{eV\,cm^{-3}}$. (Again, this excludes the Sun's contribution.) Approximating Sirius A as a blackbody with $T_A = 9900$ K and Sirius B as a blackbody with $T_B = 25\,000$ K, what fraction of $\varepsilon_C$ comes from Sirius A? What fraction of $\varepsilon_C$ comes from the white dwarf Sirius B?

2.2 A gas made of two-level atoms (with level spacing $h\nu_{u\ell}$) has excitation temperature $T_{\mathrm{exc}}$. The atomic gas is in equilibrium with a radiation field, so that the rate of photon absorption (Equation 2.18) is equal to the rate of photon emission (Equation 2.19). Show that if stimulated emission is negligible then the energy density of the radiation field at frequency $\nu = \nu_{u\ell}$ is that of a blackbody in the Wien tail, with temperature $T = T_{\mathrm{exc}}$.

2.3 The total H I column density along the line of sight to a star is $3 \times 10^{19}\,\mathrm{cm^{-2}}$; the thermal broadening of the gas is $b = 10\,\mathrm{km\,s^{-1}}$.

   (a) Calculate the damping width $\gamma_{u\ell}$ and the dimensionless parameter $a$ (Equation 2.34) for the H I Ly $\alpha$, Ly $\beta$, and Ly $\gamma$ absorption lines toward this star. Atomic data for H I, including the $A_{u\ell}$ transition probabilities, are available online from the NIST Atomic Spectrum Database (physics.nist.gov/PhysRefData/ASD/lines_form.html).

   (b) The deuterium abundance along this line of sight is $n_D/n_H = 2 \times 10^{-5}$. Do you expect the D I Lyman series lines (the analogues of the H I Lyman series) to be in the linear, flat, or square-root part of the curve of growth in the spectrum of this star? For simplicity, assume that you can adopt the H I transition probabilities for the corresponding D I transitions.

(c) Using the results above, estimate the expected absorption-line equivalent widths of H I Ly $\alpha$, Ly $\beta$, and Ly $\gamma$, and of the corresponding deuterium lines, D I Ly $\alpha$, Ly $\beta$, and Ly $\gamma$, in the spectrum of this star, assuming $n_D/n_H = 2 \times 10^{-5}$. For simplicity, assume that you can adopt the H I transition probabilities for the corresponding D I transitions.

(d) Use the NIST database you employed in part (a) to look up the D I transition probabilities for the first three Lyman series lines ($\alpha$, $\beta$, and $\gamma$). Is the assumption that you could adopt the H I transition probabilities for D I valid? How much difference does it make to use the tabulated D I probabilities instead? Be quantitative.

2.4 The "Strömgren doublet" method is used to estimate the column density of an element with a closely spaced doublet of absorption lines arising from transitions between a single ground state and an excited state split into a pair of fine-structure levels. This method is commonly applied to the $D_1$ and $D_2$ doublet of Na I. If the equivalent widths of the $D_1$ and $D_2$ lines are $W_{\lambda,1}$ and $W_{\lambda,2}$ then the ratio $W_{\lambda,2}/W_{\lambda,1}$ is a monotonic function of the optical depth $\tau_0$ of the $D_1$ line (which is the weaker of the two lines). The relation between the equivalent width ratio and the optical depth $\tau_0$ was computed for the Na I D doublet by Bengt Strömgren (1948, ApJ, 108, 242; Table 2).

(a) Observations toward a planetary nebula show two components in the Na I D lines: a strong interstellar component and a weaker circumstellar component from neutral gas surrounding the nebula (slightly blueshifted relative to the interstellar component). The measured equivalent widths for each component are given below. Using Strömgren's 1948 table, estimate the column densities of Na I and H I in each component, assuming $n_{NaI}/n_{HI} = 1 \times 10^{-9}$. Include an estimate of your uncertainties.

| Region | $W_{\lambda,2}$ | $W_{\lambda,1}$ |
|---|---|---|
| Interstellar | 386±10 mÅ | 300±10 mÅ |
| Circumstellar | 275±10 mÅ | 250±10 mÅ |

(b) Suppose that you are able to measure the doublet ratio $W_{\lambda,2}/W_{\lambda,1}$ to an accuracy of ±10%. Over what range in $\tau_0$ is this measurement accuracy sufficient to derive a meaningful estimate of $\tau_0$? What is the corresponding range in total column density $N(H I)$? For a typical matter density in the galactic plane, to what distance in parsecs can you apply this method? [Hint: a graphical approach to this problem may be helpful.]

# 3

# Warm Neutral Medium

*In enterprise of martial kind,*
*When there was any fighting,*
*He led his regiment from behind –*
*He found it less exciting.*

W. S. Gilbert (1836–1911)
"The Gondoliers," Act I [1889]

The cold neutral medium (CNM) and warm neutral medium (WNM), taken together, provide over half the mass of our galaxy's interstellar medium. In the CNM and WNM, most of the hydrogen is in the form of neutral atoms. Thus, as we've seen, the Lyman $\alpha$ line of hydrogen provides a useful probe of the properties of the CNM and WNM. However, the hydrogen Lyman $\alpha$ line has some drawbacks. For one thing, it has a wavelength $\lambda_0 = 1216\,\text{Å}$ at which the Earth's atmosphere is highly opaque.[1] Thus, observing Lyman $\alpha$ absorption within our galaxy requires the use of orbiting UV satellites, which are in scarce supply. In addition, Lyman $\alpha$ can be seen in absorption only along those lines of sight toward sources with a high ultraviolet flux.

To do a global survey of atomic hydrogen in the galaxy, we need some way of easily detecting radiation from hydrogen, regardless of its kinetic temperature or number density. Such a way was found in the year 1944 by Henk van de Hulst, then a graduate student at Utrecht University in the Netherlands. The research that van de Hulst was doing for his dissertation (on the scattering of light by dust grains) had stalled when his research advisor, Marcel Minnaert, was detained in a hostage camp. As an interim project, van de Hulst was steered toward an interesting problem in what was then the barely existing field of radio astronomy. What strong emission lines could one expect to find at the wavelengths $\sim 1\,\text{cm}$ to $20\,\text{m}$ at which the Earth's atmosphere is transparent to radio waves? Running

---

[1] Lyman $\alpha$ photons from space make it as far as the Earth's ionosphere, where they photoionize nitric oxide at an altitude $h \sim 70\,\text{km}$.

through all possibilities, van de Hulst came to the realization that the **hyperfine** line resulting from a flip of the electron spin within a hydrogen atom should be detectable at $\lambda_0 \sim 21$ cm.[2] The prediction of this line was confirmed by Ewen and Purcell in 1951, when they first detected 21 cm emission from the Milky Way.

## 3.1 Twenty-One Centimeter Emission and Absorption

Understanding the hyperfine transition requires a little quantum mechanics. An electron and a proton each have spin quantum number $s = 1/2$; in the ground state of hydrogen, the electron has no orbital angular momentum, so only the spin angular momenta of the electron and proton come into play. Because they have non-zero spins, both the electron and the proton have non-zero magnetic moments. The magnetic moment of the electron, $\mu_e$, is much larger than that of the proton, with $|\mu_e| \approx 660\mu_p$. In a classical cartoon version, the electron is a strong bar magnet zipping around in three dimensions in the magnetic field produced by the proton's weak bar magnet. Integrating over the complete probability distribution of the electron, it is found that the higher energy state occurs when the spins are parallel, and thus the magnetic moments are antiparallel. The difference in energy between the spin-parallel state and the spin-antiparallel state is $E_{u\ell} = E_u - E_\ell = 5.8743 \times 10^{-6}$ eV, corresponding to a frequency $\nu_{u\ell} = E_{u\ell}/h = 1420.4$ MHz or wavelength $\lambda_{u\ell} = c/\nu_{u\ell} = 21.106$ cm. (When speaking casually rather than calculating rigorously, astronomers round off these numbers and talk about the "1420 megahertz" or "21 centimeter" emission from hydrogen atoms.)

Properties of the hyperfine transition between the spin-parallel and spin-antiparallel state of hydrogen are shown in Figure 3.1. In addition to having a small energy difference, $E_{u\ell} = 5.87$ µeV, the two-level absorber provided by the hyperfine splitting has a tiny Einstein A coefficient: $A_{u\ell} = 2.88 \times 10^{-15}$ s$^{-1}$. Although it is energetically favorable for a hydrogen atom in the spin-parallel state to convert to the spin-antiparallel state, an isolated H atom in the higher energy state will have a mean lifetime $A_{u\ell}^{-1} \sim 11$ Myr before spontaneously emitting a 21 cm photon.

Although both the Lyman $\alpha$ transition and the 21 cm hyperfine transition are important probes of neutral hydrogen, there are obvious physical differences between these transitions. The most important difference, from an astronomical point of view, is the factor $1.7 \times 10^6$ difference in the energies of a Lyman $\alpha$ photon and a 21 cm photon. The Lyman $\alpha$ energy $E = 10.2$ eV has a temperature equivalent $E/k = 1.18 \times 10^5$ K. This is much higher than the kinetic temper-

---

[2] A **fine-structure** transition, such as the transition that produces the C II 158 µm line, is the result of spin–orbit coupling; an energy level $E_n$ will show fine-structure splitting $\Delta E \propto \alpha^2 E_n$, where $\alpha \approx 1/137$ is the fine-structure constant. A **hyperfine** transition results from spin–spin coupling and gives rise to a splitting $\Delta E \propto \alpha^2 (m_e/m_p) E_n$.

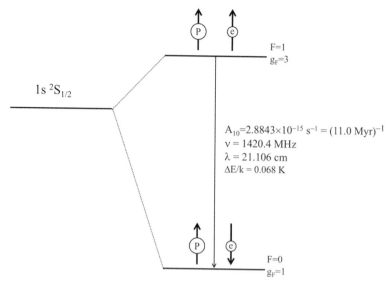

**Figure 3.1** Properties of the spin-flip transition in the ground state of the hydrogen atom. The parameter $F$ is the hyperfine quantum number; $F = 0$ for the spin-antiparallel state and $F = 1$ for the spin-parallel state.

ature of the neutral ISM; thus, collisional excitation is unimportant, and most hydrogen atoms are in the lower level of the Lyman $\alpha$ transition. Only rarely does an ultraviolet photon come along to excite the atom to the upper Lyman $\alpha$ level. By contrast, the 21 cm energy $E = 5.87 \, \mu eV$ has a temperature equivalent $E/k = 0.0682$ K. This is much lower than the kinetic temperature of the neutral ISM, as well as being much lower than the temperature of the cosmic microwave background. Thus, there is ample opportunity to populate the upper energy level of the 21 cm hyperfine transition.

In excitation equilibrium, the ratio of the populations of the 21 cm hyperfine levels would be

$$\frac{n_u}{n_\ell} = \frac{g_u}{g_\ell} e^{-h\nu/kT} = 3e^{-0.0682 \, \text{K}/T} \approx 3, \tag{3.1}$$

since the kinetic temperature $T$ everywhere in the ISM is much greater than 0.0682 K. In reality, the hyperfine levels are not exactly in excitation equilibrium. In this case, it is usual to describe the relative level populations in terms of an excitation temperature,[3] as described in Section 1.2. If we know $n_u/n_\ell$, we can define $T_{\text{exc}}$ for the 21 cm line using the relation

$$\frac{n_u}{n_\ell} \equiv \frac{g_u}{g_\ell} e^{-h\nu/kT_{\text{exc}}} = 3e^{-0.0682 \, \text{K}/T_{\text{exc}}}. \tag{3.2}$$

---

[3] Radio astronomers studying the 21 cm line often use the term **spin temperature** rather than the more general term "excitation temperature." It's the same concept, despite the difference in names.

Within the ISM, the excitation temperature $T_{exc}$ is generally much greater than 0.0682 K, so we can safely write $n_u/n_\ell \approx g_u/g_\ell \approx 3$. In any population of neutral hydrogen atoms in the $n = 1$ state, three-fourths will be in the upper hyperfine level and one-fourth will be in the lower hyperfine level. Consider a sphere carved from the warm neutral medium, with radius $r \approx 3$ pc and containing $1.2 \times 10^{57}$ hydrogen atoms. (We choose this size of sphere because it contains $1\,M_\odot$ of hydrogen, a good round number.) At the temperature of the WNM, three-fourths of the H atoms, or $N_u = 9 \times 10^{56}$, will be in the upper hyperfine level of the electronic ground state. Multiplying by the Einstein A coefficient, we find that 21 cm photons are spontaneously emitted at a rate $R \approx 2.6 \times 10^{42}\,\text{s}^{-1}$, corresponding to a luminosity $L \approx 2.4 \times 10^{25}\,\text{erg s}^{-1}$. Expressed as a fraction of the Sun's bolometric luminosity ($1\,L_\odot = 3.828 \times 10^{33}\,\text{erg s}^{-1}$), this can be written as $L \approx 6.4 \times 10^{-9}\,L_\odot$. Thus, the 21 cm emission from the WNM has a huge mass-to-light ratio compared to that of a Sun-like star. However, all the 21 cm photons come out in a small range of frequencies, which makes detection easier.

An all-sky map of the 21 cm emission is shown in Figure 3.2. As frequently happens when plotting data from 21 cm surveys, Figure 3.2 isn't simply a map of the flux at $\lambda_0 = 21.106$ cm. Instead, it is a plot of the column density of neutral hydrogen, deduced from the properties of the 21 cm line. The column density is $N_{HI} < 10^{20}\,\text{cm}^{-2}$ in some directions at high galactic latitude, away from the midplane of our galaxy's disk. For instance, the Lockman Hole, at galactic latitude $b \approx 52°$ and longitude $\ell \approx 150°$, has column density $N_{HI} \approx 6 \times 10^{19}\,\text{cm}^{-2}$. By contrast, looking through the midplane of our galaxy, at $b = 0°$,

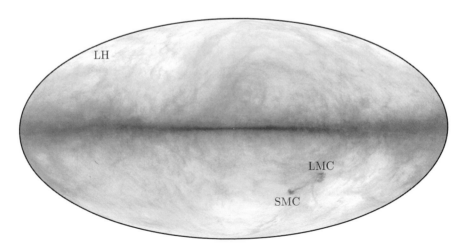

**Figure 3.2** All-sky map of the H I column density, deduced from hydrogen 21 cm emission. Column density is plotted on a logarithmic scale from $N_{HI} = 5 \times 10^{19}\,\text{cm}^{-2}$ (white) to $N_{HI} = 5 \times 10^{22}\,\text{cm}^{-2}$ (black). LMC = Large Magellanic Cloud, SMC = Small Magellanic Cloud, LH = Lockman Hole. [HI4PI collaboration 2016]

the column density can be as high as $N_{HI} \sim 10^{23}\,\text{cm}^{-3}$. Going from the detection of 21 cm photons to the deduction of a neutral hydrogen column density isn't a completely trivial task. In addition to considering spontaneous emission, we must also consider absorption and stimulated emission.

The hyperfine levels of the ground state of hydrogen (Figure 3.1) represent a simple two-level absorber. If the upper level were unpopulated then the attenuation coefficient would be purely due to the absorption of photons: $\kappa_v = n_\ell \sigma_{\ell u}$, as given in Equation 2.40. However, at the same time that absorption is populating the upper level at the rate

$$\left( \frac{dn_u}{dt} \right)_{abs} = n_\ell B_{\ell u} \varepsilon_v, \tag{3.3}$$

stimulated emission is depopulating the upper level at the rate

$$\left( \frac{dn_u}{dt} \right)_{se} = -n_u B_{u\ell} \varepsilon_v = -n_u \left( \frac{g_\ell}{g_u} B_{\ell u} \right) \varepsilon_v$$

$$= - \left( \frac{n_u}{n_\ell} \frac{g_\ell}{g_u} \right) \left( \frac{dn_u}{dt} \right)_{abs}. \tag{3.4}$$

Thus, if the two levels were populated exactly according to their statistical weights, as they would be in the limit of infinite excitation temperature, absorption and stimulated emission would exactly cancel and the net attenuation coefficient would be zero. In the case of finite excitation temperature, we can write

$$\kappa_v = n_\ell \sigma_{\ell u} - n_u \sigma_{u\ell}, \tag{3.5}$$

where $\sigma_{u\ell}$, the cross section for stimulated emission, is given by

$$\sigma_{u\ell} = \frac{g_\ell}{g_u} \sigma_{\ell u}. \tag{3.6}$$

Thus, in a system with both absorption and stimulated emission,

$$\kappa_v = n_\ell \sigma_{\ell u} \left[ 1 - \frac{g_\ell}{g_u} \frac{n_u}{n_\ell} \right]. \tag{3.7}$$

Using the definition of excitation temperature (Equation 1.19), we can write the attenuation coefficient as

$$\kappa_v = n_\ell \sigma_{\ell u} \left[ 1 - \exp \left( -\frac{h\nu_{u\ell}}{kT_{exc}} \right) \right]. \tag{3.8}$$

If the excitation temperature $T_{exc}$ is constant along the line of sight, this produces a corresponding optical depth

$$\tau_v = \sigma_{\ell u} \left[ 1 - \exp \left( -\frac{h\nu_{u\ell}}{kT_{exc}} \right) \right] N_\ell, \tag{3.9}$$

where $N_\ell = \int n_\ell ds$ is the column density of absorbers in their lower state.

When we considered the Lyman $\alpha$ line, we were looking at the case where $kT_{exc}$ was much less than $h\nu_{u\ell} = 10.2\,\text{eV}$. Thus, the correction to $\kappa_v$ from stimulated emission was negligible, and we could use the absorption-only approximation, $\kappa_\nu = n_\ell \sigma_{\ell u}$ (Equation 2.40). However, when looking at the 21 cm line, we are in the regime where $kT_{\text{exc}}$ is much greater than $h\nu_{u\ell} = 5.87$ μeV. Thus, the correction from stimulated emission is now very important. With $kT_{\text{exc}} \gg h\nu_{u\ell}$,

$$\kappa_\nu \approx n_\ell \sigma_{\ell u} \left( \frac{h\nu_{u\ell}}{kT_{\text{exc}}} \right) \ll n_\ell \sigma_{\ell u}. \tag{3.10}$$

For 21 cm absorption lines, the attenuation coefficient $\kappa_\nu$, and the resulting optical depth $\tau_\nu$ along a line of sight, is thus smaller by a factor $h\nu_{u\ell}/kT_{\text{exc}}$ than it would be in the absence of stimulated emission. Given this reduction in $\kappa_\nu$, can we expect optically thick 21 cm absorption lines along any line of sight through our galaxy? Let's go back to Equations 2.47 and 2.48, and write the optical depth as a function of the quantum mechanical properties of the hyperfine transition of the ground state of hydrogen (remembering to include the $h\nu_{u\ell}/kT_{\text{exc}}$ correction factor for stimulated emission):

$$\tau_\nu = \tau_0 e^{-v^2/b^2}, \tag{3.11}$$

where

$$\tau_0 = \frac{1}{(4\pi)^{3/2}} \left[ \frac{g_u}{g_\ell} \frac{c^2}{v_{u\ell}^2} A_{u\ell} \right] \frac{c}{b} N_\ell \left( \frac{h\nu_{u\ell}}{kT_{\text{exc}}} \right). \tag{3.12}$$

We have assumed, since the intrinsic width of the 21 cm line is so tiny, that the line width is entirely due to thermal Doppler broadening. Plugging in the appropriate properties for the 21 cm line, the optical depth at line center becomes

$$\tau_0 \approx 0.31 \left( \frac{N_{\text{HI}}}{10^{21}\ \text{cm}^{-2}} \right) \left( \frac{10\ \text{km s}^{-1}}{b} \right) \left( \frac{100\ \text{K}}{T_{\text{exc}}} \right). \tag{3.13}$$

For thermal broadening $b$ values typical of the warm neutral medium and excitation temperatures $T_{\text{exc}} \sim 100$ K, lines of sight with $N_{\text{HI}} > 10^{21}$ cm$^{-2}$ show significant absorption. (Remember from Section 2.3 that Lyman $\alpha$ becomes optically thick at a column density of only $N_{\text{HI}} \sim 10^{12}$ cm$^{-2}$.) As shown in Figure 3.2, column densities as large as this are found for lines of sight running through the disk of our galaxy.

## 3.2 Radiative Transfer of Line Emission

From where we sit in the Milky Way Galaxy, at 21 cm lines of sight at high galactic latitude are optically thin and lines of sight at low galactic latitude are optically thick. We can see what this means for the observed properties of 21 cm lines by starting with the equation of radiative transfer:

$$\frac{dI_\nu}{d\tau_\nu} = -I_\nu + S_\nu, \tag{3.14}$$

where the source function $S_\nu \equiv j_\nu / \kappa_\nu$. The emissivity accounts for those photons produced by spontaneous emission and can be written as

$$j_\nu = n_u \frac{A_{u\ell}}{4\pi} h\nu_{u\ell} \Phi_\nu. \tag{3.15}$$

For 21 cm emission, with $kT_{\text{exc}} \gg h\nu_{u\ell}$, the attenuation coefficient is

$$\kappa_\nu \approx n_\ell \sigma_{\ell u} \left( \frac{h\nu_{u\ell}}{kT_{\text{exc}}} \right) \approx n_\ell \frac{g_u}{g_\ell} \frac{c^2}{8\pi \nu_{u\ell}^2} A_{u\ell} \Phi_\nu \left( \frac{h\nu_{u\ell}}{kT_{\text{exc}}} \right), \tag{3.16}$$

where we have inserted the expression for the absorption cross section $\sigma_{\ell u}$ in terms of the quantum mechanical properties of the absorber (Equation 2.26). Dividing Equation 3.15 by Equation 3.16, we find that the source function for 21 cm emission is

$$S_\nu \approx \frac{n_u}{n_\ell} \frac{g_\ell}{g_u} \frac{2\nu_{u\ell}^2}{c^2} kT_{\text{exc}}. \tag{3.17}$$

However, since $n_u / n_\ell \approx g_u / g_\ell$ for $kT_{\text{exc}} \gg h\nu_{u\ell}$, we can safely rewrite this as

$$S_\nu \approx \frac{2\nu_{u\ell}^2}{c^2} kT_{\text{exc}}. \tag{3.18}$$

This is a particularly useful result if the excitation temperature is constant along the line of sight in question, since it means that the source function $S_\nu$ can be taken to be constant in the equation of radiative transfer:

$$\frac{dI_\nu}{d\tau_\nu} = -I_\nu + \frac{2\nu_{u\ell}^2}{c^2} kT_{\text{exc}}. \tag{3.19}$$

The equation of radiative transfer for 21 cm radiation can be written in an even simpler form if we change our variable from the intensity $I_\nu$ to the **antenna temperature** $T_A(\nu)$.

The antenna temperature is another of those "temperatures" used, slightly confusingly, to refer to parameters that aren't necessarily identical to the kinetic temperature $T$. The antenna temperature is closely related to the **brightness temperature**, which is an alternative way of stating the intensity $I_\nu$ of radiation at frequency $\nu$. The brightness temperature $T_b(\nu)$ is defined as the temperature for which an observed intensity $I_\nu$ is equal to a Planck function at the same frequency. That is,

$$I_\nu = B_\nu(T_b), \tag{3.20}$$

where $B_\nu$ is given in Equation 2.7. In the Rayleigh–Jeans tail of the Planck function, where $h\nu \ll kT_b$, the brightness temperature is given by the simple relation (Equation 2.8):

$$I_\nu = \frac{2\nu^2}{c^2} kT_b. \tag{3.21}$$

Since radio astronomers are used to working at very low frequencies, they define the antenna temperature $T_A$ as being equal to the brightness temperature of a blackbody in the Rayleigh–Jeans tail. Thus, the antenna temperature is, by definition,

$$T_A \equiv \frac{c^2}{2k} \frac{I_\nu}{\nu^2}. \qquad (3.22)$$

This may seem like an odd thing to do; why should we talk about a fictitious antenna "temperature" when we could talk about a real intensity instead? The payoff comes when we convert the equation of radiative transfer (Equation 3.19) from intensities to antenna temperatures:

$$\frac{2\nu^2 k}{c^2} \frac{dT_A}{d\tau_\nu} = -\frac{2\nu^2 k}{c^2} T_A + \frac{2\nu_{u\ell}^2 k}{c^2} T_{\text{exc}}. \qquad (3.23)$$

Within our galaxy, the 21 cm line is sufficiently narrow that we can set $\nu \approx \nu_{u\ell}$ across the entire line width, reducing the equation of radiative transfer to the exquisitely simple form

$$\frac{dT_A}{d\tau_\nu} = -T_A + T_{\text{exc}}. \qquad (3.24)$$

If we can assume a constant excitation temperature along the line of sight,

$$T_A = T_A(0)e^{-\tau_\nu} + T_{\text{exc}}[1 - e^{-\tau_\nu}]. \qquad (3.25)$$

If we can't assume a constant excitation temperature then solving the equation of radiative transfer gets uglier but is still possible.

It's time for an example. Suppose you are looking through an optically thin layer of neutral hydrogen (with $\tau_\nu \ll 1$) toward a dark background (with $T_A(0) \ll \tau_\nu T_{\text{exc}}$); for instance, you are looking at high galactic latitude in a direction where there are no bright background sources at 21 cm. In that case, you will detect an emission line from the hydrogen present along the line of sight. When $\tau_\nu \ll 1$, Equation 3.25 reduces to

$$T_A \approx \tau_\nu T_{\text{exc}}. \qquad (3.26)$$

We saw in Equation 3.12 that the optical depth $\tau_\nu$ is inversely proportional to $T_{\text{exc}}$ for the 21 cm line. Thus, the observed antenna temperature (or intensity) at a given frequency will be independent of the excitation temperature along the line of sight through an optically thin layer of atomic hydrogen. Integrating the antenna temperature over all velocities, we find

$$\int T_A dv = T_{\text{exc}} \int \tau_\nu dv = T_{\text{exc}} W_v, \qquad (3.27)$$

where $W_v = c(W_\lambda/\lambda_0)$ is the equivalent width expressed in velocity units. For a thermally broadened optically thin line, $W_v = \sqrt{\pi} b \tau_0$ and

$$\int T_A dv = \sqrt{\pi} T_{exc} b \tau_0$$

$$= \frac{c}{8\pi} \left[ \frac{g_u}{g_\ell} \frac{c^2}{v_{u\ell}^2} A_{u\ell} \right] \frac{h v_{u\ell}}{k} N_\ell, \qquad (3.28)$$

where we have made use of Equation 3.12, which gives the optical depth at line center for the 21 cm line. Not only has the excitation temperature $T_{exc}$ dropped out of Equation 3.28, so also has the broadening parameter $b$. Numerically, plugging in the appropriate properties for 21 cm emission,

$$\int T_A dv = 550 \, \text{K km s}^{-1} \left( \frac{N_{HI}}{10^{21} \, \text{cm}^{-2}} \right). \qquad (3.29)$$

This implies that, in the optically thin case, the column density of neutral hydrogen is linearly proportional to the integrated line intensity of the 21 cm line:

$$N_{HI} = 1.82 \times 10^{18} \, \text{cm}^{-2} \left( \frac{\int T_A dv}{1 \, \text{K km s}^{-1}} \right). \qquad (3.30)$$

Moreover, the proportionality constant depends solely on the quantum mechanical properties of the 21 cm spin-flip transition. You don't have to make any dubious assumptions about temperature or excitation states, just the sensible assumption that $T_{exc} \gg 68.2 \, \text{mK}$ (see the text under Equation 3.2) and the easily checkable assumption that the neutral hydrogen is optically thin. Thus, for example, the observation that the 21 cm integrated line intensity toward the Lockman Hole is $\int T_A dv = 33 \, \text{K km s}^{-1}$ leads directly to a deduced column density $N_{HI} = 6 \times 10^{19} \, \text{cm}^{-2}$.

Another useful example involves looking at a compact extragalactic radio source through an optically thin layer of neutral hydrogen within our galaxy. The radio continuum of the extragalactic source will have an absorption line due to neutral hydrogen along the line of sight. Figure 3.3 shows such an absorption feature toward BL Lacertae, a radio-loud active galactic nucleus at $d \sim 300 \, \text{Mpc}$. BL Lac, at observed frequencies $v \sim 1420 \, \text{MHz}$, produces continuum radio emission with an antenna temperature $T_C$ that can be taken as constant over frequency. Thus, the antenna temperature observed after light from BL Lac passes through the neutral hydrogen in our galaxy is

$$T_A(v) = T_C e^{-\tau_v} + T_{exc}[1 - e^{-\tau_v}]. \qquad (3.31)$$

However, the antenna temperature $T_C$ of the continuum from BL Lac is much greater than the excitation temperature $T_{exc} \sim 100 \, \text{K}$ for the hydrogen in our galaxy. Thus, we may write

$$T_A(v) = T_C e^{-\tau_v}. \qquad (3.32)$$

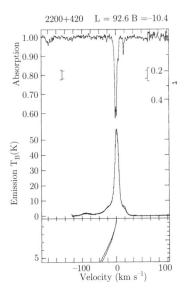

**Figure 3.3** H I 21 cm absorption and emission along the line of sight toward BL Lacertae. [Dickey et al. 1983]

In the top panel of Figure 3.3, the label on the left shows the fractional absorption, $e^{-\tau_v}$; the right-hand scale shows $\tau_v$. The absorption at line center toward BL Lac is $\tau_0 \approx 0.5$, so we may use the optically thin approximation with just a slight shiver of trepidation. The equivalent width of the absorption line is $W_v = 7\,\mathrm{km\,s^{-1}}$ in velocity units, corresponding to $W_\lambda = W_v \lambda_0/c = 0.0005\,\mathrm{cm}$.

The middle panel of Figure 3.3 shows the 21 cm emission line just to one side of BL Lac. The integrated line intensity of the emission line is $\int T_A dv \approx 930\,\mathrm{K\,km\,s^{-1}}$, leading to a deduced column density $N_{\mathrm{HI}} \approx 1.7 \times 10^{21}\,\mathrm{cm^{-2}}$ along a line of sight close to BL Lac, which is at galactic latitude $b = -10°$. Since we know the equivalent width $W_v$ from the absorption and the integrated line intensity from the emission, we can compute

$$T_{\mathrm{exc}} = \frac{\int T_A dv}{W_v} = \frac{930\,\mathrm{K\,km\,s^{-1}}}{7\,\mathrm{km\,s^{-1}}} \approx 130\,\mathrm{K}. \tag{3.33}$$

It is observations of this kind that enable astronomers to say that the excitation temperature of the hyperfine levels of neutral hydrogen in our galaxy is about 100 K. There is an assumption that goes into this calculation, however. We are assuming that the properties along the line of sight to BL Lacertae are the same as along the line of sight to the blank sky nearby.

## 3.3 Exciting Hyperfine Energy Levels

Your reaction to the heading of this section may be "What's so exciting about hyperfine levels?" This section, to clarify, is not about why hyperfine levels are exciting, but about how hyperfine levels are excited. If you have a generic two-level absorber, with the levels separated by an energy $h\nu_{u\ell} = E_u - E_\ell$, then excitation and de-excitation can be **radiative** or **collisional**.

Let's look first at radiative excitation (otherwise known as absorption) and radiative de-excitation (otherwise known as spontaneous emission and stimulated emission). Suppose that the population of absorbers is bathed with photons that have an energy density $\varepsilon_\nu$. The net rate of change for the number of absorbers in the upper energy level is, from Equations 2.18 and 2.19,

$$\frac{dn_u}{dt} = n_\ell B_{\ell u}\varepsilon_\nu - n_u(A_{u\ell} + B_{u\ell}\varepsilon_\nu). \qquad (3.34)$$

Using the relations among the Einstein coefficients (Equations 2.20 and 2.21), we find

$$\frac{dn_u}{dt} = n_\ell A_{u\ell}\frac{g_u}{g_\ell}\frac{c^3}{8\pi h\nu_{u\ell}^3}\varepsilon_\nu - n_u A_{u\ell}\left(1 + \frac{c^3}{8\pi h\nu_{u\ell}^3}\varepsilon_\nu\right). \qquad (3.35)$$

Equation 3.35 can be written more compactly if we make use of the dimensionless **photon occupation number**, defined as

$$\bar{n}_\gamma \equiv \frac{c^3}{8\pi h\nu^3}\varepsilon_\nu. \qquad (3.36)$$

For blackbody radiation at temperature $T_{\mathrm{rad}}$, the photon occupation number is

$$\bar{n}_\gamma = \frac{1}{e^{h\nu/kT_{\mathrm{rad}}} - 1}. \qquad (3.37)$$

Thus, for a blackbody, the photon occupation number ranges from $\bar{n}_\gamma \gg 1$ in the low-frequency Rayleigh–Jeans tail to $\bar{n}_\gamma \ll 1$ in the high-frequency Wien tail. Inserting the photon occupation number, Equation 3.35 becomes

$$\frac{dn_u}{dt} = n_\ell A_{u\ell}\frac{g_u}{g_\ell}\bar{n}_\gamma - n_u A_{u\ell}(1 + \bar{n}_\gamma). \qquad (3.38)$$

When the population of absorbers reaches an equilibrium state, with $dn_u/dt = 0$, the relative population of the two levels is

$$\frac{n_u}{n_\ell} = \frac{g_u}{g_\ell}\frac{\bar{n}_\gamma}{1 + \bar{n}_\gamma} \qquad (3.39)$$

in a system where excitation and de-excitation are entirely radiative. When the exciting radiation has a blackbody spectrum, then

$$\frac{n_u}{n_\ell} = \frac{g_u}{g_\ell}\exp\left(-\frac{h\nu_{u\ell}}{kT_{\mathrm{rad}}}\right). \qquad (3.40)$$

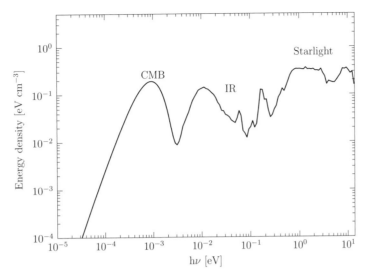

**Figure 3.4** Spectral energy distribution $\nu\varepsilon_\nu$ of light near the Sun's location. The main peaks are provided by the cosmic microwave background (CMB), infrared light from warm dust (IR), and starlight. [Data from GALPROP v56]

The levels, in this case, attain a Boltzmann distribution, the relevant temperature being that of the blackbody radiation.

The interstellar radiation field in our galaxy (Figure 3.4) comes from a variety of sources. At ultraviolet and visible wavelengths, stars are the main contributors.[4] At mid- to far-infrared wavelengths, warm dust provides most of the energy. (The emission features in the range 0.06 to 0.4 eV are from large interstellar molecules, as will be discussed in Section 6.3.) At a wavelength of 21 centimeters, however, the cosmic microwave background (CMB) is the main contributor to the radiation field. At the relevant photon energy $h\nu = h\nu_{u\ell} = 5.87\,\mu\text{eV}$, the CMB, with $T_{\text{rad}} = 2.7255\,\text{K}$, provides an energy density $\nu\varepsilon_\nu = 0.619\,\mu\text{eV cm}^{-3}$. If the hyperfine level of atomic hydrogen were excited and de-excited solely by the CMB then, with $h\nu_{u\ell}/kT_{\text{rad}} = 0.0250$, we would find $\bar{n}_\gamma = 39.48$ and

$$\frac{n_u}{n_\ell} = \frac{3}{1} \times \frac{39.48}{40.48} = 2.926. \qquad (3.41)$$

However, the CMB is not the sole source of radiation at wavelength 21 cm. There is also a significant amount of synchrotron emission, from cosmic ray electrons interacting with the interstellar magnetic field. Near the Sun's location, the energy density of synchrotron radiation at $h\nu = 5.87\,\mu\text{eV}$ is $\nu\varepsilon_\nu = 0.18\,\mu\text{eV cm}^{-3}$; this is equal to $\sim$29% of the energy density of the CMB.

---

[4] Note from Figure 3.4 that the energy density of starlight has a dip at $h\nu \sim 4\,\text{eV}$. This arises because much of the starlight energy in our galaxy comes either from hot O stars with $T > 30\,000\,\text{K}$, and thus $h\nu_{\text{peak}} > 7\,\text{eV}$, or from red giants and supergiants with $T < 5000\,\text{K}$, and thus $h\nu_{\text{peak}} < 1.2\,\text{eV}$.

Adding the local synchrotron radiation to the CMB radiation, the brightness temperature at $\nu \approx 1420\,\mathrm{MHz}$ is raised $\sim 29\%$, from $T_b = 2.7255\,\mathrm{K}$ to $T_b \approx 3.52\,\mathrm{K}$. This implies $h\nu_{u\ell}/kT_b = 0.0194$, a photon occupation number $\bar{n}_\gamma = 51.14$, and a population ratio

$$\frac{n_u}{n_\ell} = \frac{3}{1} \times \frac{51.14}{52.14} = 2.942. \tag{3.42}$$

In practice, then, the presence of the CMB places a universal floor under the excitation temperature of the hyperfine levels: $T_{\mathrm{exc}} \geq 2.73\,\mathrm{K}$ at the present time. Within our galaxy, the additional synchrotron emission hoists the floor on the excitation temperature to $T_{\mathrm{exc}} \geq 3.52\,\mathrm{K}$.

So far we have included only the effects of radiative excitation and de-excitation; let's now turn our attention to collisional excitation and de-excitation. Suppose there exists a population of identical colliders (electrons, atoms, ions, or molecules) with number density $n_c$, mass per collider $m_c$, and kinetic temperature $T$. If there were no background radiation at all ($\bar{n}_\gamma = \varepsilon_\nu = 0$) then excitation of the two-level absorbers would be purely collisional, and de-excitation would be a combination of spontaneous emission and collisional de-excitation. In equation form this amounts to

$$\frac{dn_u}{dt} = n_\ell n_c k_{\ell u} - n_u (n_c k_{u\ell} + A_{u\ell}), \tag{3.43}$$

where $k_{\ell u}$ is the **collisional rate coefficient** for excitation and $k_{u\ell}$ is the collisional rate coefficient for de-excitation.

In general, collisional rate coefficients depend on the gas temperature $T$. Consider, for example, the rate coefficient $k_{u\ell}$ for de-excitation; it can be written as $k_{u\ell} = \langle \sigma_{u\ell} v \rangle$, where $\sigma_{u\ell}$ is the energy-dependent cross section for de-excitation and $v$ is the speed of the collider relative to the excited atom. Averaged over a Maxwellian distribution of collider speeds, the collisional rate coefficient is

$$k_{u\ell}(T) = \langle \sigma_{u\ell} v \rangle = \left( \frac{8kT}{\pi m_r} \right)^{1/2} \int_0^\infty \sigma_{u\ell} \frac{E}{kT} \exp\left( -\frac{E}{kT} \right) \frac{dE}{kT}, \tag{3.44}$$

where $m_r$ is the reduced mass of the collider and the excited atom. (Compare Equation 3.44 with the similar Equation 1.26 for the radiative recombination rate $\alpha$.) For an elastic collision between two neutral atoms (a "billiard ball" collision), we would expect the collision cross section $\sigma_{u\ell}$ to be independent of the collider's energy $E$; this leads to $k_{u\ell} \propto T^{1/2}$. More generally, though, the collisions that lead to excitation are not well approximated as being elastic. If the collision cross section has energy dependence $\sigma_{u\ell} \propto E^n$ then the collisional rate coefficient has temperature dependence $k_{u\ell} \propto T^{n+1/2}$.

Although colliders of any energy can de-excite an excited atom, a collider with energy $E < h\nu_{u\ell}$ is unable to excite a two-level atom from its ground state. Thus, a gas of colliders with $\langle E \rangle = (3/2)kT \ll h\nu_{u\ell}$ will be exponentially unlikely to engage in excitation but is still able to partake in the sport of de-excitation. Thus, the collisional rate coefficients for excitation and de-excitation are related by the exponential relation

$$k_{\ell u} = \frac{g_u}{g_\ell} k_{u\ell} \exp\left(-\frac{h\nu_{u\ell}}{kT}\right). \tag{3.45}$$

In the steady state, Equation 3.43 implies that

$$\frac{n_u}{n_\ell} = \frac{n_c k_{\ell u}}{n_c k_{u\ell} + A_{u\ell}} = \frac{g_u}{g_\ell} \frac{n_c k_{u\ell} e^{-h\nu_{u\ell}/kT}}{n_c k_{u\ell} + A_{u\ell}}. \tag{3.46}$$

When spontaneous emission is extremely ineffective ($A_{u\ell} \ll n_c k_{u\ell}$), then

$$\frac{n_u}{n_\ell} \approx \frac{g_u}{g_\ell} \exp\left(-\frac{h\nu_{u\ell}}{kT}\right). \tag{3.47}$$

The levels, in this case, attain a Boltzmann distribution, the relevant temperature being the kinetic temperature $T$ of the colliders. Thus, in the pure radiative case, the excitation temperature of the absorbers will be the brightness temperature $T_b$ of the background radiation; in the pure collisional case, the excitation temperature of the absorbers will be the kinetic temperature $T$ of the colliders.

Whether electrons or atoms provide the most effective de-excitation depends on the temperature and the fractional ionization of the gas. For the hyperfine 21 cm transition of hydrogen, a good fit for H–H collisions at $T \sim 80\,\text{K}$ is

$$k_{u\ell}(\text{HH}) \approx 1.0 \times 10^{-10}\,\text{cm}^3\,\text{s}^{-1} \left(\frac{T}{80\,\text{K}}\right)^{0.78}, \tag{3.48}$$

whereas e–H collisions at $T \sim 80\,\text{K}$ have the rate coefficient

$$k_{u\ell}(\text{eH}) \approx 2.0 \times 10^{-9}\,\text{cm}^3\,\text{s}^{-1} \left(\frac{T}{80\,\text{K}}\right)^{0.5} \sim 20 k_{u\ell}(\text{HH}). \tag{3.49}$$

In the cold neutral medium, where $n_\text{H} \sim 40\,\text{cm}^{-3}$ and $n_e \sim 0.03\,\text{cm}^{-3} \sim 10^{-3} n_\text{H}$, we therefore expect hydrogen atoms to be the dominant cause of collisional excitation and de-excitation of the hydrogen hyperfine states. In addition, the rate $k_{u\ell} n_\text{H} \sim 4 \times 10^{-9}\,\text{s}^{-1}$ is six orders of magnitude greater than the Einstein A coefficient for spontaneous emission from the upper hyperfine level of hydrogen. This means that for every 21 cm photon spontaneously emitted from atomic hydrogen in the CNM, there are about a million collisional de-excitations. Because collisional de-excitation so strongly dominates over spontaneous emission in this case, the excitation temperature of the hyperfine levels is closer to the kinetic temperature of the hydrogen gas ($T \sim 80\,\text{K}$) than to the brightness temperature of the radiation at 1420 MHz ($T_b \sim 3.5\,\text{K}$).

In the most general case, both collisions and radiation can contribute to excitation and de-excitation. In general,

$$\frac{dn_u}{dt} = n_\ell \left[A_{u\ell}\frac{g_u}{g_\ell}\bar{n}_\gamma + n_c k_{\ell u}\right] - n_u \left[A_{u\ell}(1 + \bar{n}_\gamma) + n_c k_{u\ell}\right]. \tag{3.50}$$

The steady-state solution for this equation is

$$\frac{n_u}{n_\ell} = \frac{A_{u\ell}(g_u/g_\ell)\bar{n}_\gamma + n_c k_{\ell u}}{A_{u\ell}(1 + \bar{n}_\gamma) + n_c k_{u\ell}}. \tag{3.51}$$

Whether collisions or radiation play the main role in de-excitation depends on whether the number density of colliders, $n_c$, is greater than or less than the critical density $n_{\mathrm{crit}}$ at which the rate of collisional de-excitation equals the rate of radiative de-excitation. With this definition,

$$n_{\mathrm{crit}} = \frac{(1 + \bar{n}_\gamma)A_{u\ell}}{k_{u\ell}}. \tag{3.52}$$

For the 21 cm hyperfine levels, assuming $\bar{n}_\gamma = 51.1$, the critical density of colliding hydrogen atoms is

$$n_{\mathrm{crit}} = \frac{52.1(2.88 \times 10^{-15}\,\mathrm{s}^{-1})}{1.0 \times 10^{-10}\,\mathrm{cm}^3\,\mathrm{s}^{-1}} \approx 0.0015\,\mathrm{cm}^{-3} \tag{3.53}$$

at a gas temperature $T \approx 80\,\mathrm{K}$. Because the collisional rate coefficient in this case is an increasing function of temperature (Equation 3.48), the critical density is lower at higher temperatures. The critical density for the hydrogen hyperfine transition ranges from $n_{\mathrm{crit}} \approx 0.07\,\mathrm{cm}^{-3}$ at $T \approx 20\,\mathrm{K}$ to $n_{\mathrm{crit}} \approx 5 \times 10^{-4}\,\mathrm{cm}^{-3}$ at $T \approx 1000\,\mathrm{K}$. In both the CNM and WNM, the actual number density of hydrogen atoms is much greater than the critical density $n_{\mathrm{crit}}$, and collisional de-excitation dominates over radiative de-excitation of the upper hyperfine level of hydrogen. This is not necessarily the case for other levels of other atoms. For instance, at $T \approx 100\,\mathrm{K}$ the $\lambda = 158\,\mu\mathrm{m}$ fine-structure line of C II has a critical density $n_{\mathrm{crit}} \approx 3000\,\mathrm{cm}^{-3}$ for de-excitation by collisions with hydrogen atoms. This is much higher than the actual density $n_{\mathrm{H}} \sim 40\,\mathrm{cm}^{-3}$ of the cold neutral medium. Thus, the upper fine-structure level of C II undergoes primarily radiative de-excitation, emitting the $158\,\mu\mathrm{m}$ photons that permit the cold neutral medium to cool.

## Exercises

3.1   A photon with frequency $\nu$ enters a gas cloud containing a number of two-level absorbers. The excitation temperature of the population of absorbers is $T_{\mathrm{exc}}$.

   (a)   Let $P_{\mathrm{tran}}$ be the probability that the photon transits the gas cloud without being absorbed. Give an expression for $P_{\mathrm{tran}}$ in terms of $\nu$, $T_{\mathrm{exc}}$, and the optical depth $\tau_\nu$ of the cloud along the photon's path. [Hint: remember that $\tau_\nu$ includes the effects of both absorption and stimulated emission.]

   (b)   Consider a photon that successfully transited the cloud without being absorbed. Let $P_{\mathrm{stim}}$ be the probability that the transiting photon stimulated the emission of at least one photon. Give an expression for $P_{\mathrm{stim}}$

in terms of $\nu$, $T_{exc}$, and $\tau_\nu$. Comment on the conditions required for a maser, in which one entering photon produces many outgoing photons from stimulated emission.

3.2 Assume that you have both emission and absorption spectra for the 21 cm line in a particular direction, and can therefore determine $\tau_\nu$ across the line profile.

(a) Show that the measured excitation temperature $T_{exc}$ in any line of sight is actually the harmonic mean of all components with various excitation temperatures in that direction. That is, if there are $m$ separate clouds along the line of sight, each with column density $N_i$ and excitation temperature $T_i$, and total column density $N_{tot} = \Sigma N_i$, then $T_{exc}^{-1} = (1/N_{tot})[\Sigma N_i/T_i]$.

(b) Typical measured excitation temperatures for lines of sight near the galactic plane are $T_{exc} \sim 150\,\mathrm{K}$. Assume that all emission is optically thin, and all the H I is either in cold neutral clouds or in a warm neutral phase (this is the two-phase model of the ISM). What is the fraction of the H I in each phase, given $T_{exc} = 150\,\mathrm{K}$?

# 4

# Warm Ionized Medium and Ionized Nebulae

*And let my liver rather heat with wine*
*Than my heart cool with mortifying groans.*

William Shakespeare (1564–1616)
"The Merchant of Venice," I, 1

An electron that is bound to an atom, ion, or molecule can be excited to a higher energy level either radiatively or collisionally. Similarly, a bound electron can be ionized either radiatively (in which case we use the term **photoionization**) or collisionally. In the neutral portions of the ISM, the fractional ionization $x = n_e/n_H$ is much smaller than one. In the cold neutral medium $x \sim 0.001$ but in the warm neutral medium the fractional ionization can be as much as $x \sim 0.1$. Although free electrons are relatively scarce in the neutral interstellar medium, they are important in regulating the physical properties of the interstellar gas. As mentioned in Section 1.2, free electrons play a role in bringing the warm neutral medium to kinetic equilibrium; as mentioned in Section 1.3, free electrons photoejected from dust grains are a major heat source in the cold neutral medium as well as the warm neutral medium.

However, it is only in the warm ionized medium, where $x \sim 0.7$, and the hot ionized medium, where $x \sim 1$, that the ISM consists mainly of ionized hydrogen. Thus, it is only at this point, when we switch our attention to the warm ionized medium (WIM), that we will closely examine the process of ionization and the inverse process of recombination. The warm ionized medium, with $T \sim 8000\,\text{K}$ and $n \sim 0.2\,\text{cm}^{-3}$, contributes only $\sim 12\%$ of the mass of our galaxy's ISM. Despite its minority status, the WIM plays an important role in our understanding of the ISM. Moreover, H II regions and planetary nebulae, which have a similar temperature to the WIM (though a much higher density) are packed with diagnostic tools for determining densities, temperatures, chemical compositions, and ionization states.

The most important energy scale in the study of ionization is $I_H = 13.60\,\mathrm{eV}$, the ionization energy of hydrogen from its ground state. In principle, it is also possible to ionize hydrogen from an excited state; the ionization energy from the $n = 2$ level of hydrogen, for instance, is $I = 3.40\,\mathrm{eV}$. However, in a realistic interstellar radiation field, it is highly unlikely that the excited hydrogen atom will absorb a photon with $h\nu > 3.40\,\mathrm{eV}$ during the brief interval ($\sim A_{21}^{-1}$ $\sim 2$ nanoseconds) before it de-excites by emitting a Lyman $\alpha$ photon. In practice, then, we need only consider ionization from the ground state.

For atoms other than hydrogen, the **first ionization energy** is the energy required to remove the most loosely bound electron from a neutral atom in its ground state. Table 1.2 gives the first ionization energies for the most abundant elements in the ISM. The highest first ionization energy is that of helium, with $I_{He} = 24.59\,\mathrm{eV}$; the lowest is that of cesium, with $I_{Cs} = 3.89\,\mathrm{eV}$. Thus, the photoionization of atoms in their ground state will be achieved by ultraviolet photons with wavelength $\lambda < 3200\,\text{Å}$.

Because of the ubiquity and simplicity of hydrogen, we will use hydrogen as our main example when discussing ionization and recombination. However, the formulas for the ionization of hydrogen can also be applied to hydrogenic ions; that is, ions with only a single bound electron. A hydrogenic ion with atomic number $Z$ has an ionization energy $Z^2 I_H$: this comes to $54.4\,\mathrm{eV}$ for He II, $122.4\,\mathrm{eV}$ for Li III, and so forth. We will start by examining photoionization and its inverse process, radiative recombination. Then we'll look at a simple application of these processes: the Strömgren sphere.

## 4.1 Photoionization and Radiative Recombination

The photoionization of hydrogen occurs by photoelectric absorption:[1]

$$H + h\nu \rightarrow p + e^-, \tag{4.1}$$

where the photon energy $h\nu$ must be greater than or equal to the ionization energy $I_H = 13.60\,\mathrm{eV}$ in order to ionize hydrogen from its ground state. The photoionization cross section can be approximated as

$$\sigma_{\mathrm{pho}}(\nu) \approx \sigma_0 \left( \frac{h\nu}{Z^2 I_H} \right)^{-3}, \tag{4.2}$$

with an error of $\sim 20\%$ or less in the range $Z^2 I_H \leq h\nu < 50 Z^2 I_H$. The cross section at the energy threshold is

---

[1] X-ray photons with $h\nu \gg I_H$ can also photoionize H via Compton scattering. However, scattering doesn't dominate over absorption until $h\nu > 2\,\mathrm{keV}$. Thus, given the small Compton cross section ($\sigma_C \sim 10^{-24}\,\mathrm{cm}^2$ at $h\nu \sim 2\,\mathrm{keV}$) and the scarcity of X-ray photons in the ISM, Compton scattering doesn't play a significant role in photoionizing hydrogen.

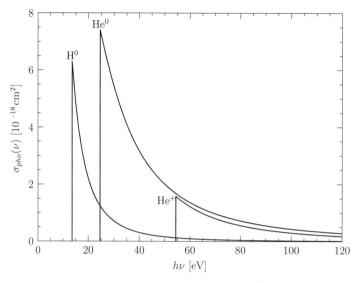

**Figure 4.1** Photoionization cross sections for hydrogen ($H^0$), neutral helium ($He^0$), and hydrogenic helium ($He^+$).

$$\sigma_0 = \frac{6.30 \times 10^{-18} \text{ cm}^2}{Z^2} = \frac{0.225 a_0^2}{Z^2}, \tag{4.3}$$

where $a_0$ is the Bohr radius. In Equations 4.2 and 4.3, we've included the atomic number $Z$ for a hydrogenic ion; if hydrogenic ions are not of interest to you, just set $Z = 1$ for hydrogen. Figure 4.1 shows the photoionization cross sections for hydrogen, neutral helium, and singly ionized (hydrogenic) helium.

If the photon energy density is $\varepsilon_\nu$ then the volumetric photoionization rate for hydrogen is

$$\frac{dn_{\text{HII}}}{dt} = n_{\text{HI}} \zeta_{\text{pho}}. \tag{4.4}$$

The **photoionization rate** $\zeta_{\text{pho}}$ (that is, the rate at which a single neutral hydrogen atom undergoes photoionization) is

$$\zeta_{\text{pho}} = \int_{\nu_0}^{\infty} \frac{\varepsilon_\nu}{h\nu} c\sigma_{\text{pho}}(\nu) d\nu, \tag{4.5}$$

where $\nu_0 = I_H/h = 3.29 \times 10^{15}$ Hz is the threshold ionization frequency. The photoionization rate thus depends on the background energy density above a photon energy of $I_H = 13.6$ eV. Since the photoionization cross section decreases fairly steeply with increasing photon energy, most photoionizing photons have energies just above 13.6 eV.

The inverse process to photoionization is radiative recombination:

$$p + e^- \rightarrow H + h\nu. \tag{4.6}$$

When hydrogen recombines, the electron doesn't necessarily recombine into the $n = 1$ ground state. If a free electron with kinetic energy $E$ triggers a radiative recombination then the emitted photon will have an energy $h\nu = E + I_H/n^2$, where $n = 1$ for the ground state, $n = 2$ for the $2s$ and $2p$ levels, $n = 3$ for the $3s$, $3p$, and $3d$ levels, and so forth.

The volumetric rate of radiative recombination can be written as

$$\frac{dn_{HII}}{dt} = -n_e n_{HII} \alpha, \tag{4.7}$$

where $\alpha$ is the radiative recombination rate first introduced in Section 1.2. The rate $\alpha$ has units of $cm^3\,s^{-1}$. For recombination to any particular electronic state of the hydrogen atom ($1s$, $2s$, $2p$, and so on), the value of $\alpha$ is

$$\alpha_{n\ell}(T) = \left(\frac{8kT}{\pi m_e}\right)^{1/2} \int_0^\infty \sigma_{rr,n\ell}(E) \frac{E}{kT} e^{-E/kT} \frac{dE}{kT}, \tag{4.8}$$

where $E$ is the kinetic energy of the free electron and $T$ is the kinetic temperature of the gas from which the electrons are drawn. Just as an electron of any energy can trigger a collisional de-excitation, an electron of any energy can trigger radiative recombination.

Since the cross section for radiative recombination, $\sigma_{rr,n\ell}$, is closely related to the cross section for photoionization, $\sigma_{pho,n\ell}$, it is possible to compute the recombination rate $\alpha_{n\ell}$ as a function of the electron gas temperature $T$. In general, $\alpha_{n\ell}$ is a decreasing function of $T$; it's easier to recombine with a slow electron than with a fast electron. In general, $\alpha_n$, summing over all applicable values of $\ell$, is a decreasing function of $n$; it's easier to recombine to a low energy level than to a high energy level. If we sum over *all* the electron energy levels of a hydrogen atom, the radiative recombination rate is $\alpha \sim 5 \times 10^{-13}\,cm^3\,s^{-1}$ at $T = 8000\,K$, typical for the warm ionized medium. Given a free electron density $n_e \sim 0.1\,cm^{-3}$, also typical for the warm ionized medium, the characteristic time scale for H II to undergo radiative recombination is

$$t_{rec} \equiv \frac{1}{\alpha n_e} \sim 0.6\,Myr. \tag{4.9}$$

The time scale $t_{rec}$ is known as the **recombination time**.

It often matters to which energy level a free electron recombines. If an electron with energy $E$ recombines to the ground state of hydrogen ($n = 1$), it produces a photon of energy $h\nu = E + I_H \geq I_H$; this emitted photon is capable of photoionizing any hydrogen atom that it encounters. In regions that are optically thick to photons at energies just above $I_H$, the emitted photon is soon destroyed in photoionizing a relatively nearby hydrogen atom. Thus, in regions that are very optically thick, it is useful to make an **on-the-spot** approximation, in which it is assumed that every photon produced by radiative recombination to the ground state of hydrogen is immediately, then and there, destroyed in photoionizing another hydrogen atom. In the on-the-spot approximation, recombination to the ground state thus has no net effect on the ionization state of the hydrogen gas.

For practical purposes, it is useful to assign regions of ionized hydrogen gas to one of two cases; these were labeled by Baker and Menzel in 1938 as "Case A" and "Case B." In **Case A**, the region of gas is optically thin to ionizing radiation with $hv \geq 13.6\,\mathrm{eV}$; every photon produced by radiative recombination is assumed to escape, even those produced by recombination to the ground state of hydrogen. In Case A, the relevant radiative recombination rate is found by summing over *all* energy levels of the hydrogen atom:

$$\alpha_{A,H} = \sum_{n,\ell} \alpha_{n\ell} \approx 4.9 \times 10^{-13}\,\mathrm{cm}^3\,\mathrm{s}^{-1} \left(\frac{T}{8000\,\mathrm{K}}\right)^{-0.71}. \tag{4.10}$$

This power-law approximation for the dependence on temperature works very well for temperatures in the range $4000\,\mathrm{K} < T < 16\,000\,\mathrm{K}$. Very hot ($T > 10^6\,\mathrm{K}$) regions are frequently optically thin to the photons emitted by radiative recombination, and thus constitute Case A recombination. At these high temperatures, the Case A recombination rate is more steeply dependent on temperature, with

$$\alpha_{A,H} \approx 5 \times 10^{-16}\,\mathrm{cm}^3\,\mathrm{s}^{-1} \left(\frac{T}{10^7\,\mathrm{K}}\right)^{-1.5}, \tag{4.11}$$

an approximation that is good in the temperature range $T > 10^6\,\mathrm{K}$.

In **Case B** radiative recombination, the region of gas is optically thick to ionizing radiation with $hv \geq 13.6\,\mathrm{eV}$; photons produced by radiative recombination to the ground state are assumed to be reabsorbed on the spot. In Case B, the relevant radiative recombination rate is found by summing over all energy levels other than the ground state:

$$\alpha_{B,H} = \alpha_{A,H} - \alpha_{1s} \approx 3.1 \times 10^{-13}\,\mathrm{cm}^3\,\mathrm{s}^{-1} \left(\frac{T}{8000\,\mathrm{K}}\right)^{-0.82}. \tag{4.12}$$

Again, this power-law dependence on temperature is a good approximation for temperatures from $4000\,\mathrm{K} < T < 16\,000\,\mathrm{K}$. The percentage of radiative recombinations that go directly to the ground state is 32% at $T = 4000\,\mathrm{K}$ but increases to 41% at $T = 16\,000\,\mathrm{K}$. Thus, the distinction between Case A and Case B becomes increasingly important at higher temperatures. H II regions and planetary nebulae, with their relatively high densities, are frequently optically thick to ionizing radiation above $13.6\,\mathrm{eV}$ and thus are examples of Case B recombination.

If we consider a gas of pure atomic hydrogen, with $n_H \equiv n_{HI} + n_{HII}$ and $n_e = n_{HII}$, we can compute the fractional ionization $x = n_e/n_H$ in the case of ionization equilibrium, with radiative recombination (Equation 4.7) exactly balancing photoionization (Equation 4.4):

$$n_{HI}\zeta_{pho} = n_e n_{HII}\alpha$$
$$(1-x)n_H\zeta_{pho} = x^2 n_H^2 \alpha. \tag{4.13}$$

In Equation 4.13, $\alpha$ is the radiative recombination rate for either Case A or Case B, whichever is appropriate. Thus, we have a quadratic equation for the fractional ionization $x$:

$$1 - x - x^2 \frac{n_H \alpha}{\zeta_{pho}} = 0. \qquad (4.14)$$

In the high-density limit $n_H \gg \zeta_{pho}/\alpha$, every proton that has been bereft of its electron soon finds a new electronic partner by radiative recombination, and thus the fractional ionization is small: $x \approx (n_H \alpha / \zeta_{pho})^{-1/2}$. In the opposite limit, $n_H \ll \zeta_{pho}/\alpha$, radiative recombinations are infrequent events, and the fractional ionization is large: $x \approx 1 - n_H \alpha / \zeta_{pho}$.

To solve for the fractional ionization $x$ in the general case, we need to know the density and temperature of the hydrogen gas. We need to know the value of $\zeta_{pho}$ for the radiation to which the gas is exposed. We need to know whether the gas is optically thin (Case A) or optically thick (Case B). In addition, we need to verify that collisional ionization actually is negligible compared to photoionization.

If you are at a random point in the ISM, it may not be intuitively obvious whether photoionization or collisional ionization is dominant. However, there is one set of circumstances in which the dominance of photoionization is obvious: in an ionized nebula, such as an H II region or planetary nebula, that surrounds a luminous source of UV photons. In that case, photoionization by light from the central source utterly dominates, and the situation is simple. Let's look at H II regions and planetary nebulae in the simplified guise of the **Strömgren sphere**.

## 4.2   Strömgren Spheres

Although receiving his Ph.D. in 1929 at the age of 21, Bengt Strömgren didn't get around to describing the Strömgren sphere until the year 1939. Strömgren, seeking the simplest possible idealization of a real H II region, assumed that pure atomic hydrogen gas, with uniform number density $n_H$, surrounds a central point source of light. The central light source produces ionizing photons, with energy $h\nu > I_H$, at a constant rate $Q_0$. For real main sequence stars, $Q_0 = 4.4 \times 10^{49}\ \mathrm{s}^{-1}$ for an ultra-hot O3 star with effective temperature $T_{eff} \approx 45\,000$ K, and $Q_0 = 7.6 \times 10^{47}\ \mathrm{s}^{-1}$ for a not-quite-so-hot O9.5 star with $T_{eff} \approx 32\,000$ K.[2]

Since this is a theoretical exercise, let's switch on the central UV source abruptly at some time $t = 0$. The mean free path for the ionizing photons as they travel through the gas of neutral atoms is

$$\lambda_{mfp} = \frac{1}{n_H \sigma_{pho}} = 0.0014\,\mathrm{pc} \left( \frac{n_H}{40\ \mathrm{cm}^{-2}} \right)^{-1} \left( \frac{\sigma_{pho}}{6 \times 10^{-18}\ \mathrm{cm}^2} \right)^{-1}, \qquad (4.15)$$

---

[2]  Since the UV spectra of stars are chopped up with absorption lines, determining $Q_0$ requires detailed modeling of stellar atmospheres. For this book, we are using the results of Martins et al. 2005.

where we have scaled to the number density of atoms appropriate for a star in the cold neutral medium and to the cross section appropriate for a photon just above the 13.6 eV threshold for ionization. Initially, just after the UV source has been turned on, the photons eat their way through the neutral medium, converting it into ionized gas. Assuming isotropic stellar radiation, the volume $V$ of ionized gas is a sphere that initially expands at a rate given by the relation

$$Q_0 = n_H \frac{dV}{dt} = n_H 4\pi R^2 \frac{dR}{dt}. \tag{4.16}$$

Integrating this equation yields a radius that increases as

$$R(t) = \left( \frac{3Q_0}{4\pi n_H} t \right)^{1/3}. \tag{4.17}$$

However, the sphere of ionized gas doesn't expand at this rate forever. The protons and electrons recombine at a rate proportional to $\alpha n_e n_{H\,II}$, and some of the UV photons produced by the central source must reionize the atoms that have recombined. Including the effects of recombination, Equation 4.16 is altered to

$$Q_0 = n_H 4\pi R^2 \frac{dR}{dt} + \alpha_{B,H} n_e n_{H\,II} \frac{4\pi}{3} R^3. \tag{4.18}$$

However, inside the ionized sphere the fractional ionization is near unity, and we may write $n_e = n_{H\,II} \approx n_H$. Thus

$$Q_0 = n_H 4\pi R^2 \frac{dR}{dt} + \alpha_{B,H} n_H^2 \frac{4\pi}{3} R^3. \tag{4.19}$$

Notice the assumptions that we have made. First, the ionized sphere is assumed to be optically thick (Case B); next, the fractional ionization is assumed to be $x = 1$. Equation 4.19 has the solution

$$R(t) = R_s \left( 1 - e^{-t/t_{rec}} \right)^{1/3}, \tag{4.20}$$

where the length scale introduced is the **Strömgren radius**

$$R_s \equiv \left( \frac{3Q_0}{4\pi \alpha_{B,H} n_H^2} \right)^{1/3} \tag{4.21}$$

$$= 5.5\,\text{pc} \left( \frac{Q_0}{10^{49}\,\text{s}^{-1}} \right)^{1/3} \left( \frac{\alpha_{B,H}}{3 \times 10^{-13}\,\text{cm}^3\,\text{s}^{-1}} \right)^{-1/3} \left( \frac{n_H}{40\,\text{cm}^{-3}} \right)^{-2/3}$$

and the time scale is the recombination time $t_{rec}$. In general, the recombination time is, from Equation 4.9, $t_{rec} = 1/(\alpha n_e)$. However, within a pure hydrogen H II region with fractional ionization $x \approx 1$, we may assume $n_e \approx n_H$. This leads to a recombination time

$$t_{rec} \approx \frac{1}{\alpha_{B,H} n_H} \tag{4.22}$$

$$\approx 2600\,\text{yr} \left( \frac{\alpha_{B,H}}{3 \times 10^{-13}\,\text{cm}^3\,\text{s}^{-1}} \right)^{-1} \left( \frac{n_H}{40\,\text{cm}^{-3}} \right)^{-1}.$$

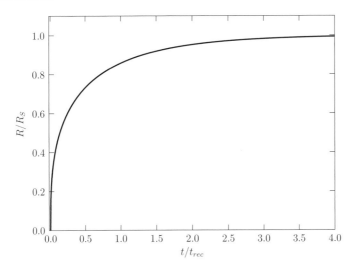

**Figure 4.2** Radius of a Strömgren sphere in units of the Strömgren length $R_s$, as a function of time in units of the recombination time $t_{rec}$.

The time required to create a Strömgren sphere by turning on a hot star in the CNM is ~3000 yr; since radiative recombination is the time reversal of photoionization, this is also the time it takes the Strömgren sphere to revert to neutral hydrogen after the central UV source has been turned off. Figure 4.2 shows the rate of growth of a Strömgren sphere as a function of time. At times $t \gg t_{\text{rec}} \sim 3000$ yr, the Strömgren sphere will consist of a sphere of almost fully ionized gas with radius $R \rightarrow R_s \sim 6$ pc, surrounded by a partially ionized boundary of thickness $\sim \lambda_{\text{mfp}} \sim 0.001$ pc $\ll R_s$, embedded in a sea of neutral hydrogen.

Although a Strömgren sphere approaches ionization equilibrium at $t > t_{\text{rec}}$, it is still far from *pressure* equilibrium. Outside the Strömgren sphere, if it is embedded in the cold neutral medium, the temperature is $T \sim 80$ K. Inside the sphere, the temperature is $T \sim 8000$ K; we'll see in the next section how heating and cooling processes fix this temperature. Also, the number density of particles inside the Strömgren sphere doubles when the hydrogen is ionized. Thus, the pressure inside the sphere will be ~200 times the pressure outside, and the ionized gas will begin to expand. However, the time scale for expansion is much longer than the recombination time $t_{\text{rec}}$. The speed at which the sphere expands is the speed of sound within its ionized gas:

$$c_s = \left( \frac{2kT}{m_{\text{H}}} \right)^{1/2} = 11.5 \text{ km s}^{-1} \left( \frac{T}{8000 \text{ K}} \right)^{1/2}. \quad (4.23)$$

The time it takes for the Strömgren sphere to expand to twice its original size is roughly the sound crossing time

$$t_{\text{sound}} = \frac{R_s}{c_s} \approx 0.5 \text{ Myr}, \quad (4.24)$$

more than a hundred times longer than the recombination time $t_{rec}$. Thus, there is a time span $t_{rec} < t < t_{sound}$ when the Strömgren sphere is close to ionization equilibrium but has not yet had time to expand significantly.

Real H II regions, of course, don't consist entirely of hydrogen. Assuming a solar abundance, out of every 1000 atoms there are on average 911 hydrogen atoms, 88 helium atoms, and one non-hydrogen, non-helium atom. Let's start by examining what happens when we acknowledge the existence of helium. (We'll turn our attention to that one-in-a-thousand heavier atom later.) In a hydrogen-only medium, the only ionization energy we needed to memorize was the ionization energy of hydrogen, $I_H = 13.6\,\mathrm{eV}$. When we add helium, we also need to know the first ionization energy of helium, $I_{He} = 24.6\,\mathrm{eV} = 1.81 I_H$, and its second ionization energy, $I_{HeII} = 54.4\,\mathrm{eV} = 4 I_H$.

Looking at the photoionization cross sections for H I, He I, and He II, as shown in Figure 4.1, we see that, above the 24.6 eV threshold for ionizing neutral helium, the photoionization cross section for helium is larger than the photoionization cross section for hydrogen. More exactly, just above 24.6 eV, $\sigma_{pho}(\mathrm{He})/\sigma_{pho}(\mathrm{H}) \approx 6.5$; just below 54.4 eV, $\sigma_{pho}(\mathrm{He})/\sigma_{pho}(\mathrm{H}) \approx 14$. Thus, the photoionization cross section for helium is $\sim 10$ times that of hydrogen in this range, while the number density of helium is $\sim 0.1$ times that of hydrogen. This implies that if we suddenly turn on a hot star in the midst of gas with the solar abundances of hydrogen and helium, the initial photons in the range $24.6\,\mathrm{eV} < h\nu < 54.4\,\mathrm{eV}$ will be about as likely to photoionize a helium atom as a hydrogen atom. This is a strong contrast to the $13.6\,\mathrm{eV} < h\nu < 24.6\,\mathrm{eV}$ range, where nearly all the photons go to ionize hydrogen; scarcer atoms like oxygen and carbon account for only a tiny fraction of the ionizations.

Suppose that you have a Strömgren sphere that contains a roughly solar mix of hydrogen and helium, with a helium to hydrogen ratio $f \equiv n_{He}/n_H \approx 0.097$. The central star emits photons with $h\nu > I_H = 13.6\,\mathrm{eV}$ at a rate $Q_0$; it emits photons with $h\nu > I_{He} = 24.6\,\mathrm{eV}$ at a rate $Q_1 < Q_0$. In the central region of the Strömgren sphere, we may assume that the hydrogen is all ionized, and the helium is all singly ionized. This will result in

$$n_{HII} = n_H$$
$$n_{HeII} = n_{He} = f n_H \tag{4.25}$$
$$n_e = n_{HII} + n_{HeII} = (1 + f) n_H.$$

Since even the hottest O stars don't produce a significant number of photons with $h\nu > 54.4\,\mathrm{eV}$, there will not be a significant amount of He III to complicate matters.

In the central region of the hydrogen–helium Strömgren sphere, the volumetric rate of hydrogen recombination (to states other than the ground state) is

$$\frac{dn_{HII}}{dt} = -\alpha_{B,H} n_e n_{HII} = -\alpha_{B,H}(1 + f) n_H^2. \tag{4.26}$$

Computing the rate of helium recombination is a little trickier. Recombination to the ground state of neutral helium produces a photon with $h\nu > 24.6\,\text{eV}$; this photon is *capable* of photoionizing neutral helium, but in practice it might end up photoionizing a neutral hydrogen atom instead. This means that, even though the Strömgren sphere is optically thick to UV photons with $h\nu > 24.6\,\text{eV}$, the appropriate radiative recombination rate to use for helium is not the Case B rate. Neither is it the Case A rate; instead, it is

$$\alpha_{\text{eff},\text{He}} = \alpha_{\text{B},\text{He}} + y\alpha_{1,\text{He}}, \tag{4.27}$$

where $\alpha_{1,\text{He}}$ is the radiative recombination rate to the ground state of helium, and $y$ is the fraction of $h\nu > 24.6\,\text{eV}$ photons that photoionize hydrogen rather than helium. At $T = 8000\,\text{K}$, $\alpha_{\text{B},\text{He}} = 3.24 \times 10^{-13}\,\text{cm}^3\,\text{s}^{-1}$ and $\alpha_{1,\text{He}} = 1.72 \times 10^{-13}\,\text{cm}^3\,\text{s}^{-1}$. This leads to an effective radiative recombination rate

$$\alpha_{\text{eff},\text{He}} \approx 3.2 \times 10^{-13}\,\text{cm}^3\,\text{s}^{-1}(1 + 0.53y). \tag{4.28}$$

This implies a radiative recombination rate for He that is slightly greater than that for H, with

$$\frac{\alpha_{\text{eff},\text{He}}}{\alpha_{\text{B},\text{H}}} \approx 1.04(1 + 0.53y). \tag{4.29}$$

The volumetric rate of helium recombination is

$$\frac{dn_{\text{HeII}}}{dt} = -\alpha_{\text{eff},\text{He}}n_e n_{\text{HeII}} = -\alpha_{\text{eff},\text{He}}f(1+f)n_{\text{H}}^2. \tag{4.30}$$

Comparing Equation 4.26 with Equation 4.30, we see that

$$\frac{dn_{\text{HeII}}}{dt} = \left(\frac{\alpha_{\text{eff},\text{He}}}{\alpha_{\text{B},\text{H}}}\right)f\frac{dn_{\text{HII}}}{dt} \approx 0.10(1 + 0.53y)\frac{dn_{\text{HII}}}{dt}, \tag{4.31}$$

assuming $f = 0.097$ and $T \approx 8000\,\text{K}$.

There's an additional complication to deal with when we mix helium into our Strömgren sphere. When a hydrogen ion recombines to an excited level, it then emits one or more photons as it cascades down to the ground level. However, the energy $h\nu$ of these photons is less than 13.6 eV, so they can't photoionize hydrogen or helium. When a helium ion recombines to an excited level, the photons emitted as the electron cascades down have energies $h\nu < 24.6\,\text{eV}$, so they can't photoionize helium. However, some of these emitted photons are energetic enough to photoionize hydrogen; this process is known as the *secondary* photoionization of hydrogen. A useful approximation is that, for every photoionization of helium by an $h\nu > 24.6\,\text{eV}$ photon from the central star, there is one secondary photoionization of hydrogen.[3]

---

[3] The actual number of secondary H photoionizations per primary He photoionization ranges from $N_{\text{sec}} \approx 1$ at low electron density to $N_{\text{sec}} \approx 0.7$ at high electron density, where collisional de-excitations become important. We'll take $N_{\text{sec}} = 1$ for computational simplicity.

Now we have the tools necessary to determine the structure of our hydrogen–helium Strömgren sphere. In ionization equilibrium, the total number $N(\mathrm{He\,II})$ of helium ions is determined by the balance between radiative recombination and photoionization:

$$N(\mathrm{He\,II})n_e\alpha_{\mathrm{eff,\,He}} = (1-y)Q_1, \qquad (4.32)$$

where $Q_1$ is the rate of production of stellar photons with $h\nu > 24.6\,\mathrm{eV}$ and $y$ is the fraction of photons in this energy range that ionize H instead of He. Similarly, the total number $N(\mathrm{H\,II})$ of hydrogen ions is determined by the balance between radiative recombination and photoionization. In the photon energy range $13.6\,\mathrm{eV} < h\nu < 24.6\,\mathrm{eV}$, every photon ionizes an H atom. In the energy range $h\nu > 24.6\,\mathrm{eV}$, a fraction $y$ of photons ionize an H atom directly and a fraction $1 - y$ ionize an H atom indirectly, via a secondary photoionization. Thus, every photon with $h\nu > 13.6\,\mathrm{eV}$ ionizes an H atom, either directly or indirectly, and the equilibrium number $N(\mathrm{H\,II})$ of hydrogen ions is

$$N(\mathrm{H\,II})n_e\alpha_{\mathrm{B,\,H}} = Q_0. \qquad (4.33)$$

Equations 4.32 and 4.33 can be combined to yield

$$\frac{N(\mathrm{He\,II})}{N(\mathrm{H\,II})} = \frac{\alpha_{\mathrm{B,\,H}}}{\alpha_{\mathrm{eff,\,He}}}(1-y)\frac{Q_1}{Q_0}. \qquad (4.34)$$

If we assume $T \approx 8000\,\mathrm{K}$, we can use the radiative recombination rates from Equation 4.29 to write

$$\frac{N(\mathrm{He\,II})}{N(\mathrm{H\,II})} \approx 0.96\frac{1-y}{1+0.53y}\frac{Q_1}{Q_0}. \qquad (4.35)$$

If all the hydrogen in the Strömgren sphere is ionized out to a radius $R_s$, and all the helium is singly ionized out to the same radius, this implies that $N(\mathrm{He\,II})/N(\mathrm{H\,II}) = n_{\mathrm{He}}/n_{\mathrm{H}} = f \approx 0.097$. Equation 4.35 tells us that this degree of ionization requires a central star with

$$\frac{Q_1}{Q_0} \approx 0.10\frac{1+0.53y}{1-y}. \qquad (4.36)$$

This critical ratio depends strongly on $y$, the fraction of photons with $h\nu > 24.6\,\mathrm{eV}$ that ionize H rather than He. If $y = 0$ then the central star must have $Q_1/Q_0 = 0.10$ to fully ionize all the helium in the Strömgren sphere. If a fraction $y = 0.5$ of the $h\nu > 24.6\,\mathrm{eV}$ photons ionize H then the required ratio is increased to $Q_1/Q_0 = 0.25$. For the conditions that prevail in real H\,II regions, the value $y \approx 0.2$ is fairly typical, implying a critical ratio $Q_1/Q_0 \approx 0.14$.

On the main sequence, a star with effective temperature $T_{\mathrm{eff}} = 37\,000\,\mathrm{K}$, corresponding to spectral class O7, will have the critical ratio $Q_1/Q_0 \approx 0.14$. Hotter stars will have $Q_1/Q_0 > 0.14$; cooler stars will have $Q_1/Q_0 < 0.14$. For ionizing stars with $T_{\mathrm{eff}} < 37\,000\,\mathrm{K}$, and thus $Q_1/Q_0 < 0.14$, the ionized helium

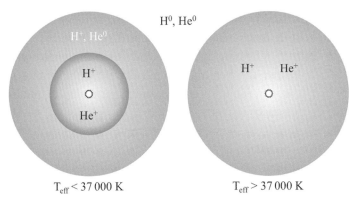

**Figure 4.3** Left: Schematic of an H II region surrounding a relatively cool star, with $T_{\text{eff}} < 37\,000\,\text{K}$. Right: Schematic of an H II region surrounding a relatively hot star, with $T_{\text{eff}} > 37\,000\,\text{K}$. Outside these regions we have neutral H and He.

sphere will have a radius $R_{\text{He}}$ smaller than the radius $R_{\text{H}}$ of the ionized hydrogen sphere. This is shown schematically in the left panel of Figure 4.3.

However, H II regions whose central star has $T_{\text{eff}} > 37\,000\,\text{K}$, and thus $Q_1/Q_0 > 0.14$, will *not* have $R_{\text{He}} > R_{\text{H}}$. Instead, H II regions around stars with $Q_1/Q_0 > 0.14$ all have $R_{\text{He}} = R_{\text{H}}$, as shown in the right panel of Figure 4.3. As $Q_1/Q_0$ increases beyond 0.14, an increasingly large fraction of newly recombined H atoms are ionized by $h\nu > 24.6\,\text{eV}$ photons rather than by lower-energy photons. This in turn drives up the value of $y$, the fraction of $h\nu > 24.6\,\text{eV}$ photons that ionize hydrogen rather than helium atoms. As a simple example, consider a magical ultraviolet star that produces photons that *all* have energy $h\nu > 24.6\,\text{eV}$; the star thus has $Q_1/Q_0 = 1$. Equation 4.36 tells us that, for $R_{\text{He}}$ to equal $R_{\text{H}}$, a fraction $y = 0.85$ of the photons must ionize a hydrogen atom directly. A fraction $1 - y = 0.15$ of the photons must ionize a helium atom (while also ionizing a hydrogen atom through a secondary reionization).[4]

## 4.3  Heating and Cooling in H II Regions

In the previous sections of this chapter, we asserted that H II regions have temperatures $\sim 8000\,\text{K}$. This is an observationally supported assertion; moreover, observations indicate that the temperatures of H II regions are remarkably independent of the effective temperature of the central star. Thus the temperature of an

---

[4] Note that these numbers imply there are 15 newly recombined He atoms for every 100 newly recombined H atoms, more than the 10 He nuclei per 100 H nuclei present at solar abundance. The reason is that when $y = 0.85$ the effective recombination rate for He is $\sim 50\%$ greater than the recombination rate for H (Equation 4.29).

H II region is not dictated by its star; instead, it is the result of a balance between the heating and cooling mechanisms in the ionized gas of the H II region.

The main source of heating in an ionized nebula is photoionization. For simplicity, let's consider the heating that results from the photoionization of hydrogen. When hydrogen is photoionized from its ground state, the free electron that is emitted carries away a kinetic energy $E = h\nu - I_H$, where $h\nu$ is the energy of the ionizing photon. If we average over all photoionizations, the mean energy of the free electrons produced is

$$\langle E \rangle = \langle h\nu \rangle - I_H. \tag{4.37}$$

The average energy $\langle h\nu \rangle$ of an ionizing photon depends on the spectrum of the central light source in the nebula. For even the hottest O stars, the required energy $h\nu > 13.6 \, \text{eV}$ for photoionization lies on the high-energy Wien tail of the Planck function. It is true that stars are not perfect blackbodies, particularly at UV frequencies, where their spectra have many absorption lines. However, we are going to engage in risky (but analytically tractable) behavior and assume that the energy density of photons produced by a star has the form appropriate for the Wien tail (Equation 2.9):

$$\varepsilon_\nu \propto \nu^3 \exp\left(-\frac{h\nu}{kT_c}\right), \tag{4.38}$$

where $T_c$ is the temperature of the central star.

The average energy $\langle h\nu \rangle$ of an ionizing photon must also be weighted by the cross section for ionization, $\sigma_{\text{pho}}(\nu) \propto \nu^{-3}$. Consequently the mean energy per ionizing photon can be written as

$$\langle h\nu \rangle = \frac{\int_{\nu_0}^{\infty} (\varepsilon_\nu/h\nu)(h\nu)\sigma_{\text{pho}} d\nu}{\int_{\nu_0}^{\infty} (\varepsilon_\nu/h\nu)\sigma_{\text{pho}} d\nu} = \frac{h \int_{\nu_0}^{\infty} (\nu^2 e^{-h\nu/kT_c})\nu \cdot \nu^{-3} d\nu}{\int_{\nu_0}^{\infty} (\nu^2 e^{-h\nu/kT_c})\nu^{-3} d\nu}, \tag{4.39}$$

where $\nu_0 = I_H/h$ is the threshold frequency for ionizing photons. If we make the substitution $x \equiv h\nu/kT_c$ and define $x_0 = h\nu_0/kT_c$ then

$$\langle h\nu \rangle = kT_c \frac{\int_{x_0}^{\infty} e^{-x} dx}{\int_{x_0}^{\infty} e^{-x} x^{-1} dx} = \frac{kT_c e^{-x_0}}{\int_{x_0}^{\infty} e^{-x} x^{-1} dx}. \tag{4.40}$$

The integral in the denominator of Equation 4.40 is the first exponential integral $E_1(x_0)$; for $x_0 \gg 1$, corresponding to photon energies $h\nu \gg kT_c$,

$$E_1(x_0) \approx \frac{e^{-x_0}}{x_0}\left[1 - \frac{1}{x_0} + O(x_0^{-2})\right]. \tag{4.41}$$

Plugging this approximation back into Equation 4.40 yields the result

$$\langle h\nu \rangle \approx h\nu_0 + kT_c \approx I_H + kT_c. \tag{4.42}$$

This means that the electrons liberated by photoionization will go darting off with mean kinetic energy

$$\langle E \rangle = \langle h\nu \rangle - I_\mathrm{H} \approx kT_c. \tag{4.43}$$

Although this result was derived for a blackbody, detailed calculations with realistic stellar spectra reveal that $\langle E \rangle \approx kT_\mathrm{eff}$ is an approximation that is good to within a factor two.

In ionization equilibrium, the photoionization rate of hydrogen is balanced by the radiative recombination rate:

$$n_\mathrm{HI}\zeta_\mathrm{pho} = n_e n_\mathrm{HII}\alpha_\mathrm{B,H}. \tag{4.44}$$

Thus, the volumetric heating rate is

$$g_\mathrm{pho} = n_\mathrm{HI}\zeta_\mathrm{pho}\langle E \rangle \approx n_e n_\mathrm{HII}\alpha_\mathrm{B,H}kT_\mathrm{eff}. \tag{4.45}$$

Since the radiative recombination rate has the temperature dependence (Equation 4.12)

$$\alpha_\mathrm{B,H} \approx 3.1 \times 10^{-13}\,\mathrm{cm}^3\,\mathrm{s}^{-1}\left(\frac{T}{8000\,\mathrm{K}}\right)^{-0.82} \tag{4.46}$$

for gas temperatures $T \sim 8000\,\mathrm{K}$, the volumetric heating rate decreases with increasing temperature.

Photoionization can act as a potent heating source. An O3 main sequence star, for instance, will cause $4 \times 10^{49}$ photoionizations per second, each producing a free electron with energy $\sim kT_\mathrm{eff} \approx 3.9\,\mathrm{eV}$.[5] However, although the photoelectrons are liberated with mean energy 3.9 eV, the free electrons in the $\sim$8000 K nebula have mean energy $(3/2)kT \approx 1.0\,\mathrm{eV}$. Some cooling mechanism must be reducing the average kinetic energy per free electron. There are three sources of cooling available in an H II region: recombination cooling, free–free (or bremsstrahlung) cooling, and line emission from collisionally excited ions. Let's look at each mechanism in turn.

**Recombination cooling** occurs when a free electron undergoes radiative recombination with a proton to form a neutral hydrogen atom. In a single recombination, the total thermal energy of an H II region is decreased by an amount equal to the kinetic energy $E$ of the recombining electron. If the number density of free electrons is $n_e$ and the number density of free protons is $n_\mathrm{HII}$, the volumetric cooling rate resulting from radiative recombination is

$$\ell_\mathrm{rr} = n_e n_\mathrm{HII}\alpha_\mathrm{B,H}\langle E_\mathrm{rr} \rangle, \tag{4.47}$$

where $\langle E_\mathrm{rr} \rangle$ is the mean kinetic energy of all recombining electrons. Note that $\langle E_\mathrm{rr} \rangle$ is *not* equal to $(3/2)kT$. The reason is that the cross section for radiative

---

[5] As a heater, an O3 main sequence star would be rated at 27 million yottawatts.

recombination is a steeply decreasing function of electron kinetic energy. Thus, the slower-moving electrons are more likely to recombine and $\langle E_{rr} \rangle$ is less than $(3/2)kT$.

The cross section for radiative recombination can be approximated as a power law in the electron energy:

$$\sigma_{rr}(E) = \sigma_0 (E/E_0)^{\gamma}, \qquad (4.48)$$

where we expect $\gamma < 0$. The average energy of a recombining electron, weighted by cross section and integrated over a Maxwellian distribution of kinetic energies, is

$$\langle E_{rr} \rangle = (2 + \gamma)kT. \qquad (4.49)$$

For Case B recombination at temperatures $T \sim 8000\,\mathrm{K}$, an exponent $\gamma = -1.32$ gives a good fit to the energy dependence of the recombination cross section. Therefore the mean energy of the recombining electrons is $\langle E_{rr} \rangle \approx 0.68kT$, and the cooling rate from radiative recombination is

$$\ell_{rr} = (2 + \gamma)n_e n_{\mathrm{HII}} \alpha_{\mathrm{B,H}} kT \approx 0.68 n_e n_{\mathrm{HII}} \alpha_{\mathrm{B,H}} kT. \qquad (4.50)$$

If radiative recombination were the only way for cooling to occur, then the temperature $T$ would be found by setting Equation 4.45 for the heating rate equal to Equation 4.50 for the cooling rate, yielding

$$T = \frac{T_{\mathrm{eff}}}{2 + \gamma} \approx 1.47 T_{\mathrm{eff}}. \qquad (4.51)$$

Thus, if radiative recombination of hydrogen were the only cooling mechanism for a Strömgren sphere, the temperature of the ionized gas in the sphere would be $\sim 47\%$ *higher* than the temperature of the central star. The reason is that radiative recombination selectively removes the lower-energy free electrons, because of their higher cross sections for recombination, and thus increases the mean kinetic energy per free electron. Thus, to match the observations ($T < T_{\mathrm{eff}}$), we are going to need an additional cooling mechanism.

**Free–free cooling**, also known as bremsstrahlung cooling, occurs when free electrons are accelerated by close encounters with free protons or other ions, and thus emit radiation. For a pure hydrogen gas, the volumetric cooling rate is

$$\ell_{ff} = \Lambda_{ff,0} \left( \frac{kT}{m_e c^2} \right)^{1/2} g_{ff} n_e n_{\mathrm{HII}}, \qquad (4.52)$$

where the normalization factor is

$$\Lambda_{ff,0} = \frac{32\pi}{3} \left( \frac{2\pi}{3} \right)^{1/2} \frac{e^6}{h m_e c^2} = 1.10 \times 10^{-22}\,\mathrm{erg\,cm^3\,s^{-1}} \qquad (4.53)$$

and $g_{ff}$ is the frequency-averaged Gaunt factor, a number of order unity that depends only weakly on temperature. At a temperature $T \approx 8000\,\mathrm{K}$, appropriate for an H II region, the Gaunt factor is $g_{ff} \approx 1.3$. Comparing the rate for radiative recombination cooling (Equation 4.50) with the rate for free–free cooling (Equation 4.52), we find that

$$\frac{\ell_{\text{ff}}}{\ell_{\text{rr}}} = \frac{\Lambda_{\text{ff},0} g_{\text{ff}}}{(2+\gamma) m_e c^2} \frac{1}{\alpha_{\text{B},\text{H}}(T)} \left(\frac{kT}{m_e c^2}\right)^{-1/2}. \tag{4.54}$$

Since both radiative recombination and free–free emission depend on two-body encounters between electrons and protons, the factors $n_e n_{\text{HII}}$ cancel. Using appropriate values for $\alpha_{\text{B},\text{H}}$ (Equation 4.46) and $g_{\text{ff}}$, the ratio becomes

$$\frac{\ell_{\text{ff}}}{\ell_{\text{rr}}} \approx 0.7 \left(\frac{T}{8000\,\text{K}}\right)^{0.32}. \tag{4.55}$$

Thus, for temperatures $T \approx 8000\,\text{K}$, the rates of free–free cooling and radiative recombination cooling are similar. Adding the free–free cooling to the radiative recombination cooling significantly decreases the temperature of an H II region. An O3 main sequence star, for instance, with $T_{\text{eff}} = 44\,900\,\text{K}$, would produce a nebula with $T \approx 66\,000\,\text{K}$ if radiative recombination were the only cooling mechanism; it would produce a nebula with $T \approx 30\,000\,\text{K}$ if free–free cooling were added as well. Although free–free cooling brings the temperature below that of the central star, it still doesn't produce the temperature $T \sim 8000\,\text{K}$ that is actually observed for H II regions. Thus, to match the observations, we are going to need another cooling mechanism.

**Collisionally excited line emission** is the third mechanism by which nebulae cool. If a free electron collisionally excites an atom or ion from a lower energy level $E_\ell$ to an excited level $E_u > E_\ell$, the energy $E_{u\ell} = E_u - E_\ell$ gained by the excited atom or ion is taken from the free electron's kinetic energy. If the excited atom or ion then undergoes radiative de-excitation, and if the emitted photon escapes from the nebula, then there is a net cooling of the electron gas. For the gas to cool from temperatures $T \sim 30\,000\,\text{K}$ to $\sim 8000\,\text{K}$, the energy levels of the excited system have to be separated by a difference $E_{u\ell} \sim kT \sim 1 \to 3\,\text{eV}$. If $E_{u\ell}$ is much lower than this value, then the photons emitted by radiative de-excitation will carry away only a small amount of energy. If $E_{u\ell}$ is much higher than this value, then only the extreme high-energy tail of the free electron distribution will be able to excite the ions or atoms. Thus, Lyman $\alpha$ emission from neutral hydrogen atoms is ineffective at cooling H II regions down to 8000 K, since the energy $E_{u\ell} = 10.2\,\text{eV}$ required to excite hydrogen from its ground state is too high. Similarly, the first excited state of neutral helium is $E_{u\ell} \approx 20\,\text{eV}$ above its ground state; this is too energetic to be collisionally excited at temperatures $T \leq 30\,000\,\text{K}$.

This is where those scarce one-in-a-thousand heavy atoms come to the fore; elements such as oxygen and nitrogen play a key role in the cooling of H II regions by collisionally excited line emission. In particular, O II, N II, and O III have electron transitions in the important $1 \to 3\,\text{eV}$ range. Although many different

**Table 4.1** Main contributors to line cooling in H II regions[a]

| Transition | $\lambda$ [Å] | $A_{u\ell}$ [$10^{-3}$ s$^{-1}$] | $n_{crit}$ [$10^4$ cm$^{-3}$] |
|---|---|---|---|
| O II $^4$S – $^2$D | 3726 | 0.164 | 0.406 |
|  | 3729 | 0.041 | 0.130 |
| N II $^3$P – $^1$D | 6548 | 0.985 | 8.86 |
|  | 6583 | 2.91 | 8.86 |
| O III $^3$P – $^1$D | 4959 | 6.97 | 69.1 |
|  | 5007 | 20.46 | 69.1 |

[a] *Data from PyNeb v1.1.12 atomic data base: critical density computed at T = $10^4$ K*

atoms and ions can potentially contribute to cooling by line emission, the main species that dominate the cooling of real H II regions are listed in Table 4.1. Among the lines that cool H II regions are our old friends, the "nebulium" lines produced by forbidden transitions in O III. All the lines in Table 4.1 are forbidden lines, with $A_{u\ell} \sim 1$ kilosecond$^{-1}$.

If a collisional excitation is to result in cooling, it must be followed by a *radiative* de-excitation. If it's followed by a collisional de-excitation then the kinetic energy that is taken from the exciting electron is merely transferred to the de-exciting electron; this is a zero-sum game that doesn't decrease the total thermal energy of the electron gas. For radiative de-excitation to dominate over collisional de-excitation, the number density of free electrons must be lower than the critical density $n_{crit}$, as defined in Section 3.3. The critical density at an electron temperature $T = 10^4$ K is given in Table 4.1 for the cooling lines of greatest interest in H II regions. The critical densities range from $n_{crit} \sim 1300$ cm$^{-3}$ for the [O II] 3729 Å line to $\sim 7 \times 10^5$ cm$^{-3}$ for the [O III] lines.

Observed H II regions have a range of densities. Our standard Strömgren sphere was assumed to be embedded in the CNM and had $n_e = n_H \sim 40$ cm$^{-3}$, resulting in a Strömgren radius $R_s \sim 6$ pc. However, a newly formed hot star in the densest regions of a molecular cloud can have $n_H \sim 10^5$ cm$^{-3}$ in its vicinity; such a star will form an **ultracompact** H II region with electron density $n_e \sim 10^5$ cm$^{-3}$ and radius $R_s \sim 0.03$ pc (Equation 4.21). When we look at ionized nebulae with higher and higher electron densities, fewer and fewer of the possible cooling lines lie above their critical density $n_{crit}$. As a consequence, cooling becomes less effective at higher densities, and the equilibrium temperature of the nebula goes up. Consider, for instance, an H II region around a star emitting blackbody radiation at a temperature $T_c = 35\,000$ K. If the gas has solar metallicity and has a density $n_H = 4000$ cm$^{-3}$ (similar to that of the Orion Nebula), then the equilibrium gas temperature of the H II region is $T = 8050$ K. However, if the density is lowered to $n_H = 100$ cm$^{-3}$, its temperature drops to $T = 6600$ K; if the density is raised to $n_H = 10^5$ cm$^{-3}$, its equilibrium temperature rises to 9050 K.

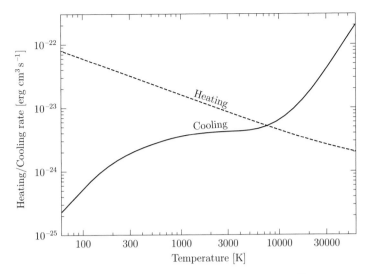

**Figure 4.4** Cooling versus heating in a nebula with $n_e = 100\,\text{cm}^{-3}$ and $Z = Z_\odot$, around a central blackbody with $T_c = 40\,000\,\text{K}$. Solid line: Cooling function $\Lambda = \ell/(n_e n_H)$. Dashed line: Heating function $g/(n_e n_H)$. The crossing point at $T \approx 7500\,\text{K}$ indicates the equilibrium temperature of this nebula. [Following Ferland 2003 and Gnedin & Hollon 2012]

The equilibrium temperature of a nebula also depends on its metallicity, the fraction of its mass provided by elements other than hydrogen and helium. Consider, for instance, our idealized H II region with density $n_H = 4000\,\text{cm}^{-3}$, metallicity $Z = Z_\odot$, and resulting gas temperature $T = 8050\,\text{K}$. If the metallicity is lowered to $Z = Z_\odot/10$, its temperature rises to $15\,600\,\text{K}$; if its metallicity is raised to $Z = 3Z_\odot$, its equilibrium temperature drops to just $T = 5400\,\text{K}$.

Changing the metallicity of an H II region by a factor 30 changes its equilibrium temperature by a factor $\sim$3; changing its density by a factor 1000 changes its temperature by a factor $\sim$1.4. This relative insensitivity to metallicity and density means that a statement like "The equilibrium temperature of this H II region is 8000 K" is quite robust, even if densities and metallicities are not perfectly known. Figure 4.4 shows a simplified plot of the cooling and heating rates for a nebula surrounding a star with effective temperature $T_{\text{eff}} = 40\,000\,\text{K}$. At different electron temperatures, different collisionally excited emission lines dominate the cooling rate. At $T \sim 1000\,\text{K}$, or $kT \sim 0.1\,\text{eV}$, the cooling is dominated by infrared fine-structure lines, such as the [S II] 18.7 μm line and, at lower temperatures, the [O I] 63 μm and [C II] 158 μm lines, which cool the CNM. Conversely, at $T \sim 25\,000\,\text{K}$, the cooling is dominated by ultraviolet lines such as Lyman $\alpha$. It is only at intermediate temperatures, $T \sim 8000\,\text{K}$, that optical forbidden lines from O II, N II, and O III dominate the cooling rate.

## 4.4 Temperature and Density Diagnostics

How do we measure the temperature and density of an H II region or a planetary nebula? Since we can't do sample-return missions, or deploy a kiloparsec-long thermometer, we must rely on determining the physical characteristics of a nebula from its spectrum.

Estimating the temperature of an astronomical object from its spectrum is a recurring theme in astronomy. For opaque objects such as stars, with (approximate) blackbody spectra, the temperature can be estimated from the continuum spectrum. However, ionized nebulae are far from being blackbodies. A spectrum of a typical H II region at visible and near-infrared wavelengths is shown in Figure 4.5. The continuum radiation is a mix of free–bound emission (from radiative recombination), free–free emission (or bremsstrahlung), and two-photon emission. To understand "two-photon" continuum emission, consider the lowest energy levels of the hydrogen atom. The ground state of hydrogen has principal quantum number $n = 1$ and orbital quantum number $\ell = 0$; in spectroscopic notation, the ground state of hydrogen is $1\,^2S$. If the principal quantum number is $n = 2$, there are two permissible values for the orbital angular momentum: $\ell = 0$ and $\ell = 1$. If $n = 2$ and $\ell = 1$ (written $2\,^2P^o$ in spectroscopic notation), then the downward transition $2\,^2P^o \rightarrow 1\,^2S$ is a permitted transition, with $A_{u\ell} = 6.26 \times 10^8\,\mathrm{s}^{-1}$ and $h\nu_{u\ell} = 10.2\,\mathrm{eV}$; this is the transition that produces Lyman $\alpha$ photons. By contrast, if $n = 2$ and $\ell = 0$ (written $2\,^2S$ in spectroscopic notation), then the downward transition $2\,^2S \rightarrow 1\,^2S$ is a forbidden transition,

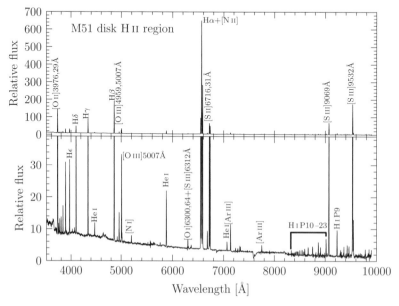

**Figure 4.5** Spectrum of a disk H II region in the Whirlpool Galaxy (M51: $d \sim 7\,\mathrm{Mpc}$). The top panel shows bright lines; the bottom panel is scaled to show fainter lines. Selected emission lines are labeled. [Data from Croxall et al. 2015]

with $A_{u\ell} = 8.22\,\text{s}^{-1}$. This forbidden transition can occur only through the emission of two photons, with total energy $h\nu' + h\nu'' = 10.2\,\text{eV}$. Although the spectral distribution of photons produced by this two-photon process peaks at 5.1 eV, it's a fairly broad peak, and photons are produced over the whole range $0 \rightarrow 10.2\,\text{eV}$. If you know enough about the temperature dependence of free–free, free–bound, and two-photon emission, you can use the continuum emission from an ionized nebula to estimate its temperature. However, since the emission lines of a nebula rise above the continuum like skyscrapers above gently rolling plains, it is usually easier to use the *emission lines* as a diagnostic of the nebula's temperature.

The key to using emission lines to estimate temperature is finding two excited states of the same ion whose energy differs by $\sim kT$, where $T$ is the temperature of the gas. For nebulae with $T \sim 10^4$ K, this implies energy differences $\Delta E \sim 1\,\text{eV}$. In practice, the lowest excited states of doubly ionized oxygen (O III), singly ionized nitrogen (N II), and doubly ionized sulfur (S III) are useful tools for estimating the temperatures of H II regions and planetary nebulae. The ions O III, N II, and S III have similar electronic structure; they are all four electrons short of a filled outer shell. Thus, their lowest energy levels have a similar form, as shown in Figure 4.6.

In each of these three ions, the ground state is compactly symbolized by the term symbol $^3P_J$, with three fine-structure sub-levels corresponding to angular momentum quantum number $J = 0$, 1, and 2. The first excited state of the ion is symbolized as $^1D$; from this excited state, a leap to the $J = 1$ or $J = 2$ sub-level of the ground state is a forbidden magnetic dipole transition, while the leap to the $J = 0$ sub-level is a more strongly forbidden electric quadrupole transition.

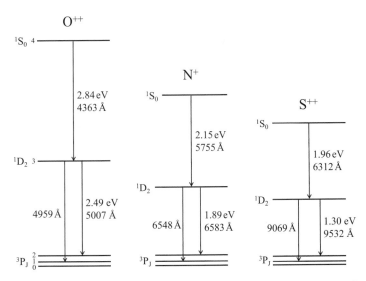

**Figure 4.6** Energy levels of the ions O III, N II, and S III; the energy separations among levels 0, 1, and 2 (typically $\Delta E \sim 0.03\,\text{eV}$) are exaggerated for clarity.

In O III, the excited $^1$D state is $\sim$2.5 eV above the ground state; radiative de-excitation from this state produces the blue-green [O III] 4959 Å, 5007 Å "nebulium" lines.[6] In N II, the $^1$D state is $\sim$1.9 eV above the ground state; radiative de-excitation from this state produces the red [N II] 6548 Å, 6583 Å doublet, which brackets the H$\alpha$ line (see Figure 4.5). In S III, the $^1$D state is $\sim$1.3 eV above the ground state; radiative de-excitation from this state produces the near-infrared doublet [S III] 9069 Å, 9532 Å, seen on the right in Figure 4.5. In general, lines produced by the transition from the $^1$D state to the $^3$P ground state are called **nebular lines**. (The reason is that these lines were first detected in planetary nebulae, the prime example being the "nebulium" lines seen by Huggins in the Cat's Eye Nebula.)

The second excited state in O III, N II, and S III is the $^1$S level. Transitions from the $^1$S state to the lower $^1$D state produce photons with wavelength $\lambda_0 = 4363$ Å for O III, $\lambda_0 = 5755$ Å for N II, and $\lambda_0 = 6312$ Å for S III. In general, lines produced by the transition from the upper $^1$S excited state to the lower $^1$D excited state are called **auroral lines**. (In this case the reason is that the $^1$S to $^1$D transition in O I produces a strong $\lambda_0 = 5577$ Å line in the Earth's aurorae; all other "auroral lines" are named by analogy with this bright line.) Note that all the nebular and auroral lines in Figure 4.6, from 4363 Å to 9532 Å, lie in the optical window where the Earth's atmosphere is transparent and thus are easily observed from the ground.

To see how observation of the relative strength of emission lines can lead to a temperature estimate, let's suppose that we are in the low-density limit, so that the free electron density is less than the critical density $n_{crit}$ for collisional de-excitation of each line. In this case, collisional excitation will be followed by radiative de-excitation; we can assume, in addition, that the radiative de-excitation will be primarily spontaneous rather than stimulated emission. To keep the notation simple, we will number the different electron energy levels from 0, for the lowest energy level, through 4, for the highest energy level. (Although there exist bound levels at still higher energies, they are nearly unpopulated at the temperatures typical of ionized nebulae, and we can safely ignore them.) The numerical labels for the different energy levels are shown in Figure 4.6.

The emissivity of the $4 \rightarrow 3$ (auroral) transition, integrated over the entire line width, is (Equation 3.15)

$$j(4 \rightarrow 3) = n_4 \frac{A_{43}}{4\pi} h\nu_{43}. \tag{4.56}$$

In excitation equilibrium, the rate of collisional excitation from the ground state to level 4 will be balanced by radiative de-excitation from level 4. Since transitions

---

[6] There is also a very weak $\lambda_0 = 4931$ Å line from the strongly forbidden transition to the $J = 0$ sub-level. It is usually too weak (when it is detected at all) to be a useful diagnostic tool.

to levels 2 and 0 from this level are strongly forbidden, we need only take into account the transitions to levels 3 and 1 when computing the equilibrium state:

$$n_0 n_e k_{04} = n_4 (A_{43} + A_{41}).$$ (4.57)

Substituting Equation 4.57 into Equation 4.56, we find that the emissivity can be written as

$$j(4 \rightarrow 3) = \frac{n_0 n_e}{4\pi} k_{04} \frac{A_{43}}{A_{43} + A_{41}} h\nu_{43}.$$ (4.58)

In Equation 4.58, the only temperature dependence is in the collisional rate coefficient $k_{04}$.

The emissivity of the $3 \rightarrow 2$ (nebular) transition is a bit more complicated, since level 3 can be populated in two ways: by collisional excitation directly from the ground state, or by collisional excitation from the ground state to level 4, followed by radiative de-excitation to level 3. This means that the emissivity is

$$j(3 \rightarrow 2) = \frac{n_0 n_e}{4\pi} \left[ k_{03} + k_{04} \frac{A_{43}}{A_{43} + A_{41}} \right] \frac{A_{32}}{A_{32} + A_{31}} h\nu_{32}.$$ (4.59)

Taking the ratio of Equation 4.58 to Equation 4.59 gives the relative strength of the $4 \rightarrow 3$ auroral line in relation to the $3 \rightarrow 2$ nebular line:

$$\frac{j(4 \rightarrow 3)}{j(3 \rightarrow 2)} = \frac{A_{43} \nu_{43}}{A_{32} \nu_{32}} \frac{(A_{32} + A_{31}) k_{04}}{(A_{43} + A_{41}) k_{03} + A_{43} k_{04}}.$$ (4.60)

Since the Einstein A coefficients and the emission frequencies are dictated by quantum mechanics, the temperature dependence in Equation 4.60 enters solely through the rate coefficients $k_{04}$ and $k_{03}$. The collisional rate coefficients for excitation and de-excitation are related by the equation

$$k_{0u} = k_{u0} \frac{g_u}{g_0} \exp \left( -\frac{h\nu_{u0}}{kT} \right).$$ (4.61)

The collisional de-excitation rate can be conveniently written in the form

$$k_{u0} = \frac{\beta}{T^{1/2}} \frac{\Omega_{u0}}{g_u},$$ (4.62)

where

$$\beta \equiv \left( \frac{2\pi \hbar^4}{k m_e^3} \right)^{1/2} = 8.629 \times 10^{-6} \, \text{cm}^3 \, \text{s}^{-1} \, \text{K}^{1/2},$$ (4.63)

and $\Omega_{u0}$ is the dimensionless **collision strength**. Equation 4.62 is convenient because when a positively charged ion, such as O III, is collisionally de-excited by a free electron, the temperature dependence of the collisional rate coefficient $k_{u0}$ is very nearly proportional to $T^{-1/2}$. This means that the collision strength $\Omega_{u0}$ is nearly independent of temperature; the constant $\beta$ is chosen so that $\Omega_{u0}$ is a factor of order unity.

By substituting Equations 4.62 and 4.61 into Equation 4.60 and making use of the fact that $h\nu_{40} = h\nu_{30} + h\nu_{43}$, we find a convenient equation relating the strength of the $4 \rightarrow 3$ auroral line to the $3 \rightarrow 2$ nebular line:

$$\frac{j(4 \rightarrow 3)}{j(3 \rightarrow 2)} = \frac{A_{43}\nu_{43}}{A_{32}\nu_{32}} \frac{(A_{32} + A_{31})\Omega_{40} \exp(-h\nu_{43}/kT)}{(A_{43} + A_{41})\Omega_{30} + A_{43}\Omega_{40} \exp(-h\nu_{43}/kT)}. \tag{4.64}$$

In Equation 4.64, all the temperature dependence, aside from the weak dependence of $\Omega_{u0}$ on $T$, is contained in the exponential $\exp(-h\nu_{43}/kT)$ factor. Thus, the line ratio is most useful for measuring temperatures $kT \sim h\nu_{43}$, which ranges from 1.96 eV for S III to 2.84 eV for O III.

For O III, to take an example, the ratio of line strengths ranges from

$$\frac{j([\text{O III}] 4363 \text{ Å})}{j([\text{O III}] 5007 \text{ Å})} = 0.0016, \tag{4.65}$$

when $T = 7000$ K ($kT \approx 0.2h\nu_{43}$), to

$$\frac{j([\text{O III}] 4363 \text{ Å})}{j([\text{O III}] 5007 \text{ Å})} = 0.033 \tag{4.66}$$

when $T = 20\,000$ K ($kT \approx 0.6h\nu_{43}$). These relatively small ratios reveal one difficulty in measuring the temperature of ionized nebulae: even when the [O III] 5007 Å line is strong, the [O III] 4363 Å line will be weak at plausible nebula temperatures. (This line is undetected, for instance, in the relatively cool H II region shown in Figure 4.5.) Moreover, the 4363 Å line lies close to a 4358 Å emission line produced by atomic mercury and thus is difficult to observe when mercury vapor lamps are a significant source of light pollution.

Emission lines can also be used to estimate the **free electron density** of an ionized nebula. For this purpose, you need an ion with two electronic transitions that are similar in energy but have very different critical densities. One such ion is singly ionized oxygen, whose energy levels are shown in Figure 4.7. In the seven-electron O II ion, the first excited state, $^2$D, is split into a pair of fine-structure sub-levels corresponding to angular momentum quantum numbers $J = 3/2$ and $5/2$. These sub-levels, $^2$D$_{3/2}$ and $^2$D$_{5/2}$, are separated by just 0.002 eV in energy; thus, transitions from the first excited level to the ground level produce a closely separated doublet, with $\lambda = 3728.8$ Å and 3726.1 Å. The critical density for the transition from the lower energy $^2$D$_{5/2}$ sub-level (level 1 in Figure 4.7) is $n_{\text{crit}, 1} \approx 1300$ cm$^{-3}$ at $T = 10^4$ K. The critical density for the transition from the higher energy $^2$D$_{3/2}$ sub-level (level 2 in Figure 4.7) is $n_{\text{crit}, 2} \approx 4100$ cm$^{-3}$ at $T = 10^4$ K.

The emissivity of the $1 \rightarrow 0$ transition, integrated over the entire line width, is

$$j(1 \rightarrow 0) = n_1 \frac{A_{10}}{4\pi} h\nu_{10}. \tag{4.67}$$

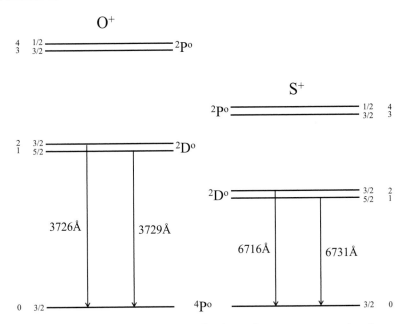

**Figure 4.7** Energy levels of the O II and S II ions; the energy separation between the doublet fine-structure sub-levels (typically $\Delta E \sim 2\,\text{meV}$ for the $^2\text{D}$ levels) is greatly exaggerated for clarity.

In ionization equilibrium, the rate of collisional excitation from the ground state will be balanced by a mix of radiative and collisional de-excitation:

$$n_0 n_e k_{01} = n_1 (A_{10} + n_e k_{10}). \tag{4.68}$$

Substituting Equation 4.68 into Equation 4.67, we find the emissivity is

$$j(1 \to 0) = n_0 n_e \frac{k_{01}}{A_{10} + n_e k_{10}} \frac{A_{10}}{4\pi} h\nu_{10}$$

$$= n_0 n_e \frac{k_{01}}{1 + n_e/n_{\text{crit},1}} \frac{h\nu_{10}}{4\pi}. \tag{4.69}$$

Similarly, we may write

$$j(2 \to 0) = n_0 n_e \frac{k_{02}}{1 + n_e/n_{\text{crit},2}} \frac{h\nu_{20}}{4\pi}. \tag{4.70}$$

Thus, the ratio of the observed strengths of the two lines in the doublet will be

$$\frac{j(2 \to 0)}{j(1 \to 0)} = \frac{\nu_{20}}{\nu_{10}} \frac{k_{02}}{k_{01}} \frac{1 + n_e/n_{\text{crit},1}}{1 + n_e/n_{\text{crit},2}}. \tag{4.71}$$

From what we know about collisional rate coefficients, encapsulated in Equations 4.61 and 4.62, we can write

$$\frac{k_{02}}{k_{01}} = \frac{\Omega_{20}}{\Omega_{10}} \exp(-h\nu_{21}/kT) \approx \frac{\Omega_{20}}{\Omega_{10}}, \tag{4.72}$$

since $h\nu_{21}$ for the O II ion is just 2 meV, far smaller than the thermal energy of ionized gas. Thus, we can write the ratio of the line strengths as

$$\frac{j(2 \to 0)}{j(1 \to 0)} \approx \frac{\Omega_{20}}{\Omega_{10}} \frac{1 + n_e/n_{\text{crit}, 1}}{1 + n_e/n_{\text{crit}, 2}}, \tag{4.73}$$

where we have set $\nu_{20} = \nu_{10}$, since the frequency of the two lines differs by less than one part per thousand.

In the low-density limit, where $n_e \ll n_{\text{crit}, 1} < n_{\text{crit}, 2}$, the ratio of line strengths is

$$\frac{j(2 \to 0)}{j(1 \to 0)} \approx \frac{\Omega_{20}}{\Omega_{10}}, \tag{4.74}$$

which is independent of the free electron density $n_e$. In the high-density limit, where $n_e \gg n_{\text{crit}, 2} > n_{\text{crit}, 1}$, the ratio of line strengths is

$$\frac{j(2 \to 0)}{j(1 \to 0)} \approx \frac{\Omega_{20}}{\Omega_{10}} \frac{n_{\text{crit}, 2}}{n_{\text{crit}, 1}} \approx \frac{\Omega_{20}}{\Omega_{10}} \frac{A_{20}/k_{20}}{A_{10}/k_{10}} \approx \frac{g_2}{g_1} \frac{A_{20}}{A_{10}}, \tag{4.75}$$

and is again independent of the free electron density $n_e$.

For the O II ion, to take an example, the ratio of line strengths ranges from

$$\frac{j([\text{O II}] \, 3726.1 \, \text{Å})}{j([\text{O II}] \, 3728.8 \, \text{Å})} = 1.50, \tag{4.76}$$

when $n_e \ll n_{\text{crit}, 1} \approx 1300 \, \text{cm}^{-3}$, to

$$\frac{j([\text{O II}] \, 3726.1 \, \text{Å})}{j([\text{O II}] \, 3728.8 \, \text{Å})} = 0.38 \tag{4.77}$$

when $n_e \gg n_{\text{crit}, 2} \approx 4100 \, \text{cm}^{-3}$. Thus, the strengths of the two lines in the doublet will always be comparable, as shown in Figure 4.8. Unfortunately, since the two lines are separated by only 2.7 Å, you will need a fairly high-resolution spectrograph, with $R \gg 3730 \, \text{Å}/2.7 \, \text{Å} \approx 1400$, to get accurate measurement of the [O II] doublet line ratio.

The [O II] doublet is not the only useful diagnostic for estimating free electron density. For instance, the [S II] 6716, 6731 Å doublet has the critical density $n_{\text{crit}, 1} \approx 3100 \, \text{cm}^{-3}$ for one of its lines and $n_{\text{crit}, 2} \approx 1200 \, \text{cm}^{-3}$ for the other; this makes it a useful diagnostic in the same density range as the [O II] doublet (Figure 4.8). The main drawback to using S II, however, is that sulfur is a rarer element than oxygen, with only one sulfur atom for every 40 oxygen atoms at solar abundance. However, the [S II] doublet lines are separated by $\sim$15 Å, making them measurable with relatively low-resolution spectrometers.

In practice, nebular temperature and density diagnostics are rarely used in isolation from each other; temperature diagnostics have a weak density dependence, and vice versa. Instead, numerical methods are employed that combine observed temperature and density diagnostic line ratios to derive self-consistent

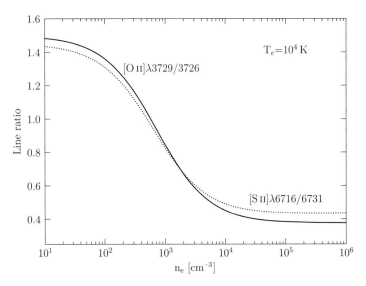

**Figure 4.8** Changes in the O II (solid) and S II (dotted) doublet line ratios as a function of free electron density. An electron temperature $T = 10^4$ K is assumed. [Calculated using PyNeb v1.1.12]

temperature and density solutions for a region; more sophisticated methods might apply multivariate or Bayesian analysis to multiple diagnostics to estimate likely values and their associated uncertainties. Single-ratio diagnostics, however, are still useful for making qualitative assessments: the relative strengths of the diagnostic emission lines can reveal at a glance whether an H II region is at the hot or cold end of the typical range for warm ionized gas or whether it is composed of relatively high- or low-density gas.

## 4.5 Dynamics of H II Regions

During World War I, the newspapers were full of news from the Western Front and the Eastern Front. The term "front," as a synonym for the boundary between opposing armies, thus changed from a technical military term to a word in everyday use. Around the year 1920, meteorologists at the University of Bergen introduced the terms "cold front" to describe the boundary between advancing cold air and retreating warm air and "warm front" to describe the boundary between advancing warm air and retreating cold air. The term "front" was then borrowed by astronomers to describe a more-or-less abrupt boundary between two regions of the ISM with very different properties. For instance, as Strömgren pointed out, an ionized nebula can be approximated as a region of highly ionized gas, separated from the surrounding neutral medium by a thin boundary region, of thickness $\lambda_{\mathrm{mfp}} \sim 0.001$ pc. Thus, an H II region is surrounded by an **ionization front**.

From Equation 4.20, which gives the size of a Strömgren sphere as a function of time, we can compute the rate of expansion of the ionization front:

$$\frac{dR}{dt} = u_s \frac{e^{-t/t_{\text{rec}}}}{(1 - e^{-t/t_{\text{rec}}})^{2/3}}, \tag{4.78}$$

where the characteristic expansion speed is

$$u_s = \frac{R_s}{3t_{\text{rec}}} \tag{4.79}$$

$$= 680 \, \text{km s}^{-1} \left(\frac{Q_0}{10^{49} \, \text{s}^{-1}}\right)^{1/3} \left(\frac{\alpha_{\text{B,H}}}{3 \times 10^{-13} \, \text{cm}^3 \, \text{s}^{-1}}\right)^{2/3} \left(\frac{n_{\text{H}}}{40 \, \text{cm}^{-3}}\right)^{1/3}.$$

The expansion speed of the ionization front at $t \ll t_{\text{rec}}$ is

$$\frac{dR}{dt} \approx u_s \left(\frac{t}{t_{\text{rec}}}\right)^{-2/3} \qquad [t \ll t_{\text{rec}}]. \tag{4.80}$$

Although formally this diverges as $t \to 0$, it's physically impossible for the ionization front to expand faster than light. The expansion speed of the ionization front at $t \gg t_{\text{rec}}$ dwindles exponentially to

$$\frac{dR}{dt} \approx u_s e^{-t/t_{\text{rec}}} \qquad [t \gg t_{\text{rec}}]. \tag{4.81}$$

Suppose the ionization front is expanding into the cold neutral medium, which has a sound speed $c_s \sim 1 \, \text{km s}^{-1}$. The front then expands at supersonic speed relative to the unwitting neutral gas when

$$t < t_{\text{rec}} \ln \left(\frac{u_s}{c_s}\right) \approx 6.5 t_{\text{rec}}. \tag{4.82}$$

According to this analysis, by the time the expanding Strömgren sphere is $\sim 20\,000$ years old, it has settled sedately into a boring near-equilibrium state, with radius $R \sim 0.999 R_s$, and a leisurely subsonic expansion of the ionized region. However, our analysis has so far ignored the extreme pressure imbalance between the hot ionized gas inside the Strömgren sphere and the cold neutral gas outside. The global expansion of the Strömgren sphere as a result of pressure forces occurs only on a relatively long time scale, the sound crossing time, $R_s/c_s \sim 0.5 \, \text{Myr}$. However, the structure of the gas in the immediate vicinity of the ionization front can become significantly altered on much shorter time scales.

Let's look at the basic hydrodynamics of an expanding H II region or supernova remnant. At a location $\vec{r}$ and time $t$, gas can be characterized by its mass density $\rho(\vec{r}, t)$, bulk velocity $\vec{u}(\vec{r}, t)$, and pressure $P(\vec{r}, t)$. Low-density gas, such as that of the ISM, can be treated as an ideal gas, with no viscosity and with the pressure given by the ideal gas law

$$P = \frac{\rho kT}{m}, \tag{4.83}$$

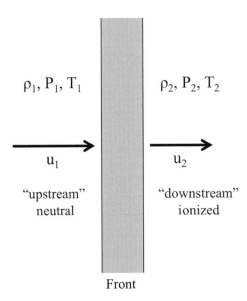

$\rho_1, P_1, T_1$        $\rho_2, P_2, T_2$

$u_1$        $u_2$

"upstream"        "downstream"
neutral        ionized

Front

**Figure 4.9** Geometry of a plane parallel ionization front.

where $T(\vec{r}, t)$ is the kinetic temperature and $m$ is the mean molecular mass of the gas particles. Consider a small patch of the ionization front between the interior of an H II region and its exterior (or between any ionized region and an adjacent neutral region). If the patch is small compared with the ionization front's radius of curvature, then we can treat the ionization front as if it has plane parallel symmetry, as shown in Figure 4.9. The density, pressure, and temperature of the neutral gas, which has not yet been overrun by the ionization front, are designated $\rho_1$, $P_1$, and $T_1$. It is convenient to use a frame of reference in which the ionization front is stationary; in this frame the bulk velocity $\vec{u}_1$ of the neutral gas points toward the ionization front. The speed $u_1$ of the neutral gas relative to the ionization front is thus equivalent to the expansion speed $dR/dt$ of a spherical ionization front relative to the surrounding neutral gas. The density, pressure, and temperature of the ionized gas are $\rho_2$, $P_2$, and $T_2$. In the ionization front's frame of reference, the bulk velocity $\vec{u}_2$ of the ionized gas points away from the ionization front.

Mass conservation tells us that

$$\frac{\partial \rho}{\partial t} = -\vec{\nabla} \cdot (\rho \vec{u}). \tag{4.84}$$

In a plane parallel system, the mass conservation equation reduces to

$$\frac{\partial \rho}{\partial t} = -\frac{\partial}{\partial x}(\rho u). \tag{4.85}$$

Momentum conservation tells us that

$$\rho \frac{\partial \vec{u}}{\partial t} + \rho(\vec{u} \cdot \vec{\nabla})\vec{u} = -\vec{\nabla}P \qquad (4.86)$$

when gravitational and magnetic forces are small compared with the pressure gradient force. In a plane parallel system, the momentum conservation equation reduces to

$$\rho \frac{\partial u}{\partial t} = -\rho u \frac{\partial u}{\partial x} - \frac{\partial P}{\partial x}. \qquad (4.87)$$

Now we are going to make a slightly riskier assumption. We're going to assume that there exists a steady-state solution in which the density $\rho$ and velocity $u$ have no explicit time dependence in the ionization front's reference frame. This assumption is not strictly true for an expanding ionization front around a hot star; we saw in Equation 4.78 that the expansion speed $dR/dt$ of the front decreases with time. However, at times $t \gg t_{\rm rec}$, the deceleration of the expansion becomes exponentially small, with

$$\frac{d^2 R}{dt^2} \approx -g_s e^{-t/t_{\rm rec}}, \qquad (4.88)$$

where the characteristic acceleration is

$$g_s = \frac{u_s}{t_{\rm rec}} = \frac{R_s}{3t_{\rm rec}^2} \qquad (4.89)$$

$$= 0.0012 \, {\rm cm\,s}^{-2} \left( \frac{Q_0}{10^{49}\,{\rm s}^{-1}} \right)^{1/3} \left( \frac{\alpha_{\rm B,H}}{3 \times 10^{-13}\,{\rm cm}^3\,{\rm s}^{-1}} \right)^{5/3} \left( \frac{n_{\rm H}}{40\,{\rm cm}^{-3}} \right)^{4/3}.$$

Thus, although a frame of reference attached to a spherically expanding ionization front is not perfectly inertial, the effective acceleration in such a frame becomes arbitrarily tiny at late times.

For a plane parallel steady-state system, the mass conservation equation becomes

$$\frac{d}{dx}(\rho u) = 0, \qquad (4.90)$$

and the momentum conservation equation becomes

$$\rho u \frac{du}{dx} + \frac{dP}{dx} = 0, \qquad (4.91)$$

which can also be written as

$$\frac{d}{dx}(\rho u^2) - u\frac{d}{dx}(\rho u) + \frac{dP}{dx} = 0. \qquad (4.92)$$

However, once we substitute Equation 4.90, this becomes

$$\frac{d}{dx}(\rho u^2 + P) = 0. \qquad (4.93)$$

Thus, the quantities $\rho u$ and $\rho u^2 + P$ must be the same at every location $x$. In particular, when we compare the properties $\rho_1$, $u_1$, and $P_1$ just upstream of the ionization front with the properties $\rho_2$, $u_2$, and $P_2$ just downstream of the ionization front, we find

$$\rho_1 u_1 = \rho_2 u_2 \qquad (4.94)$$

and

$$\rho_1 u_1^2 + P_1 = \rho_2 u_2^2 + P_2. \qquad (4.95)$$

We may also write the equation of momentum conservation as

$$\rho_1 (u_1^2 + c_1^2) = \rho_2 (u_2^2 + c_2^2), \qquad (4.96)$$

where $c_1 = (P_1/\rho_1)^{1/2}$ and $c_2 = (P_2/\rho_2)^{1/2}$ are the isothermal sound speeds in the neutral and ionized gas. If the neutral gas consists of atomic hydrogen then

$$c_1 = \left(\frac{kT_1}{m_H}\right)^{1/2} = 0.81 \text{ km s}^{-1} \left(\frac{T_1}{80 \text{ K}}\right)^{1/2}. \qquad (4.97)$$

If the ionized gas consists of almost entirely ionized hydrogen,

$$c_2 = \left(\frac{2kT_2}{m_H}\right)^{1/2} = 11.5 \text{ km s}^{-1} \left(\frac{T_2}{8000 \text{ K}}\right)^{1/2}. \qquad (4.98)$$

Thus, for typical nebulae, we expect $c_2$, the sound speed of the ionized gas inside the nebula, to be more than ten times $c_1$, the sound speed in the neutral gas outside the nebula.

Normally, when we use Equations 4.94 and 4.96 we have some idea of the properties of the ambient neutral medium ($\rho_1$, $P_1$, $T_1$), and we want to know what happens to the properties of the gas when it has been ionized. A bit of algebra applied to Equations 4.94 and 4.96 yields the relation

$$\frac{\rho_2}{\rho_1} = \frac{1}{2c_2^2}[c_1^2 + u_1^2 \pm \sqrt{f(u_1)}], \qquad (4.99)$$

where the function $f(u_1)$ is given by

$$f(u_1) \equiv (c_1^2 + u_1^2)^2 - 4c_2^2 u_1^2. \qquad (4.100)$$

More usefully, this function can be written as

$$f(u_1) = (u_1^2 - v_R^2)(u_1^2 - v_D^2), \qquad (4.101)$$

where

$$v_R \equiv c_2 + \sqrt{c_2^2 - c_1^2} \approx 2c_2 \qquad (4.102)$$

and

$$v_D \equiv c_2 - \sqrt{c_2^2 - c_1^2} \approx \frac{1}{2}\left(\frac{c_1}{c_2}\right)^2 c_2 \ll c_2. \qquad (4.103)$$

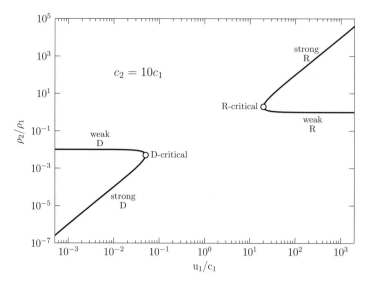

**Figure 4.10** Physically permitted values of the density contrast $\rho_2/\rho_1$ across a steady-state ionization front with $c_2 = 10c_1$.

The ratio $\rho_2/\rho_1$ must be a real number; thus, the speed of the ionization front relative to the neutral gas must be either $u_1 \geq v_R \sim 2c_2$ or $u_1 \leq v_D \ll c_2$. The racing, rapidly propagating ionization fronts with $u_1 \geq v_R$ are called **R-type fronts**. The dawdling, dilatory ionization fronts with $u_1 \leq v_D$ are called **D-type fronts**.[7]

In a real nebula, with $v_R \gg v_D$, an R-type front has $u_1 > v_R > c_2 > c_1$ and is supersonic with respect to the neutral medium. A D-type front has $u_1 < v_D < c_1 < c_2$ and is subsonic with respect to the neutral medium. For a given front propagation speed $u_1$, there are two possible values of the density ratio $\rho_2/\rho_1$, corresponding to taking the plus or minus sign in the solution for the quadratic equation (Equation 4.99). This is shown graphically in Figure 4.10, which is a plot of the density jump $\rho_2/\rho_1$ across a steady-state ionization front as a function of the propagation speed $u_1$. The front that has the larger density contrast is called the **strong** front; the front that has the smaller density contrast is called the **weak** front. Thus, there are four types of ionization front: weak R, strong R, weak D, and strong D. It turns out that strong R-type fronts are unstable, and are not seen in nature. The other types are seen around H II regions and planetary nebulae.

Consider a young, expanding H II region. At early times (Equation 4.82), the expansion speed $u_1$ of the ionization front is supersonic with respect to the sound speed $c_1$ of the surrounding neutral gas. The ionization front therefore starts out

---

[7] When Franz Daniel Kahn first wrote of R-type and D-type fronts in 1954, he chose "R" to stand for "rarefied" and "D" to stand for "dense." An R-type front runs through rarefied gas and compresses it ($\rho_2 > \rho_1$), while a D-type front runs through dense gas and expands it ($\rho_2 < \rho_1$).

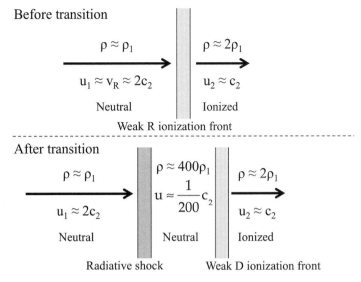

**Figure 4.11** Transition from an R-critical front (upper panel) to a shock front followed by a D-critical front (lower panel).

as a weak R-type front. In the limit that $u_1 \gg c_2 \gg c_1$, the density jump across the ionization front is

$$\frac{\rho_2}{\rho_1} \approx 1 + \frac{c_2^2}{u_1^2} \qquad \text{[weak R].} \qquad (4.104)$$

A weak R-type front compresses the gas only slightly. This period in the expansion, when the interior and exterior densities are nearly the same, is when the approximation of a Strömgren sphere is useful.

As the H II region grows, its expansion speed continuously decreases. Eventually, the speed $u_1$ drops to a value $u_1 = v_R \approx 2c_2$; for our standard Strömgren sphere, this occurs when $t \sim 3.4 t_{\text{rec}} \sim 10^4$ yr. At this moment, the front is called an "R-critical" front; the density ratio has risen to $\rho_2/\rho_1 \approx 2$ and the speed of the ionized gas relative to the front is $u_2 \approx c_2$. Once the ionization front slows still farther, the R-type front can no longer exist. However, to become a D-type front, the propagation speed $u_1$ has to drop to $u_1 = v_D \ll v_R$. What happens? Does the front instantaneously (and unphysically) decelerate? Does the entire system promptly vanish in a puff of logic?

What happens in reality is that after a time-dependent transition period, the R-critical ionization front splits into a pair of fronts, as shown in Figure 4.11. After the split, a leading **shock front** is followed by a D-critical ionization front. A shock front is the boundary between two regions of gas with different densities, pressures, and temperatures, but not necessarily different ionization states. One characteristic of a shock front is that it propagates with a speed $u_1 > c_1$; that is, it travels at supersonic speeds relative to the gas that it has not yet encountered. The shock front in the lower panel of Figure 4.11 increases the density of the gas, decreases its speed (relative to the shock front), and increases its temperature and thus its sound speed. Because of the decrease in the speed of the gas flow, and the increase in the sound speed, the gas is now subsonic and is ready to pass through a D-type ionization front. As the H II region expands still further, the leading shock front gradually weakens and the trailing D-critical front develops into a weak D-type front. Thus, a parcel of neutral gas will be first compressed by the shock front and then blasted with UV photons and ionized.

As time goes on, the shock front expands outward and decreases in amplitude until it deteriorates into a sound wave of infinitesimal amplitude, propagating outward at speed $c_1$. The ionization front that follows behind slows down until $u_1 \ll c_1$, at which point the density contrast is $\rho_2/\rho_1 \approx c_1^2/c_2^2 \approx 0.5(T_1/T_2)$ and the two sides of the ionization front approach pressure equilibrium. If the ionizing star is immortal, the final equilibrium radius, $R_{\text{final}}$, is attained when the sphere of ionized gas is in ionization equilibrium, which is given by the relation

$$R_{\text{final}} = \left( \frac{3Q_0}{4\pi \alpha_{\text{B, H}} n_{\text{final}}^2} \right)^{1/3}, \tag{4.105}$$

and is also in pressure equilibrium, given by the relation

$$2n_{\text{final}}kT_2 = n_1 kT_1. \tag{4.106}$$

Thus, we may write

$$R_{\text{final}} = \left( \frac{2T_2}{T_1} \right)^{2/3} R_s, \tag{4.107}$$

where $R_s$ is the Strömgren radius that we calculated when we were not worrying about the pressure (Equation 4.21). For our standard Strömgren sphere,

$$R_{\text{final}} \sim 34R_s \sim 200 \, \text{pc}. \tag{4.108}$$

In practice, H II regions around hot, luminous O stars cannot reach this equilibrium radius, since an O star will experience a supernova explosion before pressure equilibrium can be reached.

That's as good a transition as any to our next chapter, which deals with supernova remnants and how they lead to the existence of a hot ionized medium.

### Exercises

4.1   A photon of any energy can undergo Thomson scattering from a free electron. The electron's cross section for scattering is

$$\sigma_e = 6.6525 \times 10^{-25} \, \text{cm}^2. \tag{4.109}$$

A photon of energy $hv \approx 10.2\,\mathrm{eV}$ can be absorbed while lifting an electron from the $n = 1$ to the $n = 2$ state of a hydrogen atom. The atom's cross section for this Lyman $\alpha$ absorption is (see Equation 2.49)

$$\sigma_{\mathrm{Ly}\alpha} = \frac{\tau_0}{N_\ell} = 7.5 \times 10^{-13}\,\mathrm{cm}^2 \left(\frac{b}{1\,\mathrm{km\,s^{-1}}}\right)^{-1} \qquad (4.110)$$

at line center, assuming thermal broadening. A photon of energy $hv \geq 13.6\,\mathrm{eV}$ can be absorbed in photoionizing a hydrogen atom. The atom's photoionization cross section is (Equations 4.2 and 4.3)

$$\sigma_{\mathrm{pho}} = 6.30 \times 10^{-18}\,\mathrm{cm}^2 \left(\frac{hv}{13.6\,\mathrm{eV}}\right)^{-3}. \qquad (4.111)$$

(a)  Consider a sightline through the cold neutral medium. What distance $d$ through the CNM corresponds to an optical depth $\tau = 1$ for (i) Thomson scattering, (ii) Lyman $\alpha$ absorption at line center, and (iii) hydrogen photoionization? [Assume reasonable values for $n_{\mathrm{H}}$, $n_e$, and $T$.]

(b)  Consider a sightline through the warm ionized medium. What distance $d$ through the WIM corresponds to an optical depth $\tau = 1$ for (i) Thomson scattering, (ii) Lyman $\alpha$ absorption at line center, and (iii) hydrogen photoionization?

(c)  Consider an H II region with density $n_{\mathrm{H}}$ around a star with $Q_0 = 10^{48}\,\mathrm{s^{-1}}$. What is the Strömgren radius of the H II region, assuming pure hydrogen? If the neutral fraction is $f_n = 1 - x = 10^{-4}$ inside the H II region, what is the optical depth through the center of the H II region for (i) Thomson scattering, (ii) Lyman $\alpha$ absorption at line center, and (iii) hydrogen photoionization?

4.2  Use the table of stellar parameters for O3, O6, and O9 stars in Martins et al. (2005, A&A, 436, 1049) to write down $Q_0$ and $Q_1$, the rates of production of hydrogen- and helium-ionizing photons, respectively (use Table 5 for luminosity class III stars).

(a)  Calculate the radii of the Strömgren spheres around each of these stars assuming a density $n = 10^3\,\mathrm{cm^{-3}}$. Assume Case B recombination and a temperature $T = 10^4\,\mathrm{K}$.

(b)  For the O3 star only, imagine that it has broken free of its parent molecular cloud and is now ionizing part of the WNM with density $n = 0.3\,\mathrm{cm^{-3}}$. How big is the Strömgren sphere associated with this O star?

(c)  Using the rate of production of helium-ionizing photons, $Q_1$, for these three stars, estimate the relative sizes of the hydrogen and helium Strömgren spheres, using the discussion in Section 4.2 (Equations 4.32–4.36). Do the H II regions follow the description of the relative sizes of the H and He ionization zones in Figure 4.3?

4.3   While the Strömgren radius of an ionized nebula is usually defined in terms of hydrogen ionization, the concept can be generalized to compute a Strömgren radius $R(X^i)$ for any species $X^i$ in photoionization equilibrium with a source of $X^i$-ionizing photons:

$$Q(X^i) = \int_{\nu_0}^{\infty} \frac{L_\nu}{h\nu} d\nu = n_e n_{i+1} \alpha(X^{i+1}) \frac{4\pi}{3} R(X^i)^3, \qquad (4.112)$$

where $\nu_0$ is the ionization threshold frequency for $X^i$, and $n_{i+1}$ and $\alpha(X^{i+1})$ are the number density and radiative recombination rate for $X^{i+1}$.

Consider a B0 giant with $T_{\rm eff} = 31\,500$ K embedded in uniform-density gas composed of only hydrogen and carbon, with $n_C/n_H = 2 \times 10^{-4}$. (For simplicity, we ignore helium.) $C^0$ has an ionization threshold $h\nu_0 = 11.18$ eV, and the $C^+$ recombination coefficient is $\alpha(C^+) = 4.66 \times 10^{-13}$ cm$^3$ s$^{-1}$. For hydrogen, assume Case B recombination.

(a)   Compute the relative number of photons that can ionize $H^0$ and $C^0$, $Q(C^0)/Q(H^0)$, from this B0 star. Assume the star's ionizing spectrum is a blackbody in the Wien tail (Equation 2.9). Further, assume that photons with $h\nu > 13.6$ eV ionize only H and photons with $11.18 < h\nu < 13.6$ eV ionize only C.

(b)   Estimate the relative radii of the H II and C II Strömgren spheres around this star. Assume that in the $H^+$ ionization zone all the electrons come from $H^0$, and in the $C^+$ ionization zone all the electrons come from $C^0$.

(c)   Assume this B0 star and its ionized C region are in the WNM with $n_H = 1$ cm$^{-3}$ and that the rate of production of hydrogen-ionizing photons is $Q_0 = 3.55 \times 10^{48}$ s$^{-1}$. What is the radius of the star's carbon Strömgren sphere, assuming all its free electrons come from ionizing $C^0$?

(d)   The actual ionization state of the WNM is higher than you assumed above. Repeat your calculation in part (c) for $x \sim 10^{-2}$ (i.e., $n_e \sim 0.01 n_H$).

(e)   What are the implications of the calculations above for the ionization state of carbon in the WNM?

4.4   An H II region consists of pure hydrogen with density $n_H$; it is almost fully ionized by a central star with temperature $T_\star$.

(a)   Write down the volumetric heating rate $g$ for this H II region. Assume the central star is a blackbody in the Wien tail.

(b)   Write down the volumetric cooling rate $\ell$ for this pure hydrogen H II region, including all relevant cooling mechanisms. [Hint: don't forget Lyman $\alpha$.]

(c)   For $n_H = 100$ cm$^{-3}$ and $T_\star = 30\,000$ K, plot the heating and cooling rates as a function of the temperature $T$ of the gas (NOT that of the star). Overplot the heating curves for $T_\star = 20\,000\,\text{K}$ and $40\,000\,\text{K}$, and the cooling curves for $n_{\text{H}} = 1000\,\text{cm}^{-3}$ and $n_{\text{H}} = 10^4\,\text{cm}^{-3}$.

(d) What is the equilibrium temperature of the pure hydrogen H II region if the density is $n_{\text{H}} = 100\,\text{cm}^{-3}$ and the temperature of the star $T_\star$ is $20\,000\,\text{K}$, $30\,000\,\text{K}$, or $40\,000\,\text{K}$? (Numerical estimates are acceptable.)

(e) What is the equilibrium temperature of the pure hydrogen H II region if the temperature of the star is $T_\star = 30\,000\,\text{K}$ and the density of the region $n_{\text{H}}$ is $100\,\text{cm}^{-3}$, $1000\,\text{cm}^{-3}$, or $10^4\,\text{cm}^{-3}$?

(f) In words, explain what sets the temperature of the H II region. Is the temperature very sensitive to the temperature of the central star? Is it very sensitive to the density of the gas?

# 5

# Hot Ionized Medium

*Give me leave*
*To tell you once again that at my birth*
*The front of heaven was full of fiery shapes.*

William Shakespeare (1564–1616)
"Henry IV, Part I," III, 1

The hot ionized medium (HIM) occupies about half the volume of the interstellar medium; however, because of its low density, it provides only a few percent of its mass. The hot ionized medium is hot because it has been heated by shock fronts that result from supernova explosions. Thus, it consists of a series of bubbles; we, for instance, live in the Local Bubble, which is $\sim 50$ pc in radius. The Local Bubble is thought to have been blown by one or more supernovae that went off $\sim 2$ Myr ago. This time scale is supported by the presence in the Earth's ocean sediments of the unstable isotope $^{60}$Fe. This isotope of iron, produced in core collapse supernovae, has a lifetime $\tau \approx 3.8$ Myr. The gas in the Local Bubble has temperature $T \sim 10^6$ K, estimated from its X-ray emission, and density $n_H \sim 0.004$ cm$^{-3}$.

From our position within the Local Bubble, we can determine that the Bubble contains cooler, denser regions within itself. The Sun, for instance, is near the edge of the Local Interstellar Cloud, a region $\sim 10$ pc across with properties consistent with those of the warm ionized medium ($T \sim 8000$ K, $n_H \sim 0.2$ cm$^{-3}$). It is the Local Interstellar Cloud, sometimes also known as the "Local Fluff," that provides the Lyman $\alpha$ absorption in the direction of Alpha Centauri A and B (Figure 2.8).

In this chapter, we will examine how shock fronts can heat the gas of the ISM, with particular attention to the shocks that bound supernova remnants. Then we will look at the observations that enable us to measure temperatures of order $\sim 10^6$ K and densities of order $\sim 0.004$ cm$^{-3}$.

## 5.1 Shocking Information

A **shock front** is a thin boundary layer between two regions of gas with very different densities, pressures, and bulk velocities. A shock front will naturally tend to form from a sound wave traveling through gas. To see why this is so, consider a medium with density $\rho$ and pressure $P$. A low-amplitude sound wave traveling through this medium will be adiabatic; that is, it will not increase the entropy of the gas through which it passes. For an adiabatic process, the equation of state for the gas is

$$P(\rho) = P_0 \left(\frac{\rho}{\rho_0}\right)^\gamma, \tag{5.1}$$

where the adiabatic index $\gamma$ is 5/3 for a monatomic gas and 7/5 for a cool gas of diatomic molecules like $H_2$. The speed with which an adiabatic sound wave travels is

$$c_s = \left(\frac{\gamma P}{\rho}\right)^{1/2} \propto \rho^{(\gamma-1)/2}. \tag{5.2}$$

Thus, for $\gamma > 1$, sound travels more rapidly in a denser gas. In a sound wave of finite amplitude, as shown in the top panel of Figure 5.1, the crest of the wave, where the density and pressure are highest, travels faster than the trough of the wave, where the density and pressure are lowest. As the crest starts to catch up with the trough, as shown in the middle panel of Figure 5.1, the gradient in density between the crest and trough steepens. Although an ocean wave can be triple-valued, as shown in the bottom panel of Figure 5.1, creating a "breaker," this behavior is forbidden for a sound wave since at any given point, there can be

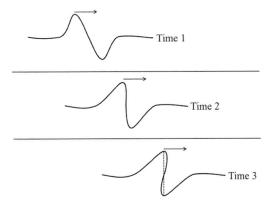

**Figure 5.1** The steepening of a sound wave to form a shock. At Time 3, the sound wave is unable to be triple-valued (solid curve); instead it develops a nearly discontinuous jump in pressure (vertical line).

only one density and one pressure. A sound wave, therefore, will steepen until a shock front forms, with a thickness comparable to the mean free path between particle collisions.

If the sound wave in Figure 5.1 has a dimensionless amplitude $\delta \equiv (P_{crest} - P_{trough})/(P_{crest} + P_{trough})$, then the distance it will travel before steepening into a shock will be $d \sim \lambda/\delta$, where $\lambda$ is the wavelength. The ordinary sounds that we hear every day will not, in practice, steepen into shocks. A loud ambulance siren three meters away will assault your ear with a sound wave of amplitude $\delta \sim 2 \times 10^{-4}$. Even if you could collimate that wave perfectly so that the amplitude didn't fall off as the inverse square of the distance, it would still travel 5000 wavelengths, or about two kilometers, before steepening into a shock front (assuming a fundamental frequency $\sim$900 Hz for the siren). However, high-amplitude pressure fluctuations with $\delta \sim 1$, such as those produced by stun grenades, lightning bolts, and supernova explosions, will rapidly steepen into shocks.

A simple plane parallel steady-state shock, like that in Figure 5.2, will obey the mass conservation law (Equation 4.90)

$$\frac{d}{dx}(\rho u) = 0, \tag{5.3}$$

the momentum conservation law (Equation 4.93)

$$\frac{d}{dx}(\rho u^2 + P) = 0, \tag{5.4}$$

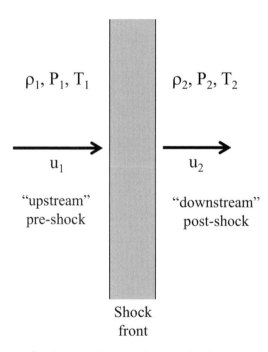

**Figure 5.2** Geometry of a plane parallel steady-state shock.

and the energy conservation law

$$\frac{d}{dx}(u\rho[u^2/2 + \epsilon] + uP) = 0, \tag{5.5}$$

where $\epsilon$ is the specific internal energy of the gas, which includes the thermal energy and also, if the gas is molecular, any rotational or vibrational energy of the molecules. In general,

$$\epsilon = \frac{1}{\gamma - 1}\frac{kT}{m} = \frac{1}{\gamma - 1}\frac{P}{\rho}, \tag{5.6}$$

where $\gamma$ is the adiabatic index and $m$ is the mass per gas particle. Notice that the energy conservation law (Equation 5.5) assumes that the gas does not gain or lose energy in the form of photons. Later in this section, we will revisit this assumption, but for the moment let's stick with simple, non-radiative shocks.

The gas properties immediately before the gas is shocked (subscript "1") and immediately afterwards (subscript "2") are linked, thanks to the conservation laws (Equations 5.3, 5.4, and 5.5), by the **Rankine–Hugoniot conditions**:[1]

$$\rho_1 u_1 = \rho_2 u_2, \tag{5.7}$$

$$\rho_1 u_1^2 + P_1 = \rho_2 u_2^2 + P_2, \tag{5.8}$$

and

$$u_1\rho_1[u_1^2/2 + \epsilon_1] + u_1 P_1 = u_2\rho_2[u_2^2/2 + \epsilon_2] + u_2 P_2. \tag{5.9}$$

Dividing Equation 5.9 by Equation 5.7, we can simplify the energy conservation equation to

$$u_1^2/2 + \epsilon_1 + P_1/\rho_1 = u_2^2/2 + \epsilon_2 + P_2/\rho_2. \tag{5.10}$$

Using the relation between specific internal energy and pressure (Equation 5.6) makes things simpler still:

$$\frac{u_1^2}{2} + \frac{\gamma}{\gamma - 1}\frac{P_1}{\rho_1} = \frac{u_2^2}{2} + \frac{\gamma}{\gamma - 1}\frac{P_2}{\rho_2}. \tag{5.11}$$

We assume an adiabatic index $\gamma$ that is the same on both sides of the shock front; this is not necessarily true if, for instance, the shock is accompanied by the dissociation of cool molecular gas ($\gamma_1 = 7/5$) to form atomic gas ($\gamma_2 = 5/3$).

A dimensionless number that is useful for describing shock fronts is the **Mach number**

$$\mathcal{M}_1 \equiv \frac{u_1}{c_1} = \left(\frac{\rho_1 u_1^2}{\gamma P_1}\right)^{1/2}, \tag{5.12}$$

[1] The Rankine–Hugoniot conditions are named after two of the nineteenth century engineers who helped to bring us the laws of thermodynamics.

where $c_1$ is the sound speed in the pre-shock gas. Using the Rankine–Hugoniot conditions (Equations 5.7, 5.8, and 5.11), we can solve for the density, pressure, and temperature ratios across the shock front:

$$\frac{\rho_2}{\rho_1} = \frac{(\gamma + 1)\mathcal{M}_1^2}{(\gamma - 1)\mathcal{M}_1^2 + 2} = \frac{u_1}{u_2} \tag{5.13}$$

$$\frac{P_2}{P_1} = \frac{2\gamma \mathcal{M}_1^2 - (\gamma - 1)}{\gamma + 1} \tag{5.14}$$

$$\frac{T_2}{T_1} = \frac{[(\gamma - 1)\mathcal{M}_1^2 + 2][2\gamma \mathcal{M}_1^2 - (\gamma - 1)]}{(\gamma + 1)^2 \mathcal{M}_1^2}. \tag{5.15}$$

Having post-shock density, pressure, and temperature values that are greater than their pre-shock values is equivalent to having a Mach number greater than one. A **strong** shock is defined as a shock with $\mathcal{M}_1 \gg 1$. For a strong shock,

$$\frac{\rho_2}{\rho_1} \approx \frac{\gamma + 1}{\gamma - 1} \tag{5.16}$$

$$P_2 \approx \frac{2}{\gamma + 1} \rho_1 u_1^2 \tag{5.17}$$

$$T_2 \approx \frac{2(\gamma - 1)}{(\gamma + 1)^2} \frac{m}{k} u_1^2. \tag{5.18}$$

Thus, no matter how strong the non-radiative shock is, the ratio $\rho_2/\rho_1$ has a finite value; for a monatomic gas, with $\gamma = 5/3$, the post-shock gas has at most four times the density of the pre-shock gas. However, a strong shock can produce very high pressures and temperatures. An interstellar shock front with propagation speed $u_1 \sim 1000 \, \text{km s}^{-1}$ (not unusual for a supernova shock wave) produces shock-heated gas with

$$T_2 \approx 1.1 \times 10^7 \, \text{K} \left( \frac{u_1}{1000 \, \text{km s}^{-1}} \right)^2, \tag{5.19}$$

assuming the shocked gas is fully ionized hydrogen. Generally speaking, shock fronts convert supersonic gas ($u_1/c_1 > 1$) into subsonic gas ($u_2/c_2 < 1$) in the shock's frame of reference. Shocks increase the density, pressure, and temperature, and decrease the bulk velocity relative to the shock front. Shocks act as entropy generators, with the increase in specific entropy $s_2 - s_1$ proportional to $\ln \mathcal{M}_1$ for strong shocks.

The hot shocked gas is out of equilibrium and will start to cool. Thus, the shock will be followed by a **radiative zone** in which the shock-heated gas cools down by emitting photons, as diagrammed in Figure 5.3. At temperatures $T > 2 \times 10^7 \, \text{K}$, cooling is dominated by free–free (bremsstrahlung) emission.

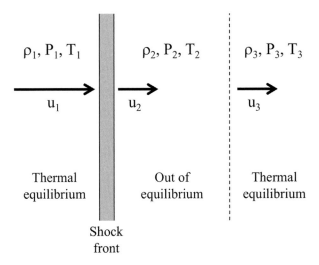

**Figure 5.3** The structure of a plane parallel radiative shock.

If we assume, for simplicity, fully ionized hydrogen gas, the volumetric cooling rate for free–free emission is (Equation 4.52)

$$\ell_{ff} \approx 5.4 \times 10^{-24} \, \text{erg cm}^{-3} \, \text{s}^{-1} \left(\frac{T}{10^7 \, \text{K}}\right)^{1/2} \left(\frac{n_H}{1 \, \text{cm}^{-3}}\right)^2, \tag{5.20}$$

assuming a frequency-averaged Gaunt factor $g_{ff} \approx 1.2$, which is appropriate for a temperature $T \sim 2 \times 10^7 \, \text{K}$. The internal energy density for fully ionized hydrogen, with number density $n = n_{HII} + n_e = 2n_H$, is

$$\varepsilon = \frac{3}{2}(2n_H)kT = 4.14 \times 10^{-9} \, \text{erg cm}^{-3} \left(\frac{T}{10^7 \, \text{K}}\right)\left(\frac{n_H}{1 \, \text{cm}^{-3}}\right). \tag{5.21}$$

This means that the free–free cooling time is

$$t_{cool} = \frac{\varepsilon}{\ell_{ff}} \approx 24 \, \text{Myr} \left(\frac{T}{10^7 \, \text{K}}\right)^{1/2} \left(\frac{n_H}{1 \, \text{cm}^{-3}}\right)^{-1}. \tag{5.22}$$

Using Equation 5.19 for the post-shock temperature in terms of the shock velocity, the cooling time can be expressed as

$$t_{cool} \sim 25 \, \text{Myr} \left(\frac{u_1}{1000 \, \text{km s}^{-1}}\right)\left(\frac{n_H}{1 \, \text{cm}^{-3}}\right)^{-1}. \tag{5.23}$$

During this time, the gas will move a distance

$$R_{cool} \sim u_2 t_{cool} \sim \frac{u_1}{4} t_{cool} \sim 6 \, \text{kpc} \left(\frac{u_1}{1000 \, \text{km s}^{-1}}\right)^2 \left(\frac{n_H}{1 \, \text{cm}^{-3}}\right)^{-1} \tag{5.24}$$

relative to the shock front. This tells us the approximate thickness of the radiative zone for a strong shock with $T_2 > 2 \times 10^7 \, \text{K}$. This is a long distance compared with (for instance) the scale height of interstellar gas in our galaxy. Thus, the hot gas produced by high-speed shocks doesn't have time to cool before the shock runs out of gas to shock.

At lower temperatures ($10^5$ K $< T < 2 \times 10^7$ K), corresponding to slower shock speeds (90 km s$^{-1}$ $< u_1 < 1300$ km s$^{-1}$), collisionally excited line emission does most of the cooling. In this temperature range the cooling rate is a decreasing function of temperature. At solar abundance, a useful approximation for the volumetric cooling rate at $10^5$ K $< T < 2 \times 10^7$ K is

$$\ell_{\text{line}} \approx 1.5 \times 10^{-22} \text{ erg cm}^{-3}\text{ s}^{-1} \left(\frac{T}{10^6 \text{ K}}\right)^{-0.73} \left(\frac{n_H}{1 \text{ cm}^{-3}}\right)^2. \tag{5.25}$$

The internal energy density for highly ionized gas of solar abundance, with number density $n \approx 2.3 n_H$ (Equation 1.2), is

$$\varepsilon \approx \frac{3}{2}(2.3 n_H)kT \approx 4.76 \times 10^{-10} \text{ erg cm}^{-3} \left(\frac{T}{10^6 \text{ K}}\right)\left(\frac{n_H}{1 \text{ cm}^{-3}}\right). \tag{5.26}$$

This results in a cooling time

$$t_{\text{cool}} = \frac{\varepsilon}{\ell_{\text{line}}} \approx 0.10 \text{ Myr} \left(\frac{T}{10^6 \text{ K}}\right)^{1.73} \left(\frac{n_H}{1 \text{ cm}^{-3}}\right)^{-1} \tag{5.27}$$

$$\approx 0.10 \text{ Myr} \left(\frac{u_1}{300 \text{ km s}^{-1}}\right)^{3.46} \left(\frac{n_H}{1 \text{ cm}^{-3}}\right)^{-1}. \tag{5.28}$$

The thickness of the radiative zone is then

$$R_{\text{cool}} \sim \frac{u_1}{4} t_{\text{cool}} \sim 8 \text{ pc} \left(\frac{u_1}{300 \text{ km s}^{-1}}\right)^{4.46} \left(\frac{n_H}{1 \text{ cm}^{-3}}\right)^{-1}. \tag{5.29}$$

These shorter time scales and length scales mean that radiative cooling is more effective at changing the structure of slower shocks.

Frequently, the equilibrium temperature $T_3$ that the gas reaches after passing through the radiative zone (see Figure 5.3) is nearly equal to the initial temperature $T_1$. In the case of an **isothermal** shock, the first two Rankine–Hugoniot conditions,

$$\rho_3 u_3 = \rho_1 u_1 \tag{5.30}$$

and

$$\rho_3 u_3^2 + \frac{\rho_3 kT_3}{m} = \rho_1 u_1^2 + \frac{\rho_1 kT_1}{m}, \tag{5.31}$$

can be combined with the additional condition $T_3 = T_1$. Making use of the isothermal sound speed in the unshocked medium, $c_T \equiv (kT_1/m)^{1/2}$, we find that the increase in density caused by shocking followed by cooling is

$$\frac{\rho_3}{\rho_1} = \left(\frac{u_1}{c_T}\right)^2 = \mathcal{M}_T^2, \tag{5.32}$$

where $\mathcal{M}_T$ is the isothermal Mach number of the shock. Thus, radiative shocks can reach arbitrarily high compression as their Mach number approaches infinity.

## 5.2 Supernova Remnants

One way to generate a shock front is to dump a large amount of energy into a tiny volume of gas over a short period of time. An explosion is an effective way of doing this. Interest in spherical shock fronts was stimulated during the 1940s by the explosion of the first nuclear fission bombs. On opposite sides of the Iron Curtain, Leonid Sedov and Geoffrey Taylor found analytic self-similar solutions for the expansion of a non-radiative spherical shock front. The **Sedov–Taylor solution** that they found can be applied to expanding supernova remnants ($E \sim 10^{51}$ erg) as well as to the explosion of fission bombs ($E \sim 10^{20}$ erg). As it happens, there are two main types of supernova: a **core collapse** supernova is powered by the gravitational collapse of the dense core of an evolved massive star, while a **thermonuclear** supernova is powered by runaway nuclear fusion in a degenerate white dwarf. For concreteness, we will use a core collapse supernova as our main example of an explosion in the ISM.

Imagine a core collapse supernova going off in a homogeneous region of gas with density $\rho_1$, pressure $P_1$, and temperature $T_1$. The explosion produces an energy $\sim 10^{53}$ erg in neutrinos; this energy is not deposited in the ISM since the neutrinos have an enormously long mean free path for interactions with low-density baryonic matter. The explosion also produces $\sim 10^{49}$ erg in photons; this energy is also mostly lost. The only energy that is effectively deposited in the local ISM is the kinetic energy of the expanding ejecta. Although core collapse supernovae have a range of properties, we can take $M_{ej} \sim 10\,M_\odot$ as a typical mass of ejected material and $u_{ej} \sim 3000\,\mathrm{km\,s^{-1}}$ as the typical initial speed of the ejecta. This implies a kinetic energy[2] $E = M_{ej}u_{ej}^2/2 \sim 10^{51}$ erg. At first, the ejecta move ballistically outward at nearly constant speed; this phase is known as the **free expansion** phase of the supernova remnant. The ejecta are not slowed down significantly until they have swept up a mass of interstellar gas comparable to the mass of the ejecta. This occurs when the radius of the supernova remnant is

$$r_{sweep} = \left(\frac{3M_{ej}}{4\pi\rho_1}\right)^{1/3} = 4.1\,\mathrm{pc}\left(\frac{M_{ej}}{10\,M_\odot}\right)^{1/3}\left(\frac{n_H}{1\,\mathrm{cm^{-3}}}\right)^{-1/3}, \tag{5.33}$$

where $n_H$ is the number density of hydrogen in the surrounding interstellar gas. Given a speed $u_{ej} \sim 3000\,\mathrm{km\,s^{-1}}$, we expect the free expansion phase to last for a time $t_{fe} = r_{sweep}/u_{ej} \sim 1000\,\mathrm{yr}$.

The initial speed of the ejecta is much greater than the sound speed,

$$c_1 = 1.29\,\mathrm{km\,s^{-1}}\left(\frac{T_1}{100\,\mathrm{K}}\right)^{1/2}, \tag{5.34}$$

---

[2] Thermonuclear supernovae have lower ejecta mass, $M_{ej} \sim 1\,M_\odot$, but higher initial ejecta speed, $u_{ej} \sim 10^4\,\mathrm{km\,s^{-1}}$, leading to a similar kinetic energy, $E \sim 10^{51}$ erg.

in the surrounding interstellar gas. Thus, as the initially dense ejecta move outward at supersonic speed, they create a shock wave ahead of them, like a supersonic bullet moving through the Earth's atmosphere. The Mach number of the shock will be

$$\mathcal{M}_1 = \frac{u_{ej}}{c_1} = 2300 \left(\frac{u_{ej}}{3000 \text{ km s}^{-1}}\right) \left(\frac{T_1}{100 \text{ K}}\right)^{-1/2} \gg 1. \qquad (5.35)$$

Immediately behind this shock front, the shock-heated interstellar gas has a high pressure, $P_2 \sim \mathcal{M}_1^2 P_1 \sim 10^7 P_1$. Meanwhile, the gas of the ejecta is becoming less dense as it expands, and also less hot, thanks to adiabatic cooling. When the pressure within the ejecta becomes small compared with the pressure of the shock-heated interstellar gas, a reverse shock propagates backward toward the center of the supernova remnant, shock-heating the ejecta.

Toward the end of the free expansion phase, in this idealized picture, a strong spherical shock is expanding outward into the ISM. Behind the shock, a layer of hot, relatively dense gas, consisting of mixed ejecta and interstellar gas, surrounds a core of hot, low-density gas consisting of ejecta that have been shock-heated by the reverse shock. However, young remnants of core collapse supernovae are frequently more complex than in this idealized view. Consider, for instance, the Crab Nebula, shown in Figure 1.7. Light from the supernova that gave birth to the Crab was seen on Earth in the year 1054; we now see the supernova remnant at an age $t \sim 1000$ yr. The mass of the luminous remnant is estimated to be $M \approx 4.6 \, M_\odot$; the bright filaments at the edge of the remnant are moving outward with speed $u \approx 1500$ km s$^{-1}$. If these numbers represent the ejecta mass $M_{ej}$ and the ejecta speed $u_{ej}$, they imply a free expansion time $t_{fe} \sim 2000$ yr and a (relatively low) kinetic energy $E \sim 10^{50}$ erg. This would be consistent with the Crab Nebula still being in the free expansion phase today. However, the currently observed proper motion of the Crab's expansion, if assumed to be constant in time, implies that the expansion started in the year 1130, as seen by an earthly observer. Since the expansion actually started $\sim 76$ yr earlier, we conclude that the expansion of the Crab Nebula has actually speeded up during its existence. The current view of the Crab is that the luminous portion of the nebula (Figure 1.7) is a pulsar wind nebula, accelerated by the magnetized pulsar at the heart of the nebula.

In a supernova remnant that does not draw significant power from a central pulsar, the end of the free expansion phase marks the transition into the **Sedov–Taylor** phase, also known as the blastwave phase. During the Sedov–Taylor phase, the amount of shocked interstellar gas is much greater than $M_{ej}$, so the properties of the shock front no longer depend on the value of $M_{ej}$. However, during this phase, the amount of energy radiated away in the form of photons is much smaller than the initial kinetic energy $E$, so the properties of the shock front do still depend on the value of $E$. The pressure $P_2$ behind the shock is much greater than the pressure $P_1$ of the surrounding gas, so the properties of the shock front don't depend on the value of $P_1$; the shock expands as if it were traveling into a pressureless medium. Finally, the properties of the shock front do depend on the value of $\rho_1$, the density of the surrounding medium; this is the mass that is going to be shocked and added to the growing supernova remnant.

The energy $E$ deposited in the interstellar gas has the dimensionality [mass]×[length]²/[time]²; the mass density $\rho_1$ of the gas has the dimensionality [mass]/[length]³. The parameters $E$ and $\rho_1$ can't be combined to form a characteristic length scale or time scale for the problem. Moreover, we can't "cheat" by using constants such as $G$, $c$, or $\hbar$; gravity is a negligibly small force in this problem, so $G$ is irrelevant, the problem is non-relativistic and non-radiative, so $c$ is irrelevant, and the problem is classical, so $\hbar$ is irrelevant. The solution for the expanding spherical shock front must therefore be a scale-free, **self-similar** solution. Let's try a solution for the radius of the shock front that takes the form

$$r_{sh}(t) = AE^\alpha \rho_1^\beta t^\eta, \tag{5.36}$$

where $A$ is a dimensionless factor of order unity.[3] The exponents $\alpha$, $\beta$, and $\eta$ are chosen so that $r_{sh}$ has the correct dimensionality. The unique solution is $\alpha = 1/5$, $\beta = -1/5$, and $\eta = 2/5$, yielding

$$r_{sh} = A\left(\frac{Et^2}{\rho_1}\right)^{1/5}. \tag{5.37}$$

The resulting expansion speed for the shock front is

$$u_{sh} = \frac{2}{5}A\left(\frac{E}{\rho_1 t^3}\right)^{1/5} = \frac{2}{5}\frac{r_{sh}}{t} \propto r_{sh}^{-3/2}. \tag{5.38}$$

Using reasonable values for a supernova remnant in its Sedov–Taylor phase,

$$r_{sh} \approx 5.3\,\mathrm{pc}\left(\frac{E}{10^{51}\,\mathrm{erg}}\right)^{0.2}\left(\frac{n_H}{1\,\mathrm{cm}^{-3}}\right)^{-0.2}\left(\frac{t}{1000\,\mathrm{yr}}\right)^{0.4} \tag{5.39}$$

$$u_{sh} \approx 2100\,\mathrm{km\,s}^{-1}\left(\frac{E}{10^{51}\,\mathrm{erg}}\right)^{0.2}\left(\frac{n_H}{1\,\mathrm{cm}^{-3}}\right)^{-0.2}\left(\frac{t}{1000\,\mathrm{yr}}\right)^{-0.6}.$$

An example of a middle-aged supernova remnant in its Sedov–Taylor phase is the Cygnus Loop, shown in Figure 1.8. The Cygnus Loop, with angular radius $\theta \approx 1.5°$, is currently expanding with a proper motion $\mu \approx 0.1\,\mathrm{arcsec\,yr}^{-1}$. (At an assumed distance $d = 735\,\mathrm{pc}$ for the Cygnus Loop, this corresponds to $r_{sh} \approx 19\,\mathrm{pc}$ and $u_{sh} \approx 350\,\mathrm{km\,s}^{-1}$.) From the results of Equation 5.38, we can compute the age of the Cygnus Loop to be

$$t = \frac{2}{5}\frac{r_{sh}}{u_{sh}} = \frac{2}{5}\frac{\theta}{\mu} \approx 2 \times 10^4\,\mathrm{yr}. \tag{5.40}$$

---

[3] For a monatomic gas, with $\gamma = 5/3$, the dimensionless factor turns out to be $A = 1.1517$.

The observed properties of the Cygnus Loop are thus consistent with its being in the Sedov–Taylor phase, given the reasonable numbers $t \approx 2 \times 10^4$ yr and $E/n_H \approx 1.3 \times 10^{51}$ erg cm$^3$.

Equation 5.39 tells us how a spherical shock front expands with time when radiative losses are insignificant and the mass of the shocked gas is large compared with the mass of the supernova ejecta. But what are the density $\rho$, bulk velocity $u$, and pressure $P$ between the shock front and the center of the supernova remnant? Sedov and Taylor each found analytic solutions for $\rho$, $u$, and $P$ during this phase. The solutions are self-similar and thus take the form

$$\rho(r,t) = \rho_1 f(x), \tag{5.41}$$

$$u(r,t) = \frac{r_{sh}(t)}{t} g(x), \tag{5.42}$$

$$P(r,t) = \frac{\rho_1 r_{sh}(t)^2}{t^2} h(x), \tag{5.43}$$

where $f$, $g$, and $h$ are dimensionless functions of the dimensionless radius $x \equiv r/r_{sh}(t)$. If we insert these values into the conservation equations for a spherically symmetric system, we find three fairly gruesome differential equations that can be solved for $f$, $g$, and $h$ as a function of $x$. Taylor and Sedov, with admirable skill, found analytic solutions to the gruesome threesome.[4]

The **Sedov–Taylor solution** for the density, bulk velocity, and pressure in the energy-conserving phase of a supernova expansion is shown in Figure 5.4. As $r \to 0$, the density $\rho$ and velocity $u$ go to zero but the pressure $P$ approaches a finite value: $P(0) \approx 0.306 P_2$ for a gas with $\gamma = 5/3$. This implies that, formally, the temperature at the origin is infinite. Because of the steep climb in density from $r = 0$ to $r = r_{sh}$, most of the mass of the supernova remnant lies in a relatively thin outer layer: about 50% of the remnant's mass lies within the outermost 17% of the remnant's volume.

The Sedov–Taylor phase lasts until the amount of energy radiated by the cooling gas is comparable to the initial energy $E$. Since free–free cooling at $T > 2 \times 10^7$ K is not very effective, as we saw in Section 5.1, the gas won't cool much until $T_2$, the temperature in the hot dense region just behind the shock, drops below 20 million kelvin. At that point, we can use Equation 5.28 for the time for cooling by collisionally excited line radiation:

$$t_{cool} \approx 0.10 \, \text{Myr} \left( \frac{u_{sh}}{300 \, \text{km s}^{-1}} \right)^{3.46} \left( \frac{n_H}{1 \, \text{cm}^{-3}} \right)^{-1}. \tag{5.44}$$

---

[4] Landau and Lifshitz, in their text *Fluid Mechanics*, call the solution "elementary but laborious." When Landau thinks something is laborious, it is definitely laborious; when Landau thinks it is elementary, others might disagree.

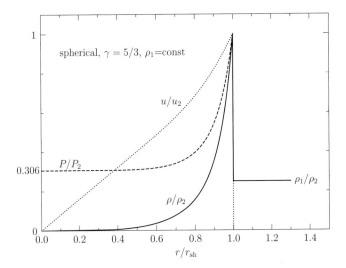

**Figure 5.4** Sedov–Taylor solution for a spherical blastwave with $\gamma = 5/3$. The density, velocity, and pressure are normalized by their immediate post-shock values $\rho_2$, $u_2$, and $P_2$. The plotted velocity $u$ is relative to the center of the blastwave.

Inserting the value for $u_{\text{sh}}$ in the Sedov–Taylor phase (Equation 5.39), we find

$$t_{\text{cool}} \approx 83\,\text{Myr} \left(\frac{E}{10^{51}\,\text{erg}}\right)^{0.69} \left(\frac{n_{\text{H}}}{1\,\text{cm}^{-3}}\right)^{-1.69} \left(\frac{t}{1000\,\text{yr}}\right)^{-2.08}, \qquad (5.45)$$

with the cooling time dropping rapidly as the shell expands and $u_{\text{sh}}$ decreases. Solving Equation 5.45 for the instant when $t = t_{\text{cool}}$, we find

$$t_{\text{cool}} \approx 4.0 \times 10^4\,\text{yr} \left(\frac{E}{10^{51}\,\text{erg}}\right)^{0.22} \left(\frac{n_{\text{H}}}{1\,\text{cm}^{-3}}\right)^{-0.55}. \qquad (5.46)$$

At this time, the size of the supernova remnant has grown to (Equation 5.39)

$$r_{\text{cool}} \approx 23\,\text{pc} \left(\frac{E}{10^{51}\,\text{erg}}\right)^{0.29} \left(\frac{n_{\text{H}}}{1\,\text{cm}^{-3}}\right)^{-0.42} \qquad (5.47)$$

and its expansion speed has dropped to

$$u_{\text{cool}} \approx 230\,\text{km s}^{-1} \left(\frac{E}{10^{51}\,\text{erg}}\right)^{0.07} \left(\frac{n_{\text{H}}}{1\,\text{cm}^{-3}}\right)^{0.13}. \qquad (5.48)$$

Although the Sedov–Taylor solution is highly simplified, it gives us a good starting point for understanding supernova remnants whose ages are in the range $1000\,\text{yr} < t < 40\,000\,\text{yr}$.

Once radiative losses become important, at $t \sim t_{\text{cool}}$, a dense shell forms behind the shock, as illustrated in Figure 5.5. The dense shell contains most of

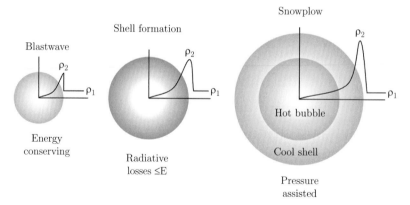

**Figure 5.5** An expanding supernova remnant makes the transition from the Sedov–Taylor, or blastwave, phase to the snowplow phase. [Following Shu 1992]

the mass of the supernova remnant, so its mass at $t \sim t_{\text{cool}}$ is

$$M_{\text{sh}} \sim \frac{4\pi}{3} \rho_1 r_{\text{cool}}^3 \sim 1600\,M_\odot \left(\frac{E}{10^{51}\,\text{erg}}\right)^{0.87} \left(\frac{n_{\text{H}}}{1\,\text{cm}^{-3}}\right)^{-0.26}. \tag{5.49}$$

The "radial momentum" of the dense shell at $t \sim t_{\text{cool}}$ is

$$\mu_{\text{sh}} \equiv M_{\text{sh}} u_{\text{sh}} \sim 7 \times 10^{43}\,\text{g cm s}^{-1} \left(\frac{E}{10^{51}\,\text{erg}}\right)^{0.94} \left(\frac{n_{\text{H}}}{1\,\text{cm}^{-3}}\right)^{-0.13}. \tag{5.50}$$

The radial momentum $\mu_{\text{sh}}$ can be thought of in physical terms if you look at a small patch of the dense shell covering a solid angle $d\Omega$. The linear momentum of the patch of shell is then $\mu_{\text{sh}} d\Omega$. Once the dense shell forms, the supernova remnant enters the **snowplow phase** of its evolution. It is called the "snowplow" phase because the expanding dense shell scoops up additional mass as it expands, just as a snowplow scoops up additional mass in the form of snow.

Although most of the mass of a supernova remnant in the snowplow phase is in the dense shell, the inner region contains gas at very high temperatures. Thus, despite its low density, the gas in the inner region has a significant pressure, which helps to push the dense shell outward. The **pressure-assisted snowplow phase** can be mathematically analyzed. Let $P_i$ be the mean pressure within the hot, low-density interior of the supernova remnant. Since hot, low-density gas can't cool effectively by free–free emission (with $t_{\text{cool}} \propto T^{1/2} n^{-1}$), its pressure drops primarily through adiabatic expansion: $PV^\gamma = \text{constant}$, where $V$ is the volume of hot, low-density gas inside the dense shell and $\gamma$ is the adiabatic index of the gas. For adiabatic cooling, then, the pressure drops as $P_i \propto V^{-\gamma} \propto r_{\text{sh}}^{-3\gamma}$. Thus, after the time $t \sim t_{\text{cool}}$ when the snowplow phase begins, the interior pressure goes as

$$P_i = P_i(t_{cool}) \left( \frac{r_{sh}}{r_{cool}} \right)^{-3\gamma}. \tag{5.51}$$

Since the pressure inside the dense shell is higher than the negligible pressure $P_1$ outside the dense shell, the pressure force creates a change in the radial momentum $\mu_{sh} = M_{sh} u_{sh}$ given by the equation

$$\frac{d}{dt}(M_{sh} u_{sh}) \approx P_i 4\pi r_{sh}^2 \approx 4\pi P_i(t_{cool}) r_{cool}^{3\gamma} r_{sh}^{2-3\gamma}. \tag{5.52}$$

Let's guess that there exists a power-law solution to pressure-assisted snowplow expansion, with $r_{sh} \propto t^b$ and thus $M_{sh} u_{sh} \propto r_{sh}^4/t \propto t^{4b-1}$. Equation 5.52 then implies $4b - 2 = (2 - 3\gamma)b$, or

$$b = \frac{2}{3\gamma + 2}. \tag{5.53}$$

Thus, $b = 2/7$ in a monatomic gas with $\gamma = 5/3$. We can then approximate the expansion of a pressure-assisted snowplow as

$$r_{sh}(t) = r_{cool} \left( \frac{t}{t_{cool}} \right)^{2/7} \tag{5.54}$$

$$\approx 58 \, pc \left( \frac{E}{10^{51} \, erg} \right)^{0.23} \left( \frac{n_H}{1 \, cm^{-3}} \right)^{-0.26} \left( \frac{t}{1 \, Myr} \right)^{0.29}. \tag{5.55}$$

The expansion speed is

$$u_{sh} = \frac{2}{7} \frac{r_{cool}}{t_{cool}} \left( \frac{t}{t_{cool}} \right)^{-5/7} \tag{5.56}$$

$$\approx 16 \, km \, s^{-1} \left( \frac{E}{10^{51} \, erg} \right)^{0.23} \left( \frac{n_H}{1 \, cm^{-3}} \right)^{-0.26} \left( \frac{t}{1 \, Myr} \right)^{-0.71}. \tag{5.57}$$

When the expansion speed, $u_{sh} \propto t^{-0.71}$, drops to the sound speed $c_1$ of the surrounding medium, the expanding shell degenerates into an expanding sound wave.[5] Thus the shock becomes a sound wave when $u_{sh} = c_1$, which occurs when

$$t_{unsh} \approx 35 \, Myr \left( \frac{E}{10^{51} \, erg} \right)^{0.32} \left( \frac{n_H}{1 \, cm^{-3}} \right)^{-0.37} \left( \frac{T_1}{100 \, K} \right)^{-0.7}. \tag{5.58}$$

The radius of the shell at this time is

$$r_{unsh} \approx 160 \, pc \left( \frac{E}{10^{51} \, erg} \right)^{0.32} \left( \frac{n_H}{1 \, cm^{-3}} \right)^{-0.37} \left( \frac{T_1}{100 \, K} \right)^{-0.2}. \tag{5.59}$$

Thus we have a picture in which supernovae blow hot, low-density bubbles, bounded by dense shells, that grow to be $\sim$300 pc in diameter over times of $\sim$35 Myr. These bubbles are the origin of the hot ionized medium.

---

[5] Old supernova remnants never die; they just fade away into silence.

## 5.3 Ionizing, Heating, and Cooling

We still need to consider the question whether supernovae blow isolated bubbles in the interstellar medium or whether the bubbles join up to form regions of hot ionized gas much larger than an individual bubble. In our galaxy, the massive stars that explode as core collapse supernovae lie close to the midplane of the disk, with a scale height $h_z \sim 50\,\mathrm{pc}$. Since this scale height is smaller than the maximum size $r_{\mathrm{unsh}}$ of a supernova shock wave, we can approximate the supernova explosions as all going off exactly in the midplane. The current supernova rate in the solar neighborhood is

$$\Sigma_{\mathrm{SNR}} \approx 2 \times 10^{-5}\,\mathrm{pc}^{-2}\,\mathrm{Myr}^{-1}. \tag{5.60}$$

If a supernova goes off at time $t = 0$, the expected number of additional supernovae that will go off at a distance $r < r_{\mathrm{unsh}}$ at a time $t < t_{\mathrm{unsh}}$ is

$$N_{\mathrm{SN}} = \Sigma_{\mathrm{SNR}} \pi r_{\mathrm{unsh}}^2 t_{\mathrm{unsh}}. \tag{5.61}$$

Using the results from Equations 5.58 and 5.59, we find

$$N_{\mathrm{SN}} \approx 55 \left(\frac{E}{10^{51}\,\mathrm{erg}}\right)^{0.95} \left(\frac{n_{\mathrm{H}}}{1\,\mathrm{cm}^{-3}}\right)^{-1.1} \left(\frac{T}{100\,\mathrm{K}}\right)^{-1.1}, \tag{5.62}$$

assuming a supernova rate equal to that in the solar neighborhood. Thus, supernova remnants in cool, low-density regions, which take a longer time to reach a bigger radius before fading away, are more likely to overlap. However, in our galaxy it is found that $nkT \sim 4 \times 10^{-13}\,\mathrm{dyn\,cm}^{-2}$ for all phases of the ISM. If we assume in Equation 5.62 that $n_{\mathrm{H}} kT = 4 \times 10^{-13}\,\mathrm{dyn\,cm}^{-2}$ (corresponding to $n_{\mathrm{H}} T = 2900\,\mathrm{cm}^{-3}\,\mathrm{K}$), we find the intriguing result

$$N_{\mathrm{SN}} \approx 1.4 \left(\frac{E}{10^{51}\,\mathrm{erg}}\right)^{0.95} \left(\frac{\Sigma_{\mathrm{SNR}}}{2 \times 10^{-5}\,\mathrm{pc}^{-2}\,\mathrm{Myr}^{-1}}\right). \tag{5.63}$$

Thus, in our galaxy, supernovae don't blow highly isolated bubbles ($N_{\mathrm{SN}} \ll 1$); nor do supernova remnants overlap to completely obliterate the cooler phases ($N_{\mathrm{SN}} \gg 1$). Instead, supernova remnants are in the interesting regime where they sometimes overlap ($N_{\mathrm{SN}} \sim 1$), forming topologically interesting structures that occupy, in total, about half the volume of interstellar space.

The hot interstellar medium (HIM) is the totality of all the hot gas in the bubbles blown by supernovae; there is some contribution as well from bubbles blown by the hot stellar winds around Wolf-Rayet stars, but most of the HIM has been shock-heated by supernova explosions. The ionized gas of the HIM differs in one crucial aspect from the ionized gas in H II regions; it is collisionally ionized by free electrons rather than being photoionized. For an atom of element X in its $i$th ionization state, collisional ionization by free electrons can be written as

$$X^i + e^- \rightarrow X^{i+1} + e^- + e^-, \tag{5.64}$$

where the kinetic energy of the colliding electron must be greater than the ionization energy $I$ of $X^i$. Collisional ionization results in a net cooling of the ionized gas, since energy $I$ is removed from the kinetic energy of the particles in Equation 5.64.

The volumetric rate at which $X^i$ is collisionally ionized by free electrons is

$$\frac{dn(X^i)}{dt} = -n_e n(X^i) k_{ci}, \tag{5.65}$$

where the rate coefficient $k_{ci}$ can be written in terms of the cross section $\sigma_{ci}(E)$ for the collisional ionization of $X^i$ by a free electron of kinetic energy $E$. In general, the collisional ionization cross section doesn't have a neat analytic dependence on $E$ and must be determined empirically from experiments with high-energy plasmas. The case of ionizing hydrogen, however, is admirably simple, and yields a rate coefficient

$$k_{ci} = 2.32 \times 10^{-8} \, \text{cm}^3 \, \text{s}^{-1} \left( \frac{kT}{I_H} \right)^{1/2} e^{-I_H/kT}. \tag{5.66}$$

Since $I_H/k = 1.58 \times 10^5$ K, we expect the collisional ionization of hydrogen to be suppressed at temperatures $T \ll 10^5$ K.

In **collisional ionization equilibrium (CIE)**, collisional ionization is balanced by radiative recombination. Since the hot ionized medium has a small neutral fraction of hydrogen atoms, and since the number density of hydrogen is small, very few of the photons emitted in radiative recombination to the ground state ($n = 1$) of hydrogen will photoionize nearby H atoms. In other words, for the hot ionized medium, the Case A recombination rate is more appropriate. The net rate of change of the number density of neutral hydrogen will be, in this case,

$$\frac{dn_{HI}}{dt} = -n_e n_{HI} k_{ci} + n_e n_{HII} \alpha_{A,H}. \tag{5.67}$$

In a state of collisional ionization equilibrium, Equation 5.67 reduces to

$$n_e n_{HI} k_{ci} = n_e n_{HII} \alpha_{A,H}. \tag{5.68}$$

The factor $n_e$ on either side cancels out; the ionization state, in collisional ionization equilibrium, is independent of the electron density, and depends only on the electron temperature. In CIE,

$$\frac{n_{HII}}{n_{HI}} = \frac{k_{ci}(T)}{\alpha_{A,H}(T)}. \tag{5.69}$$

At $T \sim 8000$ K, we may use the low-temperature approximation for $\alpha_{A,H}$ given in Equation 4.10, which yields

$$\frac{n_{HII}}{n_{HI}} \approx 4.0 \times 10^5 \left( \frac{kT}{I_H} \right)^{1.21} e^{-I_H/kT}. \tag{5.70}$$

With this approximation for the recombination rate, the ratio of collisionally ionized hydrogen to neutral hydrogen equals one at a temperature $T \approx 0.099 I_H/k \approx$ 15 700 K. A generally useful approximation is that for any collisionally ionized species, $n(X^{i+1})/n(X^i) = 1$ when $kT$ is roughly one-tenth the relevant ionization energy. At $T = 8000\,\mathrm{K} \approx 0.05 I_H/k$, the ratio of collisionally ionized hydrogen to neutral hydrogen has plummeted to $\sim 3 \times 10^{-5}$. Thus, the ionization of H II regions, which have $T \approx 8000\,\mathrm{K}$, is done almost entirely by UV photons from the hot stars at their centers.

At $T \sim 10^6\,\mathrm{K}$, typical of the hot ionized medium, we may use the high-temperature approximation for $\alpha_{A,H}$ given in Equation 4.11, which yields the following ratio of ionized to neutral hydrogen:

$$\frac{n_{HII}}{n_{HI}} \approx 9.3 \times 10^4 \left(\frac{kT}{I_H}\right)^2 e^{-I_H/kT}. \tag{5.71}$$

With this approximation for the recombination rate, the ratio of ionized to neutral hydrogen at $T = 10^6\,\mathrm{K} \approx 6.3 I_H/k$ is $\sim 3 \times 10^6$. Thus, the hot ionized medium is almost entirely collisionally ionized.

For elements other than hydrogen, computing the ionization state in CIE is more difficult. These heavier elements can recombine either through radiative recombination or through **dielectronic recombination**. In dielectronic recombination, the incoming free electron transfers some of its energy to a bound electron, elevating the bound electron to an excited state at the same time that the previously free electron is captured to an excited state. Either or both of the excited electrons can then undergo radiative de-excitation. For some ions, the cross section for dielectronic recombination can be significantly larger than that for standard radiative recombination.

Given the complexity of computing ionization states of heavy elements as a function of temperature, this is usually done numerically. Figure 5.6, for instance, shows the ion fractions for carbon and oxygen as a function of the electron temperature, assuming collisional ionization equilibrium. At $T \approx 10^6$ K, carbon is a mix of C v, C vi, and fully ionized C vii, while oxygen is nearly all in the form O vii.

The hot ionized medium is hot because it has been shock-heated. After the shock front passes through, there is a bit of additional heating from mechanisms such as photoionization, but it's a good assumption that the hot ionized medium is born hot and thereafter cools down. The cooling function in the hot ionized medium depends on both temperature and metallicity. Figure 5.7 shows the cooling function for a gas of solar metallicity in the range $10^4\,\mathrm{K} < T < 10^8\,\mathrm{K}$. (This figure uses the convention, common for gas at higher temperatures, that the cooling function $\Lambda$ is related to the volumetric cooling rate $\ell$ by the equation $\ell = n_e n_H \Lambda$.)

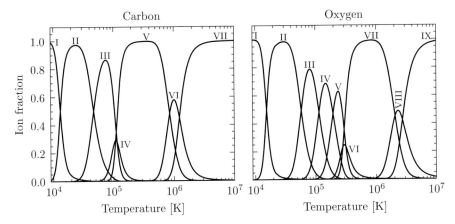

**Figure 5.6** Ion fractions for carbon (left) and oxygen (right) in collisional ionization equilibrium. [Calculated using CHIANTI v9]

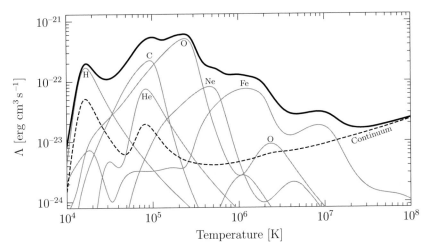

**Figure 5.7** Contributions to the total cooling function $\Lambda$ (heavy black line) from emission lines (thin gray lines) and continuum processes (dashed line), for a solar-metallicity gas in collisional ionization equilibrium. [Calculated using CHIANTI v9]

For hot gas with solar abundance, the cooling can be broken down into three different temperature ranges. At $10^4$ K $< T < 3 \times 10^4$ K, cooling occurs mainly by Lyman $\alpha$ emission from collisionally excited H atoms. At $3 \times 10^4$ K $< T < 2 \times 10^7$ K, cooling occurs mainly by permitted UV lines from collisionally excited heavy ions. The elements that provide the most effective cooling are carbon at $T \sim 8 \times 10^4$ K, oxygen at $T \sim 2.5 \times 10^5$ K, neon at $T \sim 5 \times 10^5$ K, and iron at $T \sim 10^6$ K. Finally, at $T > 2 \times 10^7$ K, continuum radiation from free–free emission, alias bremsstrahlung, is the main source of cooling, with cooling function $\Lambda_{\mathrm{ff}} \propto T^{1/2}$ (Equation 4.52).

In the hot ionized medium, heating and cooling are generally not in equilibrium. At temperatures greater than $T \approx 10^7$ K, which can be found in the hottest parts of the HIM, the cooling time from free–free emission is (Equation 5.22)

$$t_{\text{cool}} \approx 6\,\text{Gyr} \left(\frac{T}{10^7\,\text{K}}\right)^{1/2} \left(\frac{n_{\text{H}}}{0.004\,\text{cm}^{-3}}\right)^{-1}. \tag{5.72}$$

Thus, given the low density ($n_{\text{H}} \sim 0.004\,\text{cm}^{-3}$) of the hot ionized medium, the cooling time of gas with $T \geq 10^7$ K is comparable to or greater than the age of our galaxy. Even at a temperature $T \approx 10^6$ K, where the cooling time is $t_{\text{cool}} \sim 25\,\text{Myr}$ at $n_{\text{H}} \sim 0.004\,\text{cm}^{-3}$, the HIM frequently doesn't have time to cool thoroughly before another supernova shock wave comes through to heat it again.

Previously in this section, we made the implicit assumption that the hot ionized medium is in collisional ionization equilibrium. We can test whether this assumption is correct. For simplicity, consider a gas of pure hydrogen with number density $n_{\text{H}}$ and fractional ionization $x$, so that $n_{\text{HII}} = n_e = x n_{\text{H}}$ and $n_{\text{HI}} = (1 - x) n_{\text{H}}$. The balance between collisional ionization and radiative recombination (Equation 5.67) can then be written as

$$\frac{dx}{dt} = n_{\text{H}} k_{\text{ci}} \left[ x - (1 + \alpha_{\text{A,H}}/k_{\text{ci}}) x^2 \right]. \tag{5.73}$$

In collisional ionization equilibrium, Equation 5.73 yields a fractional ionization

$$x_{\text{cie}} = \frac{1}{1 + \alpha_{\text{A,H}}/k_{\text{ci}}} = \frac{1}{1 + t_{\text{ci}}/t_{\text{rec}}}, \tag{5.74}$$

where the collisional ionization time is

$$t_{\text{ci}} = \frac{1}{n_e k_{\text{ci}}} \tag{5.75}$$

$$\approx 160\,\text{yr} \left(\frac{T}{10^6\,\text{K}}\right)^{-0.5} \exp\left[0.158\left(\frac{10^6\,\text{K}}{T}\right)\right] \left(\frac{n_e}{0.004\,\text{cm}^{-3}}\right)^{-1}$$

and the much longer recombination time is

$$t_{\text{rec}} = \frac{1}{n_e \alpha_{\text{A,H}}} \approx 0.5\,\text{Gyr} \left(\frac{T}{10^6\,\text{K}}\right)^{1.5} \left(\frac{n_e}{0.004\,\text{cm}^{-3}}\right)^{-1}, \tag{5.76}$$

assuming $T \geq 10^6$ K. At $T = 10^6$ K, the fact that $t_{\text{ci}} \ll t_{\text{rec}}$ implies a fractional ionization close to unity: $x_{\text{cie}} \approx 1 - t_{\text{ci}}/t_{\text{rec}} \approx 1 - 3 \times 10^{-7}$.

Suppose, however, that a hydrogen gas with temperature $T$ and density $n_{\text{H}}$ starts in a state that is out of collisional ionization equilibrium. If at $t = 0$ the fractional ionization is $x_0 \neq x_{\text{cie}}$, how long will it take the gas to approach collisional ionization equilibrium? If we use the dimensionless variables $y \equiv x/x_{\text{cie}}$

and $\tau \equiv n_H k_{ci} t$, the balance between collisional ionization and radiative recombination (Equation 5.73) can be simply written as

$$\frac{dy}{d\tau} = y - y^2. \tag{5.77}$$

The solution for the chosen initial conditions is

$$y(\tau) = \frac{y_0}{y_0 + (1 - y_0)e^{-\tau}}, \tag{5.78}$$

where $y_0 \equiv x_0/x_{cie}$. Thus, regardless of whether the hydrogen is initially over-ionized or under-ionized relative to CIE, the time scale $t_{equil}$ for approaching CIE is

$$t_{equil} = \frac{1}{n_H k_{ci}} \sim 200 \, \mathrm{yr} \left(\frac{n_H}{0.004 \, \mathrm{cm}^{-3}}\right)^{-1} \tag{5.79}$$

when $T = 10^6$ K (and the time is even shorter at higher temperatures). The time in Equation 5.79 is sufficiently short that it is generally safe to assume that CIE holds true in the hot ionized medium.

## 5.4    Observing the Hot Ionized Medium

One reason why we have discussed H II regions at length, despite their tiny contribution to the mass of the interstellar medium, is that they have readily observable emission lines that allow us to estimate their density and temperature. Finding the density and temperature of the hot ionized medium is more difficult. In H II regions, the temperature $T \sim 10^4$ K means that we need to observe visible and near-ultraviolet lines ($h\nu \sim kT \sim 1 \, \mathrm{eV}$) to deduce the gas properties. In the HIM, the temperature $T \geq 10^6$ K implies that emission lines with $h\nu \sim kT \sim 100 \, \mathrm{eV}$ would be useful for deducing the gas properties. Unfortunately, there are practical barriers to our rushing forth and observing extreme ultraviolet and soft X-ray photons.

Photons at an energy $h\nu \approx 100 \, \mathrm{eV}$ ($\lambda \approx 120 \, \text{Å}$) are difficult to observe, even if your observatory is above the Earth's opaque atmosphere. Primarily, the reason is the large photoionization cross section of hydrogen at this photon energy. From Equation 4.2, we can write

$$\sigma_{pho} \approx 1.6 \times 10^{-20} \, \mathrm{cm}^2 \left(\frac{h\nu}{100 \, \mathrm{eV}}\right)^{-3}. \tag{5.80}$$

The mean free path of an ionizing photon through neutral gas is then

$$\lambda_{mfp} = \frac{1}{n_H \sigma_{pho}} \approx 50 \, \mathrm{pc} \left(\frac{h\nu}{100 \, \mathrm{eV}}\right)^3 \left(\frac{n_H}{0.4 \, \mathrm{cm}^{-3}}\right)^{-1}, \tag{5.81}$$

scaling to the density of the warm neutral medium. At $h\nu \approx 100 \, \mathrm{eV}$, this is a distance sufficiently short that very few 100 eV photons reach us from outside the Local Bubble. The Local Bubble is a very fine bubble indeed, but we cannot be certain that it is characteristic of all portions of the hot ionized medium.

What emission lines might we expect to see from the gas of the Local Bubble? In collisional ionization equilibrium, hydrogen is more than 99% ionized at $T \geq 2.6 \times 10^4$ K, and helium is more than 99% doubly ionized at $T \geq 1.5 \times 10^5$ K. Thus, at the high temperature of the HIM we can assume that the hydrogen and helium are almost completely ionized. To find elements that aren't thoroughly ionized at $T \geq 10^6$ K, we need to move further along the periodic table. The last two bound electrons of a heavy element require a great deal of energy to remove. The penultimate electron is difficult to remove because an ion with two electrons is helium-like, with a filled inner shell of electrons. The last electron is hard to remove because an ion with a single electron is hydrogenic, with an ionization energy $I \propto Z^2$. Carbon becomes 90% fully ionized at a temperature $T \approx 2 \times 10^6$ K, for example, while oxygen doesn't become 90% fully ionized until the temperature reaches $T \approx 5 \times 10^6$ K (see Figure 5.6).

At $T \sim 10^7$ K and higher, the only bound electrons are attached to elements with higher atomic numbers. For instance, iron doesn't become 90% fully ionized until $T \approx 4 \times 10^8$ K. At $T \sim 10^6$ K, the dominant forms of iron are Fe IX, X, and XI; at $T \sim 10^7$ K, the dominant forms are Fe XX and XXI. At temperatures near $10^6$ K, the iron ions from Fe IX to Fe XI produce a cluster of emission lines near $h\nu \approx 70$ eV ($\lambda \approx 180$ Å), as shown in Figure 5.8. For a gas with solar abundance, in collisional ionization equilibrium at $T \sim 10^6$ K, about half the radiated power will come out in the form of these iron emission lines. However, observations at $\lambda \sim 180$ Å by the *Cosmic Hot Interstellar Plasma Spectrometer* showed that

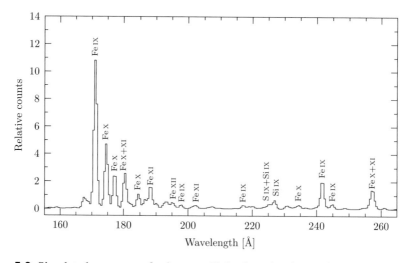

**Figure 5.8** Simulated spectrum of solar-metallicity, low-density gas in collisional ionization equilibrium at $T = 10^6$ K. Ionic species responsible for various emission lines are labeled. [Calculated using CHIANTI v9]

the iron emission lines from the Local Bubble were much weaker than originally predicted. This is speculated to be due to a combination of iron depletion onto dust grains and a somewhat lower temperature ($T \approx 7 \times 10^5$ K) than was previously assumed.

Since emission lines are lacking in the spectrum of the Local Bubble, we can fall back on absorption lines. One important pair of UV lines is due to O VI, which produces a doublet at $\lambda = 1031.9$ Å and $1037.6$ Å, corresponding to energies $h\nu = 12.01$ eV and $11.95$ eV. These lines are a major cooling mechanism for interstellar gas at $T \sim 2.5 \times 10^5$ K (see Figure 5.7). Since O VI is a fragile lithium-like ion, with three bound electrons, it never provides a majority of the oxygen ions in hot gas. The fraction of oxygen that takes the form of O VI is a maximum at $T \approx 3.0 \times 10^5$ K, where O VI contributes about 25% of the total oxygen abundance (~35% is O V and ~35% is O VII). Despite the fragility of the O VI ion, the lines of the O VI doublet have a large absorption cross section, and thus are observable even when the column density of O VI is small.

To see the 12 eV lines of O VI in absorption, we need a background light source with $kT_{eff} \geq 12$ eV, or $T_{eff} \geq 1.4 \times 10^5$ K. Main sequence stars don't run this hot, so we have to look at nearby hot white dwarfs.[6] This technique works best when applied to white dwarfs of spectral type DA, which have only hydrogen lines in their spectra. If the white dwarf has O VI absorption lines from its own photosphere, measuring the interstellar O VI absorption becomes much more difficult.

The number of hot DA white dwarfs inside the Local Bubble is limited; Figure 5.9 shows a map of 39 that were studied with the *FUSE* spacecraft. Of the 39 white dwarfs observed, 24 had O VI lines detected at the $2\sigma$ level along the line of sight. The distribution of O VI in the Local Bubble appears to be patchy. The mean density of O VI, along the lines of sight where it is detected, lies in the range $N_{OVI}/d \approx (0.7 \rightarrow 13) \times 10^{-8}$ cm$^{-3}$, with adjacent lines of sight often having very different densities. For the lines of sight with the strongest detections, the broadening parameter ranges from $b = 15$ km s$^{-1}$ to $b = 36$ km s$^{-1}$. Using the relation for the broadening parameter (Equation 2.31),

$$b = 0.32 \,\text{km s}^{-1} \left(\frac{T}{100 \text{ K}}\right)^{1/2} \left(\frac{m}{16 m_H}\right)^{1/2}, \tag{5.82}$$

we find the temperature

$$T = 2.5 \times 10^5 \,\text{K} \left(\frac{b}{16 \text{ km s}^{-1}}\right)^2 \left(\frac{m}{16 m_H}\right). \tag{5.83}$$

---

[6] The very hottest white dwarfs known have $kT_{eff} \sim 20$ eV, so looking for the 70 eV absorption lines of ionized Fe in white dwarf spectra would be "challenging," to use a favorite astronomical euphemism.

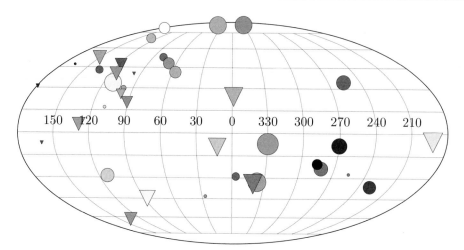

**Figure 5.9** Location in galactic coordinates of nearby hot DA white dwarfs. The circles show regions of O VI detection; the triangles are non-detections at the $2\sigma$ level. The diameters of the symbols are inversely proportional to their distances. The grayscale of the symbols runs logarithmically from $N_{\mathrm{OVI}} = 10^{12.4}\,\mathrm{cm}^{-2}$ (white) to $N_{\mathrm{OVI}} = 10^{13.6}\,\mathrm{cm}^{-2}$ (black). [Savage & Lehner 2006]

Thus, the narrowest O VI absorption lines are consistent with thermal Doppler broadening near the temperature of peak O VI abundance, $T \approx 3.0 \times 10^5$ K, with the broader lines representing higher temperatures, turbulent gas, or variation in bulk flow.

## Exercises

5.1   Technologically advanced (but ethically dubious) space aliens have found a method of converting mass $M$ to energy $E$ with 100% efficiency, using Einstein's relation $E = Mc^2$.

(a)   The star Proxima Centauri ($M_{\mathrm{pc}} = 0.123\,\mathrm{M}_\odot$, $R_{\mathrm{pc}} = 0.143\,\mathrm{R}_\odot$) is at a distance $d = 1.301$ pc from the Earth. The space aliens convert 5% of Proxima Centauri's mass into energy, and use that energy to eject the remaining 95% spherically outward at an initial speed $u_{\mathrm{ej}}$. What is $u_{\mathrm{ej}}$ in kilometers per second?

(b)   The gas between the solar system and Proxima Centauri is part of the Local Interstellar Cloud, with mean density $n_{\mathrm{H}} = 1\,\mathrm{cm}^{-3}$ and temperature $T = 8000$ K. Assuming uniform density and temperature, how long will it take the blastwave from the explosion in part (a) to reach the solar system? If the aliens beam us a taunting radio message at the same time they blow up Proxima Centauri, how long will we have to prepare for the arrival of the blastwave?

(c) What will be the speed $u_{sh}$ of the blastwave as it passes through the solar system?

(d) What will be the pressure $P_2$ and the temperature $T_2$ right behind the blastwave as it passes through the solar system?

(e) What will be the total thermal energy of the portion of the blastwave that strikes the Earth? Is this greater than or less than the thermal energy of the Earth's atmosphere? [Hint: in addition to knowing the Earth's radius and average atmospheric temperature, you will need to know the mass of its atmosphere.]

5.2 A lightning bolt can be approximated as a linear explosion, releasing an energy per unit length $\epsilon$ in a gaseous medium of mass density $\rho$.

(a) If the shock created by the lightning is a cylindrical blastwave, what is the radius of the expanding cylindrical shell as a function of time? [There will be a dimensionless factor of order unity in your solution, comparable with $A$ for the spherical Sedov–Taylor blastwave. It's not necessary to compute the exact value of this number; just set it equal to one.]

(b) If $\epsilon = 3 \times 10^{12}\,\mathrm{erg\,cm^{-1}}$, a typical value for lightning, and if $\rho = 0.001\,\mathrm{g\,cm^{-3}}$, typical for the Earth's lower atmosphere, find the radius at which the pressure behind the shock is equal to the ambient pressure of the Earth's lower atmosphere.

5.3 Converting a helium-like ion (with two bound electrons) to a hydrogenic ion (with one bound electron) requires an energy $I_{pen}$ that we may call the "penultimate ionization energy," since it removes the penultimate (or next-to-last) bound electron.

(a) For helium, with only two electrons to begin with, $I_{pen}$ is equal to the first ionization energy (Table 1.2). At what temperature $T_{pen}$ does half the helium become hydrogenic $He^+$ rather than neutral $He^0$? For a hydrogenic ion, the Lyman $\alpha$ transition has energy $h\nu_0 \propto Z^2$ and Einstein A coefficient $A_{21} \propto Z^4$, where $Z$ is the atomic number of the ion. At the temperature $T_{pen}$ that you have just calculated, what is the thermal broadening $b$ of the Lyman $\alpha$ line of $He^+$? What is the intrinsic broadening $\gamma_{21}/2\pi$ of the same line?

(b) Lithium has a penultimate ionization energy $I_{pen} = 75.64\,\mathrm{eV}$. At what temperature $T_{pen}$ does half the lithium become hydrogenic $Li^{++}$? At this temperature, what is the thermal broadening of the Lyman $\alpha$ line of $Li^{++}$? What is the intrinsic broadening of the same line? [Assume the lithium is all $^7Li$, the most abundant isotope.]

# 6

# Interstellar Dust

*Ah, make the most of what we yet may spend,*
*Before we too into the Dust descend.*

> Omar Khayyám (1046–1131)
> "The Rubaiyat," quatrain XXIV
> [Edward FitzGerald translation, 1889]

Until now, we have evaded the discussion of interstellar dust, aside from recognizing its contribution of photoelectrons to the interstellar gas. However, since dust and gas strongly influence each other, particularly in the densest regions of the interstellar medium, it's time to acknowledge further the existence of dust. The discovery of interstellar dust was a belated, drawn-out process. A look at the Milky Way and its central dust lane would make the presence of dust obvious, you might think. However, when William Herschel looked at the dust features in Scorpius, near the galactic center, he did not exclaim, "Here is truly a dark cloud of dust!" but rather "Here is truly a hole in Heaven!" It wasn't until the late nineteenth century that astronomers speculated in print about opaque material blocking the light from distant stars. In 1894, the astronomer Arthur Cowper Ranyard looked at E. E. Barnard's photograph of the star Theta Ophiuchi (Figure 6.1)[1] and wrote, "the dark vacant areas or channels running north and south of the bright star ... seem to me to be undoubtedly dark structures, or obscuring matter in space, which cut out the light from the nebulous or stellar regions behind them." To our experienced eyes, the dark structures near Theta Ophiuchi are undeniably dust features, but contemporaries of Ranyard had some doubts. Although Slipher's observations of the Pleiades in 1912 provided evidence for circumstellar dust, the existence of a more widespread distribution of interstellar dust wasn't universally accepted until the work of Robert Trumpler in 1930.

---

[1] The image of Theta Oph shown in Figure 6.1 was actually taken by Barnard a few years after his 1894 photograph. In Barnard's opinion, it "shows the peculiarity of this part of the sky considerably better than the previous photograph."

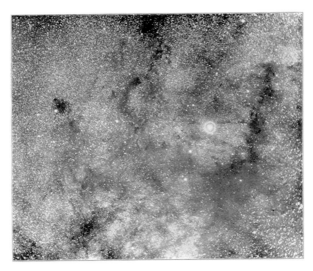

**Figure 6.1** Photograph by E. E. Barnard of the dark structures near Theta Ophiuchi (the bright star with diffraction ring to right of center). North = left, east = down. [Barnard 1899]

Trumpler was making a study of open clusters within our galaxy. He used bright main sequence stars in each cluster as standard candles, using the measured flux $f$ and assumed luminosity $L$ of the stars to compute a luminosity distance $r_L = (L/4\pi f)^{1/2}$, assuming no extinction along the line of sight to the cluster. He also assumed that clusters with similar richness and similar central concentration had the same physical diameter ($d = 3$ pc for strongly concentrated poor clusters, for instance, and $d = 6.5$ pc for low-concentration rich clusters). He then used the clusters as standard yardsticks, using the measured angular diameter $\theta$ and assumed physical diameter $d$ to compute an angular-diameter distance $r_A = d/\theta$. Figure 6.2 shows the angular-diameter distance and luminosity distance computed by Trumpler for each open cluster in his sample.

If Trumpler assumed that open clusters were dimmed solely by the inverse square law of flux, then the luminosity distances that he estimated were larger than the angular-diameter distances, with the deviation increasing with distance. However, he found a much better fit if he assumed that the light was attenuated exponentially, with an attenuation coefficient $\kappa_\lambda$ that was constant in space. That is,

$$I_\lambda = I_{\lambda,0} e^{-\kappa_\lambda r}, \tag{6.1}$$

where Trumpler found $\kappa_\lambda \approx 2 \times 10^{-22}\,\mathrm{cm}^{-1} \approx 0.6\,\mathrm{kpc}^{-1}$ at $\lambda = 4300\,\text{Å}$ and $\kappa_\lambda \approx 1 \times 10^{-22}\,\mathrm{cm}^{-1} \approx 0.3\,\mathrm{kpc}^{-1}$ at $\lambda = 5600\,\text{Å}$. The wavelength dependence of the attenuation, with shorter wavelengths having a larger $\kappa_\lambda$, means that distant stars are *reddened* as well as dimmed. Trumpler realized that this could be caused by Rayleigh scattering from "fine cosmic dust particles."

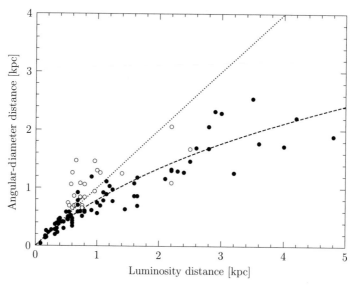

**Figure 6.2** Angular-diameter distance versus luminosity distance for Trumpler's sample of 100 open clusters. Solid dots represent clusters with better photometry. Straight dotted line: no dust. Curved dashed line: attenuation coefficient $\kappa = 0.6 \, \text{kpc}^{-1}$. [Data from Trumpler 1930]

## 6.1 Observed Properties of Dust

Although sample-return missions have given us samples of interplanetary dust to play with, the properties of more distant interstellar dust must be deduced indirectly, primarily by the effect of dust on electromagnetic radiation. Dust particles can **scatter** light. When we look at a reflection nebula, such as that surrounding the Pleiades (Figure 1.3), we are seeing light from the central stars that has been scattered by dust into our line of sight. Dust particles can also **absorb** light. The relative amount of scattering and absorbing depends on the properties of the dust grains. When we look at a distant star through the intervening dust, the excess dimming of the star is caused by a combination of scattering and absorbing. Usually, astronomers refer to the net result of scattering and absorbing as **extinction**. Extinction can be dependent on the polarization of the light being extinguished, so dust can **polarize** light from distant stars.

When dust absorbs light, it becomes warmer, so dust grains can **emit** light in the form of thermal radiation. Most of this emission is at wavelengths from a few microns (near-infrared) to the sub-millimeter range (far-infrared). By observing how dust scatters, absorbs, polarizes, and emits light as a function of wavelength, we can deduce many of the properties of the dust grains. Let's start in the footsteps of Trumpler and begin by examining extinction.

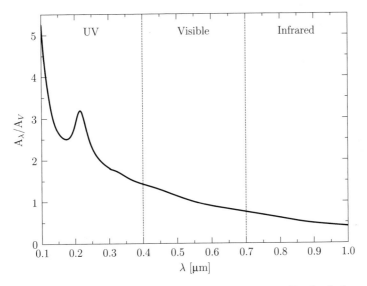

**Figure 6.3** Dust extinction as a function of wavelength, normalized relative to $A_V$, the extinction in the $V$ band. [Data from Cardelli et al. 1989 and O'Donnell 1994]

The observed flux of a star, $F_\lambda$, can be written as

$$F_\lambda = F_{\lambda,0} e^{-\tau_\lambda}, \tag{6.2}$$

where $F_{\lambda,0}$ is what the flux would be in the absence of dust extinction and $\tau_\lambda$ is the wavelength-dependent optical depth. At photon energies $h\nu > 13.6\,\mathrm{eV}$ most of the optical depth is due to photoionization of neutral hydrogen, not dust extinction. Thus, studies of dust extinction focus on photon energies $h\nu < 13.6\,\mathrm{eV}$ or, equivalently, wavelengths $\lambda > 912\,\text{Å}$. This means that dust extinction studies have fallen into the hands of optical observers; as such, they are subject to the use of **magnitudes** to describe flux. Dust extinction is commonly expressed not in terms of the optical depth $\tau_\lambda$, but in terms of $A_\lambda$, the number of magnitudes of extinction. At a wavelength $\lambda$,

$$A_\lambda = 2.5 \log_{10}\left(\frac{F_{\lambda,0}}{F_\lambda}\right). \tag{6.3}$$

The relation between $A_\lambda$ and $\tau_\lambda$ is thus

$$A_\lambda = 2.5 \log_{10}\left(e^{\tau_\lambda}\right) \approx 1.086\tau_\lambda. \tag{6.4}$$

The extinction curve $A_\lambda$ as a function of $\lambda$ can be determined by comparing the spectrum of a nearby star, with negligible extinction, to that of a more distant star of the same spectral type. Figure 6.3 shows a fairly standard extinction curve for dust in our galaxy. It is normalized relative to the extinction in the $V$ band, which has effective wavelength $\lambda \approx 5500\,\text{Å}$.

The reddening by dust is expressed in terms of a **color excess**, for instance,

$$E(B - V) = A_B - A_V \qquad (6.5)$$

is the commonly used $B - V$ color excess. The effective wavelength of the $B$ band is $\lambda \approx 4400\,\text{Å}$. Color excess is expressed as the extinction $A$ at the shorter wavelength minus $A$ at the longer wavelength. Since extinction is usually greater at shorter wavelengths, the color excess is usually a positive number. However, as Figure 6.3 shows us, extinction doesn't fall monotonically with wavelength. Most obviously, there is an excess extinction feature at $\lambda \approx 2175\,\text{Å}$. The $2175\,\text{Å}$ bump is attributed to the presence of carbon-rich particles, the height of the bump being strongly dependent on metallicity. In the Large Magellanic Cloud (with metallicity $Z \sim 0.5 Z_\odot$), the $2175\,\text{Å}$ bump is weaker than in our galaxy; in the Small Magellanic Cloud (with $Z \sim 0.1 Z_\odot$), the bump is nearly absent.

If the interstellar extinction curve were the same everywhere then by knowing $A_V$ you always would be able to read off $E(B - V)$. The ratio of $A_V$ to $E(B - V)$ is called the ratio of total to selective extinction:

$$R_V \equiv \frac{A_V}{E(B - V)}. \qquad (6.6)$$

Observationally, $R_V$ is seen to range between 2 and 6 for different lines of sight through our galaxy. The value of $R_V$ that is most frequently adopted in the astronomical literature is $R_V = 3.1$, which is typical of the lower-density portions of the ISM. However, in dense molecular clouds, the value $R_V = 5$ is more often adopted. In the limit $R_V \to \infty$, the extinction curve is perfectly horizontal between $B$ and $V$. This is characteristic of "gray" absorbers, which extinguish all wavelengths of light equally. Gray absorbers are sometimes called "bricks," since an opaque brick is equally effective at absorbing all wavelengths of light much smaller than the brick. The fact that $R_V = 2 \to 6$ for interstellar dust, implying 1.5 to 1.2 magnitudes of $B$ band extinction for every magnitude of $V$-band extinction, tells us that dust isn't made of bricks.

Within our galaxy, the amount of dust extinction along a line of sight is found to be correlated with the total column density of hydrogen. When all the hydrogen is taken into account, the molecular $H_2$ as well as the atomic H, the ratio of column density to $B - V$ color excess is

$$\frac{N_H}{E(B - V)} \approx 5.8 \times 10^{21}\,\text{cm}^{-2}\,\text{mag}^{-1}. \qquad (6.7)$$

For lines of sight with $R_V \approx 3.1$, this implies

$$\frac{N_H}{A_V} \approx 1.9 \times 10^{21}\,\text{cm}^{-2}\,\text{mag}^{-1}. \qquad (6.8)$$

The correlation between dust extinction and hydrogen column density is useful because if you have measured the extinction toward a star, you can make an estimate of the hydrogen column density. Conversely, since

$$\frac{A_V}{N_{\mathrm{H}}} \approx 5.3 \times 10^{-22} \, \mathrm{cm}^2 \, \mathrm{mag}, \tag{6.9}$$

if you have a map of hydrogen column density, you can convert it to a dust map. This is handy if you want a quick estimate of the extinction toward extragalactic sources. In more modern dust studies, it is common to normalize to the Cousins $I$ band (effective wavelength $\lambda = 8000 \, \mathring{\mathrm{A}}$), where the relation between dust extinction and hydrogen column density is found to be

$$\frac{A_I}{N_{\mathrm{H}}} \approx \left( 2.96 - 3.55 \left[ \frac{3.1}{R_V} - 1 \right] \right) \times 10^{-22} \, \mathrm{cm}^2 \, \mathrm{mag}. \tag{6.10}$$

Thus, even at high galactic latitudes, where the column density of hydrogen is $N_{\mathrm{H}} \sim 10^{20} \, \mathrm{cm}^{-2}$, there is still some foreground extinction when one is looking at extragalactic sources: $\sim 0.05$ magnitudes in the $V$ band and $\sim 0.03$ magnitudes in the Cousins $I$ band.

Since some extinction is caused by absorption rather than scattering, dust grains are heated by starlight. They then cool by the emission of infrared light. In our galaxy, the infrared portion of the interstellar radiation field (see Figure 3.4) peaks at a photon energy $h\nu \approx 0.009 \, \mathrm{eV}$, corresponding to a wavelength $\lambda \approx 140 \, \mu\mathrm{m}$. This represents thermal (but not blackbody) emission from dust grains with temperature $T \sim h\nu/k \sim 100 \, \mathrm{K}$. Dust grains are inefficient radiators at long wavelengths, as we'll see in more detail in Section 6.2, which is why their spectrum is not that of a blackbody.

Along a typical line of sight through our galaxy, the ratio of dust infrared emission to the hydrogen column density tells us the infrared luminosity per hydrogen nucleus:

$$\frac{I_{\mathrm{IR}}}{N_{\mathrm{H}}} = 5 \times 10^{-24} \, \mathrm{erg} \, \mathrm{s}^{-1}. \tag{6.11}$$

To give you a sense of scale, the Sun's luminosity per hydrogen nucleus is $\sim L_\odot m_{\mathrm{H}}/M_\odot \sim 3 \times 10^{-24} \, \mathrm{erg} \, \mathrm{s}^{-1}$. In some hydrogen-centered sense, the ISM is about as efficient at converting starlight into infrared light as the Sun is at creating sunlight to begin with. Indeed, when you look at the spectrum of our galaxy (Figure 3.4), the amount of energy in dust-emitted infrared light is comparable to the amount of energy in starlight.

## 6.2   Optical Properties of Grains

Much of what we deduce about the physical properties of dust grains comes from observing how they absorb and scatter light. Consider a single grain of dust. It has a wavelength-dependent cross section for absorption, $\sigma_{abs}(\lambda)$, and a wavelength-dependent cross section for scattering, $\sigma_{sca}(\lambda)$. From these, we can compute an **extinction cross section,**

$$\sigma_{ext}(\lambda) \equiv \sigma_{abs}(\lambda) + \sigma_{sca}(\lambda). \tag{6.12}$$

For a population of identical dust grains with number density $n$, the extinction cross section is related to the attenuation coefficient by the relation

$$\kappa_\lambda = n\sigma_{ext}(\lambda). \tag{6.13}$$

We will assume a spherical dust grain; most grains are not particularly close to spherical, but we'll start with spheres for the sake of simplicity. The geometric cross section of a spherical grain is $\pi a^2$, where $a$ is the grain radius. In general, neither the absorption cross section nor the scattering cross section is equal to $\pi a^2$. Thus, it is customary to define dimensionless **efficiency factors,**

$$Q_{abs} \equiv \frac{\sigma_{abs}}{\pi a^2} \tag{6.14}$$

and

$$Q_{sca} \equiv \frac{\sigma_{sca}}{\pi a^2}. \tag{6.15}$$

The efficiency factor can be either greater than or less than one. However, in the limit of a big shiny sphere, we expect $Q_{sca} \sim 1$; in the limit of a big matte black sphere, we expect $Q_{abs} \sim 1$.

Whether a dust grain is better at absorbing or scattering depends on the **index of refraction** for the substance of which it's made. In general, the index of refraction is a complex number:

$$\tilde{n} = n_r + i\,n_i. \tag{6.16}$$

The real part of the index of refraction, $n_r$, is sometimes simply called the "refractive index," since it determines how much a beam of light is bent, or refracted, when entering or exiting the substance. The imaginary part, $n_i$, is sometimes called the "absorption index," since it determines how strongly the substance absorbs light. The dimensionless numbers $n_r$ and $n_i$ are real and non-negative. (A quick warning: this sign convention is common but not universal. Some publications use an alternative definition in which $n_i \leq 0$ rather than $n_i \geq 0$.)

For transparent substances, the imaginary part of the index of refraction is much smaller than one. For instance, at $\lambda = 5500$ Å, pure water ice has $\tilde{n} = 1.31 + i\,2.3 \times 10^{-9}$. For highly reflective substances, the imaginary part of the index of refraction is much greater than the real part. For instance, at $\lambda = 5500$ Å, gold has $\tilde{n} = 0.32 + i\,2.60$. The index of refraction can be strongly dependent on wavelength. For instance, as shown in Figure 6.4, silicon has $n_i \approx 5$ at $0.28\,\mu m$

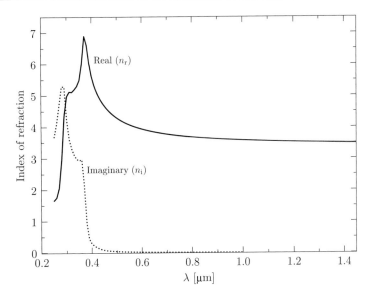

**Figure 6.4** Real (solid line) and imaginary (dotted line) components of the index of refraction for silicon at $T = 300$ K. [Data from Green 2008]

but $n_i \approx 5 \times 10^{-4}$ at 1 μm. Thus, silicon goes from being opaque in the ultraviolet to transparent in the near infrared.[2]

Suppose we have a solid slab of material with index of refraction $\tilde{n}$. If an electromagnetic wave with electric field strength

$$E \propto e^{i(kx - \omega t)} \tag{6.17}$$

travels through the material, the dispersion relation between wavenumber $k$ and frequency $\omega$ will be

$$k = \tilde{n}\frac{\omega}{c}. \tag{6.18}$$

Thus, if the imaginary part of $\tilde{n}$ is greater than zero, the power of the electromagnetic wave will decrease as it propagates through the material, with

$$|E|^2 \propto e^{-2n_i \omega x / c}. \tag{6.19}$$

The attenuation coefficient for light passing through this chunk of material will then be

$$\kappa = 2n_i\frac{\omega}{c} = \frac{4\pi n_i}{\lambda_{vac}}, \tag{6.20}$$

---

[2] In fact, in the wavelength range $1.2 \to 7$ μm silicon lenses are often used. Silicon has the great advantage, compared with other substances transparent in the near infrared, of having a low cost, a low mass density, and a high refractive index $n_r$.

where $\lambda_{vac} \equiv 2\pi c/\omega$ is the vacuum wavelength of the light. Thus, visible light traveling through pure water ice will have $1/\kappa \sim 15$ m; the same light traveling through gold will have $1/\kappa \sim 15$ nm.

Of course, when light travels through interstellar space it seldom encounters large slabs of solid material (especially not large slabs of gold). Instead, it encounters small dust grains, which are frequently tiny compared to the wavelength $\lambda$ of the light. In the limit $2\pi a \ll \lambda$, the efficiency factors $Q$ have a particularly simple form. For absorption,

$$Q_{abs} = 4 \left( \frac{2\pi a}{\lambda} \right) \text{Im} \left[ \frac{\tilde{n}^2 - 1}{\tilde{n}^2 + 2} \right]. \tag{6.21}$$

The attenuation coefficient is, in this long-wavelength limit,

$$\kappa_{abs} = n_{gr} Q_{abs} \pi a^2 \propto n_{gr} a^3, \tag{6.22}$$

where $n_{gr}$ is the number density of dust grains. Thus, the absorption coefficient is proportional to the fraction of the volume of space that is occupied by dust. For scattering, the efficiency factor is

$$Q_{sca} = \frac{8}{3} \left( \frac{2\pi a}{\lambda} \right)^4 \left| \frac{\tilde{n}^2 - 1}{\tilde{n}^2 + 2} \right|^2. \tag{6.23}$$

This yields the relation $\sigma_{sca} \propto a^6 \lambda^{-4}$ that is characteristic of Rayleigh scattering.

As long as there is any absorption at all ($n_i \neq 0$), absorption will dominate over scattering in the limit that the grain radius $a$ goes to zero. Consider the observed extinction curve at UV, visible, and near-IR wavelengths (Figure 6.3). Once the 2175 Å bump is removed, the curve shows a wavelength dependence much closer to $\lambda^{-1}$ than $\lambda^{-4}$. This is consistent with the dominant source of extinction being absorption by particles with $a \leq \lambda_{vis}/2\pi \sim 0.1\,\mu\text{m}$.

At shorter wavelengths, $2\pi a \geq \lambda$, the efficiency factors $Q_{abs}$ and $Q_{sca}$ aren't as easy to compute. One approach is to use the results of **Mie scattering**, named after the physicist Gustav Mie,[3] who investigated the scattering of plane electromagnetic waves by homogeneous spheres. The general solution that Mie found consists of an infinite series. However, the astronomer Henk van de Hulst found a useful approximation that applies in the limit when the sphere is large ($2\pi a \gg \lambda$) and is only moderately refractive and absorptive at the wavelength of interest ($|\tilde{n} - 1| < 1$). In this limit, the solution for the pure scattering case yields an efficiency factor

$$Q_{ext} = Q_{sca} \approx 2 - \frac{4}{\varrho} \sin \varrho + \frac{4}{\varrho^2} (1 - \cos \varrho), \tag{6.24}$$

---

[3] Mie's name is pronounced "mee," as in "It's all about Mie."

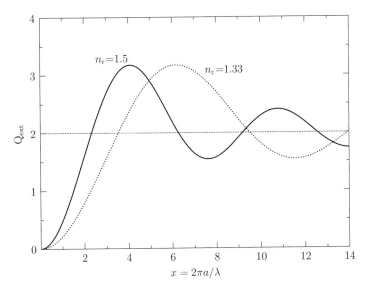

**Figure 6.5** Efficiency factor $Q_{ext}$ calculated for two values of the real index of refraction, $n_r = 1.5$ (solid line) and $n_r = 1.33$ (dotted line).

where

$$\varrho \equiv 2\left(\frac{2\pi a}{\lambda}\right)|n_r - 1|. \tag{6.25}$$

When the refractive index is greater than one, the portion of the plane wave that passes through the center of the sphere will be slowed relative to the portion that passes through the vacuum outside the sphere. Thus, the parameter $\varrho$ represents the phase delay of the wave passing through the center of the sphere.

Figure 6.5 shows the van de Hulst approximation for $Q_{ext}$ in the pure scattering case. The first maximum in $Q_{ext}$ occurs at $\varrho \approx 4.1$; this corresponds to $\lambda \approx 1.5a$ when $n_r = 1.5$ (a refractive index typical of glass at visible wavelengths) and $\lambda \approx 1.0a$ when $n_r = 1.33$ (typical of pure ice at visible wavelengths). At the maxima in $Q_{ext}$, the twice-refracted light passing through the center of the sphere interferes destructively with the light that is diffracted around the sphere.

In the short-wavelength limit, when $\varrho$ becomes very large, the value of $Q_{ext}$ settles down to $Q_{ext} \to 2$. The fact that the extinction cross section approaches *twice* the geometric cross section is called the "extinction paradox." For more realistic spheres, which absorb as well as scatter, it is also found that $Q_{ext} \to 2$. You might expect that the extinction cross section for a large opaque sphere (a "bowling ball") would be exactly equal to the geometric cross section in the limit $\lambda \ll a$. However, diffraction must be taken into account. According to Babinet's principle, the scattering produced by an opaque object is the same as that produced by an aperture of the same size and shape cut into an opaque screen.

Thus, the diffraction pattern of an opaque sphere of radius $a$ is the same Airy pattern as that produced by a hole of radius $a$ cut into a screen. In the limit of opaque bowling balls ($a = 10.86\,\text{cm} \gg \lambda$), light is dimmed with an efficiency factor $Q_{abs} = 1$ from absorption and $Q_{sca} = 1$ from scattering.

## 6.3  Composition, Shape, and Size of Grains

Even with the limited information available to us, we can make some deductions about the composition of interstellar dust. It must be made of refractory materials that are solid rather than gaseous under typical interstellar conditions. In addition, it must be made of relatively common refractory materials. The total mass of interstellar gas in our galaxy is $M_{gas} \approx 7 \times 10^9\,M_\odot$, of which a mass $M_H \approx 5 \times 10^9\,M_\odot$ is hydrogen. From studies of optical extinction and far-infrared emission, it is estimated that the mass of interstellar dust in our galaxy is $\sim 0.01 M_H$, or $M_{dust} \approx 5 \times 10^7\,M_\odot$. Since the total mass of metals at solar abundance is $M_{met} = 0.022 M_H$, we require about half the heavy elements in the ISM to be condensed into solid dust grains. Thus, we don't expect grains of beryllium or uranium or other scarce elements to contribute significantly to the interstellar dust. Most dust grains must be made of refractory elements near the top of the list of abundant elements (Table 1.2).

From these abundance arguments, we expect dust grains to be made of the relatively common refractory elements carbon (C), magnesium (Mg), silicon (Si), and iron (Fe), perhaps in chemical combination with the more volatile elements hydrogen (H), oxygen (O), and nitrogen (N). One possibility is that grains could be elemental. There could be metallic grains of magnesium or iron. There could be carbon grains, either in the form of amorphous carbon or of crystalline carbon (graphite or diamonds). There could be silicon grains, in the form of either amorphous silicon or crystalline silicon. Another possibility is that grains could be made of **silicates**, a family of minerals containing Si, O, and some combination of metals (in the everyday sense of the word). We expect that silicates in interstellar dust will be made primarily with the abundant metals Fe and Mg. Yet another possibility is that grains could be made of carbonates, a family of minerals containing C, O, and some combination of metals. Grains might also be made of silicon carbide (SiC). Finally, in the cooler regions of the interstellar medium, dust grains might be made of ices, such as $H_2O$, $CO_2$, $CH_4$, and $NH_3$.

We can estimate what elements are present in solid form by seeing which elements are missing in gaseous form. Along the well-studied line of sight to Zeta Ophiuchi, 140 parsecs away, it is found that many elements are depleted relative to their abundances in the solar photosphere. Although nitrogen is present at its solar abundance, it is found that $N_C/N_H$ is $\sim$35% of its solar value, and $N_O/N_H$ is $\sim$55% of its solar value. The more refractory elements are even more severely

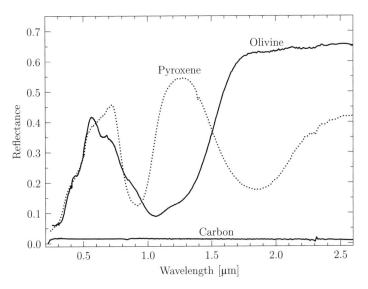

**Figure 6.6** Reflectance of olivine (upper solid line), pyroxene (dotted line), and carbon (lower solid line) at visible and near-infrared wavelengths. [Following Paton et al. 2011]

depleted in the gas phase. Toward Zeta Ophiuchi, magnesium is at $\sim$11%, silicon is at $\sim$5%, and iron is at $\sim$0.4% of their solar values relative to hydrogen. These depletion values are consistent with 30% by mass of the dust being made of carbon and carbonaceous materials and 70% being made of silicates. The two main types of silicates in dust are **pyroxene**, with chemical formula $Mg_x Fe_{1-x} SiO_3$, and **olivine**, with chemical formula $Mg_{2x} Fe_{2-2x} SiO_4$. In these formulas, the value of the magnesium fraction $x$ can run from 0 to 1. Pyroxene, for instance, runs the gamut from $MgSiO_3$, called *enstatite*, to $FeSiO_3$, called *ferrosilite*. Olivine runs from $Mg_2 SiO_4$, called *forsterite*, to $Fe_2 SiO_4$, called *fayalite*.

Pyroxene and olivine are very common minerals on the Earth. For instance, the igneous rock basalt, which makes up most of the oceanic crust, is rich in pyroxene, and can also contain significant amounts of olivine.[4] Pyroxene and olivine are also extraterrestrially common. They are frequently found in stony meteorites; the magnesium-rich forms enstatite and forsterite are more common than the iron-rich forms ferrosilite and fayalite. Forsterite has been found in comet dust returned by the *Stardust* probe. Pyroxene and olivine can be distinguished from each other at large distances because of their distinctly different reflection spectra, shown in Figure 6.6. The spectroscopic signatures of olivine and pyroxene have also been seen in the dusty gas shells around protostars and asymptotic giant branch stars.

---

[4] When Othello speaks of a world made "of one entire and perfect chrysolite," he is referring to gem-quality olivine, a translucent yellow-green stone. Pyroxene can also be gem quality; jadeite, for instance, is a variety of pyroxene that contains aluminum and sodium rather than magnesium and iron.

Determining the composition of dust grains from their optical properties can be tricky, since the index of refraction of many materials will be altered by the inclusion of small amounts of impurities or by exposure to cosmic rays. For instance, pure water ice has $\tilde{n} = 1.31 + i3.1 \times 10^{-9}$ at visible wavelengths; "dirty" water ice, of the type you might find in a comet, has $\tilde{n} = 1.33 + i0.09$. Pure diamond has $\tilde{n} = 2.42$, with the imaginary part equal to zero in the limit of a perfect crystal at visible wavelengths; nanodiamonds taken from the Allende meteorite have $\tilde{n} = 2.03 + i0.0015$. Pure pyroxene or olivine has $\tilde{n} \approx 1.7 + i5 \times 10^{-5}$. However, the observed properties of interstellar silicate dust grains are better fitted by a semi-fictional substance called "astrosilicate," which has $\tilde{n} \approx 1.7 + i0.03$ at visible wavelengths. Astrosilicate is amorphous silicate with the chemical formula $MgFeSiO_4$, assumed to have impurities that give it an $n_i$ value much greater than that of pure silicates.

Although some optical properties of silicates (such as $n_i$) depend sensitively on the impurities present, there are other more robust predictions that we can make. For instance, silicates have a strong infrared absorption/emission feature at $\lambda \approx 9.7\,\mu m$, resulting from the Si–O stretching mode. They have an additional feature at $\lambda \approx 18\,\mu m$, resulting from the O–Si–O bending mode. These broad features are seen in absorption in the spectrum of the massive young stellar object RAFGL7009S (Figure 6.7), which is embedded behind $\sim 40$ magnitudes of dust at visible wavelengths. The spectrum of RAFGL7009S also shows absorption features due to various ices, indicating the presence of icy grains, or at least icy

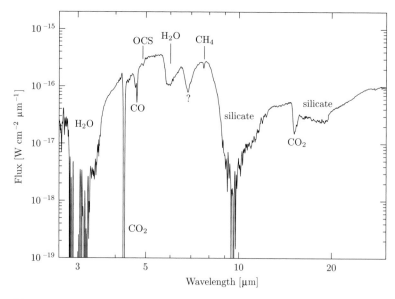

**Figure 6.7** Infrared spectrum of the young stellar object RAFGL7009S (IRAS 18316-0602), located within a dense molecular cloud at a distance $d \sim 3\,kpc$. [Data from Dartois et al. 1998]

mantles on silicate cores. Water ice has a 3.0 μm feature from the O–H stretching mode and a 6.0 μm feature from the H–O–H bending mode, both of which are seen in the spectrum of RAFGL7009S. Other features result from ices of methane, carbon dioxide, carbon monoxide, and carbonyl sulfide (OCS).

Although a spherical dust grain, like a spherical cow, is occasionally a useful approximation, true interstellar dust grains are not spherical. For one thing, to produce the observed polarization of starlight, dust grains must be non-spherical. However, even the approximation of an ellipsoidal grain is not correct. It is likely that interstellar dust, like the interplanetary dust strewn about by comets, has a fractal-like structure with a large porosity.

Just as dust grains have a variety of shapes, they also have a variety of sizes. It is customary, as a starting point, to regard interstellar dust grains as having a power-law distribution of radii:

$$\frac{dn_{gr}}{da} \propto a^{-\beta}. \tag{6.26}$$

Since dust grains are non-spherical, we are here using effective radius $a$, with $4\pi a^3/3 = V$ where $V$ is the grain volume. The power law is assumed to stretch from a minimum grain radius $a_{min}$ to a maximum grain radius $a_{max}$. Mathis, Rumpl, and Nordsieck (MRN), in their pioneering 1977 paper, found that $\beta = 3.5$ gave a good fit to the observed extinction curve from $\lambda = 0.11$ μm to 1 μm. To achieve this fit, MRN had to assume that dust grains were made of two different materials: one of the materials had to be graphite, but the other could be silicon carbide (SiC), magnetite ($Fe_3O_4$), iron, olivine, or pyroxene. The value of $a_{max}$ found by MRN was $a_{max} \approx 0.25$ μm. Unfortunately, the MRN fit is highly insensitive to $a_{min}$.

The study of very small dust grains and of very large molecules is aided by observations of their infrared emission. (The boundary line between a small grain and a large molecule is a fuzzy one; typically a dust grain with $a \sim 0.001$ μm $\sim 10$ Å, corresponding to $N \sim 500$ atoms, is called a "very small grain." Smaller structures are usually described as individual molecules or clusters of molecules.) Infrared emission features, such as those in Figure 6.8, reveal the existence of large molecules, known as **polycyclic aromatic hydrocarbons (PAHs)**, in regions where dust grains are found. Polycyclic aromatic hydrocarbons are "hydrocarbons" because they are molecules made of hydrogen and carbon atoms. They are "polycyclic" because they contain three or more benzene rings. They are "aromatic" because August Hofmann, a nineteenth century chemist, realized that molecules containing benzene rings are often smelly.[5]

In Figure 6.8, the broad emission peak on the right, centered at $\lambda \approx 140$ μm, represents thermal emission by silicate and graphite dust grains. Note that the

---

[5] The compounds cinnamaldehyde and vanillin, whose scents you can guess from their names, contain benzene rings.

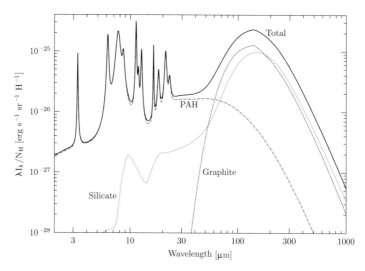

**Figure 6.8** Solid curve: total emission from dust grains and polycyclic aromatic hydro-carbons (PAHs). Dotted line, silicate grains; dashed line, large graphite grains; dot-dashed line, PAHs. [Following Li & Draine 2001]

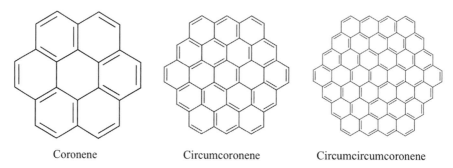

**Figure 6.9** Left: coronene [$C_{24}H_{12}$]. Middle: circumcoronene [$C_{54}H_{18}$]. Right: circum-circumcoronene [$C_{96}H_{24}$]. There is a carbon atom at each intersection; the C–C bonds are ~1.4 Å long. [ChEBI Database]

emission spectrum for silicate grains (the dotted line in Figure 6.8) has broad spectral emission features at $\lambda = 9.7\,\mu m$ and $18\,\mu m$, from the stretching and bending modes in tiny silicate grains. The narrower emission features at the left of Figure 6.8, ranging from $\lambda \approx 3\,\mu m$ to $\lambda \approx 20\,\mu m$, are produced by the stretching and bending modes of the C–H bonds in PAHs, and by vibrational modes of their carbon skeleton.

Figure 6.9 shows, as an example, three PAHs of increasing size: coronene, circumcoronene, and circumcircumcoronene. Simple extrapolation will let you visualize the structure of circum$^N$coronene in a case where $N > 2$. In the limit that $N \to \infty$, flat PAHs such as the coronene family are known as *graphene*.

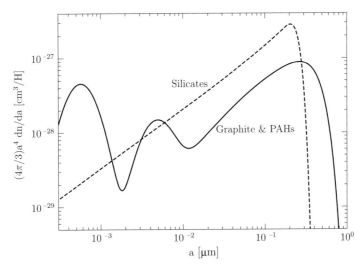

**Figure 6.10** Distribution of dust grain sizes, expressed as the volume of dust per logarithmic interval in the radius $a$. Solid line: graphite grains and PAHs. Dashed line: silicate grains. [Following Weingartner & Draine 2001]

Crystalline graphite consists of regularly stacked sheets of graphene held together by weak van der Waals forces, with an interplanar spacing of $\sim 3.4$ Å.

Weingartner and Draine (WD) modeled the distribution of grain sizes going down as far as the PAH range; they constrained their model using observations of extinction at near-UV, visible, and near-IR wavelengths, as well as using the observations of infrared emission. The distribution of grain and PAH sizes for the model of Weingartner and Draine is shown in Figure 6.10. In the distribution of carbon grain sizes, the bump at $a \sim 5 \times 10^{-4}$ μm, corresponding to PAHs with $\sim 100$ carbon atoms, is required to produce the observed PAH features in the $\lambda = 3 \rightarrow 20$ μm range; the broad peak at $a \sim 0.3$ μm is required to match the observed extinction curve.[6] The distribution of silicate grain sizes, in the WD model, is close to a power law up to a maximum grain size $a_{max} \sim 0.2$ μm. Silicate grains and carbon grains provide comparable amounts of extinction at near-UV and visible wavelengths; at near-IR wavelengths, carbon grains dominate the extinction. Carbon is also responsible for the 2175 Å bump.

A "typical" grain size may be taken as the half-mass grain size, $a_{0.5}$, defined so that half the mass of the dust is in grains of effective radius $a_{0.5}$ or greater. In the WD model, the half-mass grain size is the same for both carbon grains and silicate grains: $a_{0.5} \approx 0.12$ μm, about the size of the smallest known bacteria or the largest known viruses. The total volume of carbon grains per H atom in interstellar space is $V_{car} = 2.1 \times 10^{-27}$ cm$^3$; given a bulk density for graphite of

---

[6] The bump at $a \sim 0.005$ μm may be an artifact, so don't take it too seriously.

$\rho = 2.2\,\mathrm{g\,cm^{-3}}$, this corresponds to $4.6 \times 10^{-27}\,\mathrm{g}$ of carbon grains per H atom. The total volume of silicate grains per H atom is $V_{\mathrm{sil}} = 3.7 \times 10^{-27}\,\mathrm{cm^3}$; given a bulk density $\rho = 3.7\,\mathrm{g\,cm^{-3}}$ for silicates, this corresponds to $13.7 \times 10^{-27}\,\mathrm{g}$ of silicate grains per H atom. Thus, the WD model for dust requires $0.011\,\mathrm{g}$ of dust per gram of hydrogen; 75% of the dust mass comes from silicate grains, while the other 25% comes from carbon grains. The mass of PAHs is estimated to be $\sim 3\%$ of the total dust mass.

## 6.4  Heating and Cooling Grains

The heating and cooling of a large dust grain, with $a > 0.01\,\mu\mathrm{m}$, can be treated as a classical problem. For these large grains, the emission or absorption of a single photon doesn't significantly change the thermal energy of the grain. Except in the darkest dust-shrouded regions of molecular clouds, dust grains are primarily heated by the absorption of light. The absorption efficiency factor $Q_{\mathrm{abs}}$, as we have seen, satisfies $Q_{\mathrm{abs}} \ll 1$ in the limit that $\lambda \gg 2\pi a$. For a typical grain size $a \sim a_{0.5} \sim 0.12\,\mu\mathrm{m}$, this means that absorption is inefficient for wavelengths $\lambda \gg 8000\,\mathrm{\mathring{A}}$. Thus, grains are ordinarily heated by starlight at UV, visible, and near-IR wavelengths; mid-IR light and far-IR light are ineffective at heating dust. The rate of energy gain for a single spherical grain is then

$$G = \pi a^2 \langle Q_{\mathrm{abs}} \rangle_\star \, \varepsilon_\star c, \tag{6.27}$$

where $\varepsilon_\star$ is the energy density of starlight. In the Sun's neighborhood, $\varepsilon_\star \approx 1 \times 10^{-12}\,\mathrm{erg\,cm^{-3}} \approx 0.6\,\mathrm{eV\,cm^{-3}}$. The factor $\langle Q_{\mathrm{abs}} \rangle_\star$ is the absorption efficiency factor for the grain, averaged over wavelength after weighting by the spectrum of starlight. For grain sizes near the typical grain size,

$$\langle Q_{\mathrm{abs}} \rangle_\star \approx 0.18 \left( \frac{a}{0.1\,\mu\mathrm{m}} \right)^{0.6} \qquad \text{[silicate]} \tag{6.28}$$

and

$$\langle Q_{\mathrm{abs}} \rangle_\star \approx 0.8 \left( \frac{a}{0.1\,\mu\mathrm{m}} \right)^{0.85} \qquad \text{[graphite].} \tag{6.29}$$

Equation 6.28 is good for classical silicate grains with $a < 1.7\,\mu\mathrm{m}$; Equation 6.29 is good for classical graphite grains with $a < 0.13\,\mu\mathrm{m}$. If you have to deal with larger silicate or graphite "bricks," you can set $Q_{\mathrm{abs}}$ equal to one.

Classical dust grains are ordinarily heated by absorbing starlight; they are cooled by emitting thermal radiation. The rate of energy loss for a single spherical grain is

$$L = 4\pi a^2 \langle Q_{\mathrm{abs}} \rangle_T \, \sigma_{\mathrm{sb}} T^4, \tag{6.30}$$

where $T$ is the temperature of the dust grain, $\sigma_{sb}$ is the Stefan–Boltzmann constant, and $\langle Q_{abs}\rangle_T$ is the emission efficiency for the grain, averaged over a Planck function with temperature equal to the dust temperature $T$. For dust temperatures $T < 100\,\text{K}$,

$$\langle Q_{abs}\rangle_T \approx 5.2 \times 10^{-4} \left(\frac{a}{0.1\,\mu\text{m}}\right)\left(\frac{T}{20\,\text{K}}\right)^2 \qquad \text{[silicate]} \qquad (6.31)$$

and

$$\langle Q_{abs}\rangle_T \approx 3.2 \times 10^{-4} \left(\frac{a}{0.1\,\mu\text{m}}\right)\left(\frac{T}{20\,\text{K}}\right)^2 \qquad \text{[graphite]}. \qquad (6.32)$$

The rate of energy loss $L$ for a dust grain is thus strongly dependent on the grain temperature: $L \propto T^6$.

The equilibrium temperature $T_{eq}$ for dust grains in the classical limit can be found by setting the cooling rate (Equation 6.30) equal to the heating rate (Equation 6.27):

$$4\pi a^2 \langle Q_{abs}\rangle_T \sigma_{sb} T^4 = \pi a^2 \langle Q_{abs}\rangle_\star \varepsilon_\star c. \qquad (6.33)$$

Solving for $T$, we find that heating and cooling are in equilibrium at a temperature

$$T_{eq} \approx 16.3\,\text{K} \left(\frac{a}{0.1\,\mu\text{m}}\right)^{-1/15}\left(\frac{\varepsilon_\star}{0.6\,\text{eV cm}^{-3}}\right)^{1/6} \qquad \text{[silicate]} \qquad (6.34)$$

and

$$T_{eq} \approx 22.2\,\text{K} \left(\frac{a}{0.1\,\mu\text{m}}\right)^{-1/40}\left(\frac{\varepsilon_\star}{0.6\,\text{eV cm}^{-3}}\right)^{1/6} \qquad \text{[graphite]}. \qquad (6.35)$$

Because the cooling rate is highly sensitive to temperature, the equilibrium temperature is highly insensitive to the rate at which the starlight is heating the grains: if the energy density $\varepsilon_\star$ of the starlight is doubled, the grain temperature increases by just 12%. There is also little dependence of the grain temperature on grain radius. Thus, the estimate $T_{eq} \sim 20\,\text{K}$ is good for all large grains, no matter whether they are made of silicate or carbon, except in the immediate vicinity of stars. If dust grains were perfect blackbodies, with $\langle Q_{abs}\rangle_\star = \langle Q_{abs}\rangle_T = 1$, the temperature of the dust grains would be

$$T_{bb} = \left(\frac{\varepsilon_\star c}{4\sigma_{sb}}\right)^{1/4} \approx 3.4\,\text{K} \left(\frac{\varepsilon_\star}{0.6\,\text{eV cm}^{-3}}\right)^{1/4}. \qquad (6.36)$$

Thus, the gross inefficiency of dust grains at emitting mid-IR light elevates their temperature by a factor of $\sim 6$ above what it would be if they were perfect emitters and absorbers.

Grains much smaller than $a \sim 0.01\,\mu\text{m}$ cannot be treated classically. The number of atoms in such a small grain is

$$N \sim 4 \times 10^5 \left(\frac{a}{0.01\,\mu\text{m}}\right)^3. \qquad (6.37)$$

For small values of $N$, the absorption of a single visible or UV photon can significantly increase the grain's temperature; in this case, heating and cooling can no longer be thought of as being in perpetual equilibrium. The energy $E$ required to heat a dust grain by an amount $\Delta T$ is

$$E = C_V \Delta T, \tag{6.38}$$

where $C_V$ is the heat capacity of the grain. In the high-temperature limit, the heat capacity of a grain with $N$ atoms is

$$C_V = 3Nk. \tag{6.39}$$

The temperature increase produced by absorption of a photon of energy $h\nu$ is then

$$\Delta T = \frac{h\nu}{3Nk} \approx 0.08 \, \text{K} \left( \frac{h\nu}{2 \, \text{eV}} \right) \left( \frac{N}{10^5} \right)^{-1}. \tag{6.40}$$

A large dust grain with $a \approx a_{0.5} \approx 0.12 \, \mu\text{m}$ contains $N \sim 7 \times 10^8$ atoms. Thus, when it absorbs a photon of visible light ($h\nu \approx 2 \, \text{eV}$), it will have only a tiny temperature increase: $\Delta T \sim 10 \, \mu\text{K} \sim 5 \times 10^{-7} T_{\text{eq}}$. By contrast, when a very small grain with $N \sim 500$ absorbs a photon of visible light, it will experience a temperature jump $\Delta T \sim 16 \, \text{K} \sim T_{\text{eq}}$. In the extreme limit, a PAH with $N \sim 50$ can have a temperature jump as large as $\Delta T \sim 1000 \, \text{K}$ when it absorbs an ultraviolet photon.

At a random point in interstellar space, a large dust grain will have $T \approx 20 \, \text{K}$ and will radiate steadily in the far infrared, with typical wavelength $\lambda \sim 120 \, \mu\text{m}$. By contrast, a very small grain or PAH will spend most of its time at $T < 20 \, \text{K}$ but occasionally will reach temperatures as high as $T \sim 1000 \, \text{K}$. Thus, very small grains and PAHs can emit thermally at wavelengths as short as $\lambda \sim 3 \, \mu\text{m}$.

## 6.5  Making and Breaking Grains

Although we've been cheerfully discussing the properties of dust grains, we have dodged the question of how dust forms in the first place. Even in a fairly dense molecular cloud core, with $n_{\text{H}} \sim 3 \times 10^4 \, \text{cm}^{-3}$, at solar abundance you would expect only one Si atom and seven C atoms per cubic centimeter. How can you coax such widely separated atoms into forming a dense solid grain?

To get an estimate of how fast dust grains can grow, imagine a solid spherical grain made of element X. The number density of X atoms in the interstellar gas is $n_{\text{X}}$, and the rms thermal velocity of an X atom is

$$v_{\text{X}} = \left( \frac{3kT}{m_{\text{X}}} \right)^{1/2} = 1.57 \, \text{km s}^{-1} \left( \frac{T}{100 \, \text{K}} \right)^{1/2} \left( \frac{m_{\text{X}}}{m_{\text{H}}} \right)^{-1/2}, \tag{6.41}$$

where $T$ is the gas temperature and $m_X$ is the mass of an X atom. If there exists a "seed" grain of radius $a$, it will accrete additional X atoms from the surrounding gas, and the grain's mass will grow at the rate

$$\frac{dM_{gr}}{dt} = (n_X v_X)(\pi a^2) m_X p_s, \tag{6.42}$$

where $p_s$ is the "sticking probability"; that is, the probability that an X atom striking the grain will stick instead of bouncing off. The mass of our spherical grain is

$$M_{gr} = \frac{4\pi}{3} a^3 \rho_{gr}, \tag{6.43}$$

where $\rho_{gr}$ is the bulk density of element X. Using this relation between mass and radius, we can rewrite the mass accretion rate (Equation 6.42) as a radius growth rate:

$$\frac{da}{dt} = \frac{n_X v_X \pi a^2 m_X p_s}{4\pi \rho_{gr} a^2} = \frac{n_X v_X m_X p_s}{4\rho_{gr}}. \tag{6.44}$$

Thus, the growth rate $da/dt$ is independent of the grain radius $a$, unless the sticking probability $p_s$ is a function of grain size. For carbon atoms in the cold neutral medium ($T \approx 80\,\text{K}$, $n_H \approx 40\,\text{cm}^{-3}$),

$$\frac{da}{dt} \approx 0.3\,\mu\text{m Gyr}^{-1} \left( \frac{n_C/n_H}{2.7 \times 10^{-4}} \right), \tag{6.45}$$

assuming perfect stickiness, with $p_s = 1$.

Our back-of-envelope calculation suggests that grains can grow to their observed size during times shorter than the age of our galaxy, as long as atom–grain collisions are sticky. It becomes much easier to grow grains, however, if there exist seed grains produced in regions that are denser than typical regions of the ISM. Observationally, it is seen that these "dust nurseries" exist in the dense, cool stellar winds of asymptotic giant branch (AGB) stars. In these winds, the densities are as much as $\sim 10^9\,\text{cm}^{-3}$, and the temperatures can drop below the condensation temperatures of many heavy elements ($\sim 1000\,\text{K}$).

For instance, Mira variable stars are AGB stars with substantial emission at mid-infrared wavelengths. In *oxygen-rich* Mira variables, most of the carbon is locked up in gaseous CO, and the main condensates are silicates. Oxygen-rich Mira variables often have a strong silicate emission feature at $\lambda \approx 9.7\,\mu\text{m}$, as shown in Figure 6.11 for Mira itself and two other oxygen-rich Mira variable stars.

In *carbon-rich* Mira variables, the principal condensates are PAHs and graphite grains. Silicon takes the form not of a silicate but of silicon carbide (SiC). In carbon-rich Mira variables, the mid-IR thermal dust continuum is frequently accompanied by an emission feature at $\lambda \sim 11\,\mu\text{m}$, resulting from a vibrational mode of SiC. An example of a carbon-rich Mira variable is the star R Fornacis, whose mid-IR spectrum is shown in Figure 6.12. Silicon carbide is useful stuff

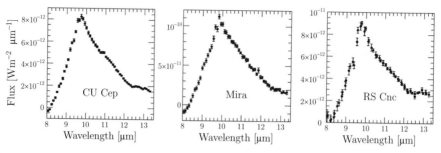

**Figure 6.11** Continuum-subtracted spectra of oxygen-rich Mira variable stars. [Speck et al. 2000]

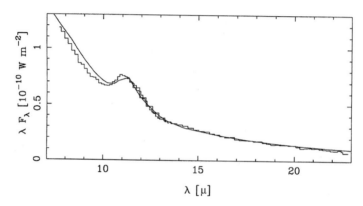

**Figure 6.12** Spectrum of the carbon-rich Mira variable R Fornacis. [Groenewegen et al. 1998]

here on Earth; under its alias of "carborundum," it's a popular abrasive. However, nearly all the silicon carbide on Earth is of synthetic origin. In the interstellar medium, silicon carbide grains are rare compared with grains of silicates and graphite. Even in the winds of carbon-rich Mira variables, because of the scarcity of silicon relative to carbon, much of the carbon condenses out in the form of PAHs rather than silicon carbide grains.

A typical mass-loss rate for a Mira variable star is $\dot{M} \sim 10^{-6}\,M_\odot\,\mathrm{yr}^{-1}$. Given a dust-to-gas mass ratio of $\sim 0.01$, this implies that a Mira variable star can produce as much as $\sim 10^{-8}\,M_\odot\,\mathrm{yr}^{-1}$ of dust during its $\sim 10^6\,\mathrm{yr}$ time scale for copious mass loss. The newly formed $\sim 0.01\,M_\odot$ of stardust is spread abroad into the ISM, where it undergoes competing processes of growth and destruction.

Dust grains can grow by accreting individual atoms, as discussed above. They can also grow as the result of low-speed collisions between grains. On the other hand, they can be shattered into smaller fragments by high-speed collisions. Grains can be vaporized by shock heating in supernova blastwaves. Moreover, what the gas gives, the gas can also take away. That is, although slow atoms in a cool gas can be accreted onto a dust grain, fast ions in a hot gas can erode the grain, kicking off atoms from the grain's surface. This process of erosion by high-speed ions is called **sputtering**.

The molecular bonds holding together the atoms in a dust grain typically have a bond energy $U_0 \sim 5\,eV$. However, the ions that do the sputtering must have an initial kinetic energy $E_i$ that is significantly greater than the bond energy $U_0$. The reason is that sputtering involves shooting low-mass "bullets" (usually $^1H$ and $^4He$ ions) at higher-mass target atoms, ranging from $^{12}C$ in a graphite grain to $^{56}Fe$ in a metallic iron grain. At ion energies $E_i < 1\,keV$, the interaction between the bullet ion and the target atom is well approximated as an elastic collision between their nuclei, and elastic collisions are not efficient at transferring energy from low-mass bullets to high-mass targets.

At ion energies below $\sim 1\,keV$, sputtering typically occurs when an ion, with mass $m_i$ and speed $v_i$, enters a dust grain's interior and undergoes an elastic collision with an atom of mass $m_a > m_i$. The maximum possible rebound speed of the ion, in the case of a head-on collision, is

$$v_{reb} = \frac{m_a - m_i}{m_a + m_i} v_i < v_i. \tag{6.46}$$

The outward-moving ion then strikes an atom at the grain surface, giving the surface atom a maximum possible speed

$$v_a = \frac{2m_i}{m_a + m_i} v_{reb} = \frac{2m_i(m_a - m_i)}{(m_a + m_i)^2} v_i. \tag{6.47}$$

This corresponds to an energy for the surface atom

$$\frac{1}{2} m_a v_a^2 = \frac{4 m_a m_i (m_a - m_i)^2}{(m_a + m_1)^4} \times \frac{1}{2} m_i v_i^2. \tag{6.48}$$

For this energy to be greater than the bond energy $U_0$, the incoming ion must have a kinetic energy

$$E_i = \frac{1}{2} m_i v_i^2 > b U_0 \tag{6.49}$$

where

$$b = \frac{(m_a + m_i)^4}{4 m_a m_i (m_a - m_i)^2}. \tag{6.50}$$

This factor varies from $b = 4.9$ for hydrogen striking a graphite grain to $b = 15.6$ for hydrogen striking an iron grain.[7]

---

[7] At ion energies $E_i > 1\,keV$, the interactions between the ion and the grain's atoms become more complicated but the transfer of energy from the light ion to the heavy atoms is still not highly efficient.

If a dust grain is surrounded by gas at a temperature $T$, the gas ions will cause sputtering if

$$T > \frac{bU_0}{k} \sim 6 \times 10^5 \, \text{K} \left( \frac{bU_0}{50 \, \text{eV}} \right). \tag{6.51}$$

For temperatures in the range $10^6 \, \text{K} < T < 10^9 \, \text{K}$, the sputtering rate for a graphite, silicate, or iron grain can be approximated as

$$\frac{da}{dt} \sim -1 \, \mu\text{m} \, \text{Myr}^{-1} \left( \frac{n_{\text{H}}}{1 \, \text{cm}^{-3}} \right). \tag{6.52}$$

At temperatures $T > 10^9 \, \text{K}$, sputtering becomes less effective, as the ion penetrates deeper into the grain and deposits its energy in the interior rather than near the surface. The sputtering rate of Equation 6.52 yields a grain lifetime

$$t_{\text{sput}} = -\frac{a}{da/dt} \approx 0.1 \, \text{Myr} \left( \frac{a}{0.1 \, \mu\text{m}} \right) \left( \frac{n_{\text{H}}}{1 \, \text{cm}^{-3}} \right)^{-1}. \tag{6.53}$$

In a supernova remnant, $n_{\text{H}} \sim 1 \, \text{cm}^{-3}$ and $T \geq 10^6 \, \text{K}$, giving a lifetime of 0.1 Myr for a typical dust grain. In the intracluster medium of the Coma Cluster, $n_{\text{H}} \sim 3 \times 10^{-3} \, \text{cm}^{-3}$ and $T \sim 10^8 \, \text{K}$, giving a lifetime of 30 Myr for a typical dust grain. Thus, we don't expect the intracluster medium to be a dusty place.

## Exercises

6.1 If the dust extinction $A_\lambda$ were a power law in the wavelength, $A_\lambda \propto \lambda^{-\alpha}$, what would be $R_V$ as a function of $\alpha$? What value of $\alpha$ would give $R_V = 3.1$?

6.2 In a region of interstellar space, the gas has a number density $n_{\text{H}}$, and there is a number density $n$ of dust grains. The distribution of dust radii $a$ is given by a power law,

$$\frac{1}{n_{\text{H}}} \frac{dn}{da} = \frac{A_0}{a_{\text{max}}} \left( \frac{a}{a_{\text{max}}} \right)^{-\beta}, \tag{6.54}$$

in the radius range $0 < a < a_{\text{max}}$.

(a) Let $V_0$ be the volume of dust grains per hydrogen nucleus. Find an expression for $V_0$ in terms of $A_0$, $a_{\text{max}}$, and $\beta$.

(b) If $V_0 = 6 \times 10^{-27} \, \text{cm}^3$, $a_{\text{max}} = 0.5 \, \mu\text{m}$, and $\beta = 3.5$ (all in approximate agreement with observations), what is the numerical value of $A_0$?

(c) Sputtering by hydrogen nuclei erodes grains at a rate

$$\frac{da}{dt} = -k n_{\text{H}}. \tag{6.55}$$

If sputtering is the *only* process at work changing grain size, and if the initial distribution of grain sizes (at $t = 0$) is given by Equation 6.54, derive an analytic formula for $V(t)$, the volume of dust grains per hydrogen nucleus as a function of time.

(d)  If $k = 1\,\mu\text{m}\,\text{Myr}^{-1}\,\text{cm}^3$ and $n_\text{H} = 1\,\text{cm}^{-3}$, how long will it take until $V(t) = 0$? How long will it take until $V(t) = V_0/2$?

6.3  A dust grain is destroyed by sublimation when its temperature becomes hotter than the *sublimation temperature* of the grain material: $\sim1200\,\text{K}$ for silicate grains and $\sim2000\,\text{K}$ for graphite grains. Consider the fate of silicate and graphite grains in the Orion Nebula. Most of the ionization in the Orion Nebula is provided by a single star, $\theta^1\text{C}$ Orionis, which has a luminosity $L_\star \approx 2 \times 10^5\,\text{L}_\odot$.

(a)  Assuming the dust grains have a radius $a = 0.1\,\mu\text{m}$, how close can silicate grains and graphite grains get to $\theta^1\text{C}$ Orionis and still survive? This defines the "sublimation radii" of the grains in the Orion Nebula.

(b)  The rate of hydrogen-ionizing photons from $\theta^1\text{C}$ Orionis is $Q_0 \approx 10^{49}\,\text{s}^{-1}$ and the density of the Orion Nebula is $n_\text{H} \approx 2000\,\text{cm}^{-3}$. Compare the sublimation radii for dust grains computed in part (a) with the Strömgren radius for the Orion Nebula.

# 7

# Molecular Clouds

*Here a star, and there a star,*
*Some lose their way.*
*Here a mist and there a mist,*
*Afterwards – day!*

Emily Dickinson (1830–1886)
"Our Share of Night to Bear" [1890]

Given the ubiquity of hydrogen in the interstellar medium, and the inability of helium to form chemical bonds, we expect the molecular gas in the ISM to consist primarily of $H_2$. Although the relative abundance of other types of molecule varies with location, CO is usually the second most abundant molecule in the ISM, with one CO molecule for every $\sim20\,000$ $H_2$ molecules. Since many other molecules of astrophysical importance are also diatomic, let's start with a discussion of molecules containing two atoms. The dissociation energy $D_0$ of a diatomic molecule is the energy required to break it into its component atoms, starting from the molecule's ground state. Typical dissociation energies, listed in Table 7.1, are $D_0 \sim 6\,\text{eV}$. Except for the robust, triple-bonded carbon monoxide (CO) molecule, the most abundant diatomic molecules have dissociation energies that are less than the ionization energies of the 10 most abundant atoms (see Table 1.2).

A molecule of hydrogen, with $D_0 = 4.5\,\text{eV}$, is not extremely tightly bound. An ultraviolet photon can **photodissociate** it. In a gas with temperature $T > D_0/k \sim 5 \times 10^4$ K, collisions with other gas particles can **collisionally dissociate** it. Thus, we expect molecular hydrogen to survive for long periods of time only in cool regions of the ISM that are shielded from UV radiation.

In a diatomic molecule, two atoms, of masses $m_1$ and $m_2$, have a distance $r$ between their nuclei. The potential energy $V(r)$ between the two atomic nuclei has a minimum at some separation $r_0$. A sketch of the potential $V(r)$ for the hydrogen molecule is given in Figure 7.1. In the molecule's ground state, the distance $r_0$ is

**Table 7.1** Properties of some diatomic molecules[a]

| Molecule | Dissociation energy $D_0$ [eV] | Equilibrium separation $r_0$ [Å] | Rotational energy scale $B_0$ [meV] | Vibrational energy scale $\hbar\omega_0$ [eV] | Electric dipole $\mu_0$ [debye] |
|---|---|---|---|---|---|
| $H_2$ | 4.48 | 0.741 | 7.554 | 0.546 | 0.000 |
| CO | 11.11 | 1.128 | 0.239 | 0.269 | 0.112 |
| OH | 4.41 | 0.970 | 2.345 | 0.463 | 1.66 |
| CH | 3.47 | 1.120 | 1.793 | 0.355 | 1.46 |
| CN | 7.72 | 1.172 | 0.236 | 0.256 | 1.45 |
| CS | 7.39 | 1.535 | 0.102 | 0.159 | 1.98 |

[a]*Data from Computational Chemistry Comparison and Benchmark Database [cccbdb.nist.gov]*

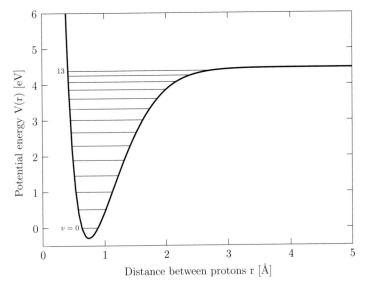

**Figure 7.1** Potential of the $H_2$ molecule as a function of $r$, the distance between the protons. The bound vibrational energy levels ($v = 0 \rightarrow 13$) are depicted as horizontal lines. [Data from Sharp 1971]

typically $\sim 1$ Å; more accurate values are given in Table 7.1 for astrophysically interesting molecules. In the immediate neighborhood of $r_0$, the potential $V(r)$ can be approximated as a parabola:

$$V(r) \approx V(r_0) + \frac{1}{2}k_s(r - r_0)^2, \qquad (7.1)$$

which is the equation for a harmonic oscillator with spring constant $k_s$. Thus, the classical cartoon version of a diatomic molecule would consist of two spherical masses connected by a spring. The fundamental frequency of the spring's vibration is $\omega_0 = (k_s/m_r)^{1/2}$, where

$$m_r \equiv \frac{m_1 m_2}{m_1 + m_2} \qquad (7.2)$$

is the reduced mass of the molecule. Typical values of $\hbar\omega_0$ for diatomic molecules are $\hbar\omega_0 \sim 0.3$ eV; more accurate values are given in Table 7.1 for astrophysically interesting molecules. The reduced mass of $H_2$ is $m_r = m_H/2$, the lowest of any molecule. Thus, hydrogen molecules have a particularly high fundamental frequency of vibration compared to other diatomic molecules. Quantum mechanically, the vibrational energy of a diatomic molecule must be

$$E_{vib} = \hbar\omega_0(v + 1/2), \tag{7.3}$$

where $v = 0, 1, 2, \ldots$ is the vibrational quantum number. Equation 7.3, which implies an equal spacing ($\Delta E_{vib} = \hbar\omega_0$) between the vibrational energy levels, assumes that the potential has a parabolic shape. As shown in Figure 7.1, this approximation is true only for small values of $v$. In reality, as $v$ becomes larger, the inter-level spacing $\Delta E_{vib}$ becomes smaller.

In addition to vibrating, a diatomic molecule can also rotate. Classically, if the molecule has rotational angular momentum $L$, its kinetic energy of rotation will be $E_{rot} = L^2/(2I)$, where $I = m_r r_0^2$ is the moment of inertia of the molecule. Quantum mechanically, the rotational energy of a diatomic molecule must be

$$E_{rot} = \frac{\hbar^2}{2I}J(J + 1), \tag{7.4}$$

where $J = 0, 1, 2, \ldots$ is the rotational quantum number. The energy normalization $B_0 \equiv \hbar^2/(2I)$ is typically $B_0 \sim 1$ meV for diatomic molecules; more accurate values are given in Table 7.1 for astrophysically interesting molecules. Molecular hydrogen, with its small reduced mass and small moment of inertia, has a particularly high value of $B_0$ compared to other diatomic molecules. In Figure 7.1, each vibrational energy level should be thought of as being split into numerous rotational sub-levels.

Diatomic molecules can emit and absorb radiation because of electronic transitions, similar to those in atoms. These transitions typically produce visible and UV photons. Diatomic molecules can also emit and absorb radiation through **vibrational transitions**. These typically produce near-infrared photons with $h\nu \sim 0.3$ eV and $\lambda \sim 4$ μm. Moreover, diatomic molecules can emit and absorb radiation through **rotational transitions**. These typically produce microwave and far-infrared photons with $h\nu \sim 1$ meV and $\lambda \sim 1$ mm.

One problem with studying the distribution of molecular gas in our galaxy is that $H_2$, the most abundant molecule, consists of two identical hydrogen atoms bonded together. Thus, an $H_2$ molecule has no permanent electric dipole moment, and cannot produce the strong dipole radiation that is created, for example, by rotating OH and CO molecules. Because $H_2$ lacks an electric dipole, the $J = 1$ rotational level is prevented from radiating by the laws of quantum mechanics. Thus, the lowest permitted rotational transition is $J = 2 \rightarrow 0$, which has energy $h\nu = 0.044$ eV and wavelength $\lambda = 28.2$ μm; this is much larger than the energy $h\nu \sim 0.001$ eV typical for rotational transitions in more massive diatomic molecules with non-zero electric dipole moments. The Einstein A coefficient for the $J = 2 \rightarrow 0$ transition of molecular hydrogen is $A_{20} = 2.95 \times 10^{-11}\,\text{s}^{-1} \sim 1\,\text{kyr}^{-1}$.

Compare this with the lowest permitted rotational transition for CO, which has an electric dipole moment $\mu_0 = 0.112\,\text{debye}$.[1] The lowest permitted rotational transition for this lopsided diatomic molecule is $J = 1 \rightarrow 0$. The energy of this transition, $E_{10} = E_1 - E_0 = 4.767 \times 10^{-4}\,\text{eV}$, corresponds to a frequency $\nu_{10} = E_{10}/h = 115.27\,\text{GHz}$ and wavelength $\lambda_{10} = c/\nu_{10} = 2.601\,\text{mm}$. (When speaking rather than calculating, astronomers round off these numbers and talk about the "115 gigahertz" or "2.6 millimeter" emission from CO molecules.) The Einstein A coefficient for the $J = 1 \rightarrow 0$ transition of CO is $A_{10} = 7.20 \times 10^{-8}\,\text{s}^{-1} \sim 2\,\text{yr}^{-1}$. Thus, compared with a hydrogen molecule, a CO molecule has its lowest accessible rotational level at an energy only 0.011 times as high, with a corresponding transition probability 2400 times larger.

Given the weak quadrupole emission of the symmetric $H_2$ molecule, it is not surprising that molecules other than $H_2$, despite their relative scarcity, were the first to be discovered in the interstellar medium. As mentioned in Chapter 2, interstellar molecules were first detected in 1937, in the form of CH electronic absorption bands at $\lambda = 4300\,\text{Å}$. Absorption bands of CN and $CH^+$ were discovered soon after. By the 1960s, radio emission lines of hydroxyl (OH), ammonia ($NH_3$), water ($H_2O$), and formaldehyde ($H_2CO$) were discovered at centimeter and millimeter wavelengths. Detection of $H_2$ in the interstellar medium didn't occur until 1970, when George Carruthers used a sounding rocket to detect the Lyman electronic absorption band of $H_2$, which lies in the ultraviolet. Molecular hydrogen is also detected by its emission in the near infrared, where the rotational–vibrational transitions of $H_2$ lie. For instance, the $(v = 1, J = 3) \rightarrow (v = 0, J = 1)$ transition is at $\lambda = 2.12\,\mu\text{m}$.

## 7.1 Interstellar CO

Although molecular hydrogen can be detected (with some difficulty), much of what we know about molecular gas in our galaxy comes from observations of "tracer" molecules such as CO. The 2.6 mm emission from the $J = 1 \rightarrow 0$ rotational transition of CO is the most important line used in studies of molecular gas. The strength of CO lines can be used, with various assumptions, to estimate the density and temperature of CO. Leaping from CO to $H_2$ requires knowledge of the local metallicity, as well as the degree to which C and O have been depleted onto grains.

---

[1] One debye is equivalent to the electric dipole of an electron and proton placed $\sim$0.21 Å apart. The debye is thus a convenient unit for expressing the electric dipole moment of a single molecule.

Let's start with a look at the equation of radiative transfer, to see how absorption and emission lines from the CO ($J = 1 \rightarrow 0$) transition tell us about the physical properties of molecular gas. The energy of this transition, $E_{10} = 0.4767\,\text{meV}$, corresponds to an equivalent temperature $E_{10}/k = 5.532\,\text{K}$. At this photon energy, the dominant background radiation is the cosmic microwave background (see Figure 3.4). The good news is that the 2.6 mm line of CO is radiatively excited by blackbody radiation, which is analytically tractable. The bad news is that we can't use the simplifying assumption that we are in the Rayleigh–Jeans tail, as we did when studying the 21 cm spin-flip transition. The intensity of the cosmic microwave background is (Equation 2.7)

$$I_\nu = B_\nu(T_{\text{rad}}) = \frac{2h\nu^3}{c^2} \frac{1}{e^{h\nu/kT_{\text{rad}}} - 1}, \tag{7.5}$$

where $T_{\text{rad}} = 2.7255\,\text{K}$. At $\lambda = 2.6\,\text{mm}$, the photon occupation number is (Equation 3.37):

$$\bar{n}_\gamma = \frac{1}{e^{5.532\,\text{K}/2.7255\,\text{K}} - 1} = 0.151. \tag{7.6}$$

If radiative excitation and de-excitation were the only way CO could transit between the $J = 0$ and $J = 1$ rotational states, then the excitation temperature would be $T_{\text{exc}} = T_{\text{rad}}$. However, collisional excitation and de-excitation are also possible.

Conditions that are favorable for the formation of CO are usually also favorable for the formation of $H_2$. The main colliders that excite and de-excite the rotational states of CO are therefore hydrogen molecules rather than hydrogen atoms or free electrons. At a temperature $T_{\text{gas}} \approx 15\,\text{K}$, the collisional rate coefficient for the de-excitation of CO by $H_2$ is

$$k_{10} \approx 4.1 \times 10^{-11}\,\text{cm}^3\,\text{s}^{-1} \left(\frac{T_{\text{gas}}}{15\,\text{K}}\right)^{0.2}. \tag{7.7}$$

The collisional rate coefficient for excitation is then (Equation 3.45)

$$k_{01} = \frac{g_1}{g_0} \exp\left(-\frac{h\nu_{10}}{kT_{\text{gas}}}\right) k_{10} \sim 8 \times 10^{-11}\,\text{cm}^3\,\text{s}^{-1} \tag{7.8}$$

at $T_{\text{gas}} \approx 15\,\text{K}$. (The statistical weight of a rotational state is $g_J = 2J + 1$.) The critical density $n_{\text{crit}}$ at which collisional de-excitation equals radiative de-excitation is, from Equation 3.52,

$$n_{\text{crit}} = \frac{(1 + \bar{n}_\gamma)A_{10}}{k_{10}} \approx 2000\,\text{cm}^{-3} \left(\frac{T_{\text{gas}}}{15\,\text{K}}\right)^{-0.2}. \tag{7.9}$$

The central dense cores of molecular clouds have $n > n_{\text{crit}} \sim 2000\,\text{cm}^{-3}$; thus, collisional excitation dominates, and the excitation temperature will be $T_{\text{exc}} \sim T_{\text{gas}} \sim 15\,\text{K}$. In the outer, more diffuse regions of molecular clouds, $n < n_{\text{crit}}$ and radiative excitation dominates, leading to $T_{\text{exc}} \sim T_{\text{rad}} \sim 2.7\,\text{K}$.

The fact that the excitation temperature lies in the range $2.7\,\mathrm{K} < T_{\mathrm{exc}} < 15\,\mathrm{K}$ is a slight nuisance, since it ensures that the excitation temperature always lies within a factor three of $h\nu_{10}/k = 5.532\,\mathrm{K}$. We can't use the $kT_{\mathrm{exc}} \ll h\nu_{u\ell}$ approximation that was useful when we studied Lyman $\alpha$ lines. Neither can we use the $kT_{\mathrm{exc}} \gg h\nu_{u\ell}$ approximation that was useful when we studied 21 cm lines. We must use the equation of radiative transfer in its full glory when we study the emission and absorption lines of the CO ($J = 1 \to 0$) transition. Thus, in the rest of this section, we will follow the physical arguments of Section 3.2 about 21 cm radiation, but without the simplifying assumption that $kT_{\mathrm{exc}} \gg h\nu_{u\ell}$.

Along a line of sight passing through a molecular cloud, the radiative transfer equation is given as usual by

$$\frac{dI_\nu}{d\tau_\nu} = -I_\nu + S_\nu, \tag{7.10}$$

where $S_\nu = j_\nu/\kappa_\nu$ is the source function. The emissivity, accounting for spontaneously emitted 2.6 mm photons, is (Equation 3.15)

$$j_\nu = n_1 \frac{A_{10}}{4\pi} h\nu_{10} \Phi_\nu, \tag{7.11}$$

where $n_1$ is the number density of CO molecules in the $J = 1$ state, and $\Phi_\nu$ is the line profile of the CO ($J = 1 \to 0$) line. The net attenuation coefficient, taking into account both absorption and stimulated emission, is (Equation 3.8)

$$\kappa_\nu = n_0 \sigma_{01} \left[ 1 - \frac{g_0}{g_1} \frac{n_1}{n_0} \right], \tag{7.12}$$

where $n_0$ is the number density of CO molecules in the $J = 0$ rotational state and $\sigma_{01}$ is the frequency-dependent cross section for absorption. From the definition of the excitation temperature (Equation 1.19), we can write

$$\frac{n_1}{n_0} = \frac{g_1}{g_0} \exp\left(-\frac{h\nu_{10}}{kT_{\mathrm{exc}}}\right) \tag{7.13}$$

and thus

$$\kappa_\nu = n_0 \sigma_{01} \left[ 1 - \exp\left(-\frac{h\nu_{10}}{kT_{\mathrm{exc}}}\right) \right]. \tag{7.14}$$

In terms of the Einstein A coefficient, the absorption cross section can be written, from Equation 2.26, as

$$\sigma_{01} = \frac{g_1}{g_0} \frac{c^2}{8\pi \nu_{10}^2} A_{10} \Phi_\nu. \tag{7.15}$$

Thus, the attenuation coefficient $\kappa_\nu$ takes the form

$$\kappa_\nu = n_0 \frac{g_1}{g_0} \frac{c^2}{8\pi \nu_{10}^2} A_{10} \Phi_\nu \left[ 1 - \exp\left(-\frac{h\nu_{10}}{kT_{\mathrm{exc}}}\right) \right]. \tag{7.16}$$

Combining Equations 7.11 and 7.16, the source function can be written as

$$S_\nu \equiv \frac{j_\nu}{\kappa_\nu} = \frac{2h\nu_{10}^3}{c^2}\frac{1}{\exp(h\nu_{10}/kT_{\text{exc}})-1}. \tag{7.17}$$

If the excitation temperature is constant over the entire region of emission, we can integrate the equation of radiative transfer (Equation 7.10) as

$$I_\nu = I_\nu(0)e^{-\tau_\nu} + S_\nu[1 - e^{-\tau_\nu}]. \tag{7.18}$$

Since the background radiation is the cosmic microwave background, with

$$I_\nu(0) = B_\nu(T_{\text{rad}}) = \frac{2h\nu^3}{c^2}\frac{1}{\exp(h\nu/kT_{\text{rad}})-1}, \tag{7.19}$$

we can write

$$I_\nu = \frac{2h\nu^3}{c^2}\frac{e^{-\tau_\nu}}{\exp(h\nu/kT_{\text{rad}})-1} + \frac{2h\nu_{10}^3}{c^2}\frac{1-e^{-\tau_\nu}}{\exp(h\nu/kT_{\text{exc}})-1}. \tag{7.20}$$

Since the thermal broadening is small, with (Equation 2.31)

$$\frac{b}{c} = 3.2 \times 10^{-7}\left(\frac{T}{15\,\text{K}}\right)^{1/2}\left(\frac{m}{28m_{\text{H}}}\right)^{-1/2}, \tag{7.21}$$

we can set $\nu = \nu_{10}$ over the entire line.

When we turn our radio antenna toward a molecular cloud, we detect an intensity

$$I_\nu(\text{on}) = \frac{2h\nu^3}{c^2}\left[\frac{e^{-\tau_\nu}}{\exp(h\nu/kT_{\text{rad}})-1} + \frac{1-e^{-\tau_\nu}}{\exp(h\nu/kT_{\text{exc}})-1}\right], \tag{7.22}$$

where $\tau_\nu$ is the optical depth of the cloud. When we "chop" our antenna over to a blank region of sky, we detect nothing but CMB:

$$I_\nu(\text{off}) = \frac{2h\nu^3}{c^2}\frac{1}{\exp(h\nu/kT_{\text{rad}})-1}. \tag{7.23}$$

The difference between the two observations is

$$\Delta I_\nu = I_\nu(\text{on}) - I_\nu(\text{off})$$
$$= \frac{2h\nu^3}{c^2}\left[\frac{1}{\exp(h\nu/kT_{\text{exc}})-1} - \frac{1}{\exp(h\nu/kT_{\text{rad}})-1}\right](1-e^{-\tau_\nu}). \tag{7.24}$$

To write things more compactly, observers of the CO ($J = 1 \to 0$) line, and of similar millimeter lines, express the observed $\Delta I_\nu$ in terms of an antenna temperature (Equation 3.22)

$$T_{\text{A}} \equiv \frac{c^2}{2k}\frac{\Delta I_\nu}{\nu^2}. \tag{7.25}$$

For an optically thick cloud ($\tau_\nu \gg 1$),

$$T_A \approx \frac{h\nu}{k} \left[ \frac{1}{\exp(h\nu/kT_{\text{exc}}) - 1} - \frac{1}{\exp(h\nu/kT_{\text{rad}}) - 1} \right]. \tag{7.26}$$

When the intensity of the CO line, with $h\nu/k = 5.532\,\text{K}$, is compared with that of the cosmic microwave background, for which $T_{\text{rad}} = 2.725\,\text{K}$, Equation 7.26 becomes

$$\frac{T_A}{5.532\,\text{K}} = \frac{1}{\exp(5.532\,\text{K}/T_{\text{exc}}) - 1} - 0.151. \tag{7.27}$$

We can then solve for the unknown $T_{\text{exc}}$ in terms of the observable $T_A$:

$$\frac{5.532\text{K}}{T_{\text{exc}}} = \ln \left[ 1 + \frac{1}{(T_A/5.532\,\text{K}) + 0.151} \right]. \tag{7.28}$$

In the limit that the observed antenna temperature is $T_A = 0$, we deduce that $T_{\text{exc}} = T_{\text{rad}}$, and the CO line is entirely radiatively excited. In the limit that $T_A \gg 5.532\,\text{K}$, we deduce that $T_{\text{exc}} \approx T_A \gg T_{\text{rad}}$, and the CO line is almost entirely collisionally excited. For observed molecular clouds in our galaxy, it is often found that $T_A \sim 5\,\text{K}$ at the center of the CO ($J = 1 \rightarrow 0$) line, implying $T_{\text{exc}} \sim 8\,\text{K}$.

The optical depth of the thermally broadened CO line can be written as

$$\tau_\nu = \int \kappa_\nu ds = \tau_0 e^{-v^2/b^2} \tag{7.29}$$

where the optical depth at line center is

$$\tau_0 = \frac{1}{(4\pi)^{3/2}} \frac{g_1}{g_0} \frac{c^2}{\nu_{10}^3} A_{10} \left[ 1 - e^{-h\nu_{10}/kT_{\text{exc}}} \right] \frac{c}{b} N_0, \tag{7.30}$$

given a column density $N_0$ of CO in the $J = 0$ rotational state. Plugging in the appropriate values for the CO ($J = 1 \rightarrow 0$) rotational transition, we find

$$\tau_0 \approx 0.90 \left( \frac{T_{\text{gas}}}{15\,\text{K}} \right)^{-1/2} \left[ 1 - e^{-5.532\,\text{K}/T_{\text{exc}}} \right] \left( \frac{N_0}{10^{14}\,\text{cm}^{-2}} \right). \tag{7.31}$$

Thus, lines of sight with a CO $J = 0$ column density greater than $N_0 \sim 2 \times 10^{14}\,\text{cm}^{-2}$ will be optically thick. Is this a large or small column density, in the context of actual molecular clouds? Let's see.

For concreteness, assume a solar carbon abundance ($n_C/n_H \approx 2.7 \times 10^{-4}$) and that a fraction $F_{\text{CO}} \leq 1$ of the carbon atoms in molecular clouds are in CO molecules, as opposed to being solitary atoms or ions, being part of other types of molecule, or being locked up in dust grains. The value of $F_{\text{CO}}$ varies from place to place; in the interior of a dusty cloud, $F_{\text{CO}} \sim 0.25$ is common. The fraction of the CO molecules that have rotational quantum number $J$ is

$$f_J = (2J + 1) \exp \left( -\frac{B_0 J(J+1)}{kT_{\text{exc}}} \right) \Big/ Q(T_{\text{exc}}), \tag{7.32}$$

where

$$Q(T_{\text{exc}}) \equiv \sum (2J+1) \exp\left(-\frac{B_0 J(J+1)}{kT_{\text{exc}}}\right) \tag{7.33}$$

is the partition function. This distribution among rotational levels implies that at excitation temperatures greater than

$$T_{\text{exc}} = \frac{2}{\ln 3}\frac{B_0}{k} = 5.06\,\text{K}, \tag{7.34}$$

more CO molecules will be in the first excited state ($J = 1$) than in the rotational ground state ($J = 0$). For example, at $T_{\text{exc}} = 8\,\text{K}$, typical for molecular clouds, $f_0 \approx 0.31$ for the ground state, $f_1 \approx 0.46$ for the first excited state, and $f_2 \approx 0.19$ for the second excited state. We thus expect the column density of CO molecules in their ground state to be

$$N_0 = 2.1 \times 10^{-5} N_{\text{H}} \left(\frac{n_C/n_H}{2.7 \times 10^{-4}}\right)\left(\frac{F_{\text{CO}}}{0.25}\right)\left(\frac{f_0}{0.31}\right), \tag{7.35}$$

and a CO $J = 0$ column density $N_0 = 2 \times 10^{14}\,\text{cm}^{-2}$ corresponds to a hydrogen column density $N_{\text{H}} \sim 10^{19}\,\text{cm}^{-2}$. (Remember that Lyman $\alpha$ becomes optically thick at $N_{\text{H}} \sim 10^{12}\,\text{cm}^{-2}$, while the hyper-forbidden 21 cm line becomes optically thick at $N_{\text{H}} \sim 10^{21}\,\text{cm}^{-2}$.) For a not-very-dense molecular cloud, with $n_{\text{H}} \sim 100\,\text{cm}^{-3}$, this corresponds to a path length $d = N_{\text{H}}/n_{\text{H}} \sim 10^{17}\,\text{cm} \sim$ 0.03 pc. Even the smallest molecular clouds, which appear as dark nebulae like Barnard 68 (Figure 1.4), have diameters $d \sim 0.1$ pc. Thus, we expect molecular clouds to be *optically thick* at the line center of CO ($J = 1 \to 0$) radiation.

## 7.2 From CO to H$_2$

The opacity of molecular clouds at $\lambda = 2.6\,\text{mm}$ has an upside and a downside. The upside is that we can use the relatively simple Equation 7.26 to go from the observable antenna temperature to the excitation temperature. The downside is that molecular clouds are "brick walls" at $\lambda \sim 2.6\,\text{mm}$; we can't see what's happening to the CO in their central regions. All the 2.6 mm emission that we see comes from a thin skin (a photosphere, of sorts) at the surface of the molecular cloud.

One way to penetrate the cloud's interior is to observe the emission from $^{13}\text{CO}$ rather than the more abundant $^{12}\text{CO}$.[2] The local ratio of $^{13}\text{C}$ to $^{12}\text{C}$ is 0.0108, or about one carbon-13 atom for every 92 carbon-12 atoms. This is enough to render all but the densest and largest molecular clouds transparent at the $^{13}\text{CO}$ ($J = 1 \to 0$) transition. The frequency of the $^{13}\text{CO}$ ($J = 1 \to 0$) transition

---

[2] The molecules $^{13}\text{CO}$ and $^{12}\text{CO}$ are **isotopologues**; that is, they have the same basic chemical formula, but at least one atom (in this case, carbon) is represented by different isotopes (in this case, $^{12}\text{C}$ and $^{13}\text{C}$).

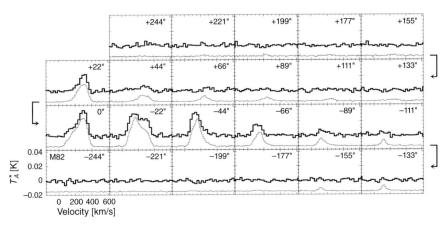

**Figure 7.2** $^{13}$CO emission lines (thin) and $^{12}$CO emission lines divided by 10 (thick) for the $J = 1 \rightarrow 0$ transition, along the apparent major axis of the galaxy M82. At the $d \approx 3.5$ Mpc distance to M82, the 22 arcsec separation between the panels corresponds to $\sim$370 pc. [Paglione et al. 2001]

is shifted to 110.20 GHz, 4.4% lower than the value for $^{12}$CO. The antenna temperature of $^{13}$CO is typically lower than that of $^{12}$CO; however, if the $^{12}$CO emission from a region is strong then $^{13}$CO is detectable. Figure 7.2, for instance, shows the $^{12}$CO and $^{13}$CO emission lines along the apparent long axis of the galaxy M82, a starburst galaxy whose copious star formation is fed by a large store of molecular gas.

The key to comparing the two isotopic lines is that we may usually assume that $^{12}$CO is optically thick for a given molecular cloud and that $^{13}$CO is optically thin. We can therefore compute $T_{\text{exc}}$ from the optically thick $^{12}$CO. A crucial assumption that we then make is that $T_{\text{exc}}$ is the same for the $^{13}$CO as for the $^{12}$CO. This is generally a safe assumption, since the two isotopes of carbon are intermingled in the ISM. However, if there is significant chemical fractionation, the $^{13}$CO and $^{12}$CO may live in different regions, with different degrees of excitation.

If we assume that we know $T_{\text{exc}}$ from our observations of $^{12}$CO, the antenna temperature that we measure for the optically thin $^{13}$CO line is (compare with Equation 7.24):

$$T_A(^{13}\text{CO}) \approx \frac{h\nu}{k} \left[ \frac{1}{\exp(h\nu/kT_{\text{exc}}) - 1} - \frac{1}{\exp(h\nu/kT_{\text{rad}}) - 1} \right] \tau_\nu, \qquad (7.36)$$

where $h\nu/k = 5.140$ K for the $^{13}$CO ($J = 1 \rightarrow 0$) transition and $T_{\text{rad}} = 2.725$ K. The integrated line intensity is then

$$\int T_A(^{13}\text{CO}) d\nu \approx 5.140 \,\text{K} \left[ \frac{1}{\exp(5.140 \,\text{K}/T_{\text{exc}}) - 1} - 0.179 \right] \int \tau_\nu d\nu. \qquad (7.37)$$

The optical depth, integrated over the thermally broadened line, is

$$\int \tau_v dv = \sqrt{\pi} b\tau_0 = \frac{1}{8\pi} \frac{g_1}{g_0} \frac{c^3}{v^3} A_{10}[1 - e^{-hv/kT_{\text{exc}}}]N_0(^{13}\text{CO}), \tag{7.38}$$

where $N_0(^{13}\text{CO})$ is the column density of $^{13}\text{CO}$ in its rotational ground state with $J = 0$. With the appropriate frequency and Einstein A coefficient ($6.29 \times 10^{-8}\,\text{s}^{-1}$ for $^{13}\text{CO}$, as opposed to $7.20 \times 10^{-8}\,\text{s}^{-1}$ for $^{12}\text{CO}$), the integrated optical depth is

$$\int \tau_v dv \approx 0.152\,\text{km s}^{-1} \left(1 - e^{-5.140\,\text{K}/T_{\text{exc}}}\right) \left[\frac{N_0(^{13}\text{CO})}{10^{15}\,\text{cm}^{-2}}\right]. \tag{7.39}$$

Substituting Equation 7.39 into Equation 7.37, we find

$$\int T_A(^{13}\text{CO})dv \approx 0.34\,\text{K km s}^{-1} \left[\frac{N_0(^{13}\text{CO})}{10^{15}\,\text{cm}^{-2}}\right], \tag{7.40}$$

when $T_{\text{exc}} \approx 8$ K. This means that if you measure a value for $\int T_A dv$, you can compute a column density of $^{13}\text{CO}$ in its ground state:

$$N_0(^{13}\text{CO}) \approx 2.9 \times 10^{14}\,\text{cm}^{-2} \left[\frac{\int T_A(^{13}\text{CO})dv}{1\,\text{K km s}^{-1}}\right] \tag{7.41}$$

when $T_{\text{exc}} \approx 8$ K.

Since a fraction $f_0 \sim 0.31$ of the $^{13}\text{CO}$ will be in its rotational ground state, we must multiply by 3.2 to go from $N_0(^{13}\text{CO})$ to $N(^{13}\text{CO})$, the total column density of $^{13}\text{CO}$. Since the $^{13}\text{C}$ to $^{12}\text{C}$ ratio is $\sim 1/92$, we must multiply by 92 to go from $N(^{13}\text{CO})$ to $N(^{12}\text{CO})$. Since a fraction $F_{\text{CO}} \sim 1/4$ of $^{12}\text{C}$ is in carbon monoxide molecules, we must multiply by 4 to go from $N(^{12}\text{CO})$ to $N_C$. Since the $n_C$ to $n_H$ ratio is $\sim 2.7 \times 10^{-4}$, we must multiply by 3700 to go from $N_C$ to $N_H$. Finally, if nearly all the hydrogen is in its molecular state, we must multiply by 1/2 to go from $N_H$ to $N(\text{H}_2)$. Thus, translating from an observed $^{13}\text{CO}$ line strength to a column density of molecular hydrogen is a multi-step task, with each step introducing new uncertainty:

$$N(\text{H}_2) \approx \frac{1}{2} \times 3700 \times 4 \times 92 \times 3.2 \times N_0(^{13}\text{CO}) \approx 2.2 \times 10^6 N_0(^{13}\text{CO}), \tag{7.42}$$

and thus

$$N(\text{H}_2) \approx 6 \times 10^{20}\,\text{cm}^{-2} \left[\frac{\int T_A(^{13}\text{CO})dv}{1\,\text{K km s}^{-1}}\right]. \tag{7.43}$$

This is a useful relation; if, for instance, you measure the integrated antenna temperature for the $^{13}\text{CO}$ line over an entire molecular cloud, you can estimate the total hydrogen mass of the cloud. However, our estimate of approximately two million $\text{H}_2$ molecules for every $^{13}\text{CO}$ molecule in the ground state is the result of a teetering tower of assumptions.

To prop up the tower of assumptions, it is useful to look at other ways of estimating the hydrogen mass of a molecular cloud. A dense cloud with $T_{gas} \approx 15\,\mathrm{K}$ and $n \gg 100\,\mathrm{cm}^{-3}$ will have a much higher gas pressure than the $P \sim 4 \times 10^{-13}\,\mathrm{dyn\,cm}^{-2}$ of the surrounding ISM. Since we do not see molecular clouds systematically expanding, we might suppose that they are in hydrostatic equilibrium, the inward force of gravity being balanced by the outward force from the thermal pressure gradient. However, when we look at individual molecular clouds in our galaxy, the line width $b$ that we measure for molecular line emission is often much broader than the thermal value $b \sim 0.1\,\mathrm{km\,s}^{-1}$ expected at $T_{gas} \approx 15\,\mathrm{K}$. For instance, a 2004 study by Heyer and Brunt found

$$b = \sqrt{2}\sigma_v \approx 1.2\,\mathrm{km\,s}^{-1}\left(\frac{d}{1\,\mathrm{pc}}\right)^{0.65}, \tag{7.44}$$

where $d$ is the size of a molecular cloud. The biggest molecular clouds in their study, with $d \sim 30\,\mathrm{pc}$, had $b \sim 12\,\mathrm{km\,s}^{-1}$. The large broadening is similar for molecules of different molecular mass, suggesting that the broadening is primarily due to turbulent motions within the cloud.

Since molecular clouds are regions where the turbulent kinetic energy dominates over the thermal energy, we must modify our picture of the cloud's static equilibrium state. The inward force of gravity is still present, but it is balanced by *turbulent* pressure, not by ordinary thermal pressure. If we assume a spherical cloud of mass $M_c$ and radius $r_c$ then, for self-gravity to be balanced by the turbulent pressure, the one-dimensional turbulent velocity must be

$$\sigma_v^2 = \frac{GM_c}{5r_c} = \frac{4\pi}{15}G\rho_c r_c^2. \tag{7.45}$$

Equation 7.45 is derived from the steady-state virial theorem applied to a spherical cloud of uniform density $\rho_c$. The average surface density of the spherical cloud is

$$\Sigma = \frac{M_c}{\pi r_c^2} = \frac{4}{3}r_c\rho_c = \frac{5\sigma_v^2}{\pi G r_c}. \tag{7.46}$$

In terms of the observed broadening parameter $b = \sqrt{2}\sigma_v$,

$$\Sigma = \frac{5}{2\pi}\frac{b^2}{G r_c} \approx 0.039\,\mathrm{g\,cm}^{-2}\left(\frac{b}{1\,\mathrm{km\,s}^{-1}}\right)^2\left(\frac{r_c}{1\,\mathrm{pc}}\right)^{-1}. \tag{7.47}$$

If the cloud is 71% hydrogen by mass, and if almost all the hydrogen is in the form of H$_2$, with molecular mass $m = 2m_H$, the average column density of molecular hydrogen is

$$N(\mathrm{H}_2) = \frac{0.71\Sigma}{2m_H} = 0.71\frac{5}{4\pi}\frac{1}{Gm_H}\frac{b^2}{r_c} \tag{7.48}$$

$$\approx 8.3 \times 10^{21}\,\mathrm{cm}^{-2}\left(\frac{b}{1\,\mathrm{km\,s}^{-1}}\right)^2\left(\frac{r_c}{1\,\mathrm{pc}}\right)^{-1}.$$

Given the empirical $b \propto r_c^{0.65}$ dependence of line broadening on cloud size seen by Heyer and Brunt, we expect the column density of molecular clouds to depend only weakly on their radius:

$$N(H_2) \sim 1 \times 10^{22}\, cm^{-2} \left(\frac{r_c}{1\, pc}\right)^{0.3}. \tag{7.49}$$

The number density of hydrogen molecules inside the cloud is

$$n(H_2) = \frac{3N(H_2)}{4r_c} \approx 2000\, cm^{-3} \left(\frac{b}{1\, km\, s^{-1}}\right)^2 \left(\frac{r_c}{1\, pc}\right)^{-2}, \tag{7.50}$$

and the total cloud mass is

$$M_c = \frac{5b^2 r_c}{2G} \approx 600\, M_\odot \left(\frac{b}{1\, km\, s^{-1}}\right)^2 \left(\frac{r_c}{1\, pc}\right), \tag{7.51}$$

with the caution that these results assume a spherical, uniform density cloud and are useful only to within a factor two.

Computing the hydrogen column density from Equation 7.48 and the virial mass from Equation 7.51 is useful, but it requires knowing the radius $r_c$ of a molecular cloud. The angular size of a cloud is difficult to resolve outside our galaxy, and the distance to a molecular cloud is difficult to measure inside our galaxy. It would be useful if we had a reliable way to convert from something relatively easy to measure, such as the integrated line intensity of $^{12}CO$,

$$W_{CO} \equiv \int T_A(^{12}CO)dv, \tag{7.52}$$

to something more difficult to measure, like the column density $N(H_2)$ of molecular hydrogen. In short, we want to know the **X-factor**, defined as

$$X_{CO} \equiv \frac{N(H_2)}{W_{CO}}. \tag{7.53}$$

Since a typical molecular cloud is optically thick at the $J = 1 \to 0$ rotational transition of $^{12}CO$, we can use the radiative transfer solution for an opaque object:

$$I_\nu = B_\nu(T_{exc}) = \frac{2h\nu^3}{c^2} \frac{1}{\exp(h\nu/kT_{exc}) - 1}. \tag{7.54}$$

Expressed in terms of the antenna temperature, this is

$$T_A \equiv \frac{c^2}{2k\nu^2} I_\nu = \frac{h\nu}{k} \frac{1}{\exp(h\nu/kT_{exc}) - 1}. \tag{7.55}$$

Since the observed emission lines from $^{12}CO$ have $b \ll c$, we can assume that $\nu \approx \nu_{10}$ over the entire line width. In this case,

$$W_{CO} \equiv \int T_A dv \approx \sqrt{\pi} b T_A(\nu_{10}) \approx \frac{h\nu_{10}}{k} \frac{\sqrt{\pi} b}{\exp(h\nu_{10}/kT_{exc}) - 1}. \tag{7.56}$$

Combined with the virial relation for the column density (Equation 7.49), this becomes

$$X_{\rm CO} = 0.71 \frac{5}{4\pi\sqrt{\pi}} \frac{1}{Gm_{\rm H}} \frac{k}{h\nu_{10}} \left[\exp(h\nu_{10}/kT_{\rm exc}) - 1\right] \frac{b}{r_{\rm c}}. \qquad (7.57)$$

We can use the turbulent virial equation (Equation 7.45) to write

$$\frac{b}{r_{\rm c}} = \left(\frac{8\pi}{15}G\rho_{\rm c}\right)^{1/2} \approx 1.4 \times 10^{-14}\,{\rm s}^{-1} \left(\frac{n}{1000\,{\rm cm}^{-3}}\right)^{1/2}. \qquad (7.58)$$

Thus, we expect $X_{\rm CO}$ to be inversely proportional to the freefall time of a cloud, with denser clouds having larger X-factors.

Plugging in the correct values for the properties of the $^{12}$CO $J = 1 \to 0$ transition, we compute an X-factor

$$X_{\rm CO} \approx 3.6 \times 10^{20} \frac{{\rm cm}^{-2}}{{\rm K\,km\,s}^{-1}} \left(\frac{n_{\rm H}}{1000\,{\rm cm}^{-3}}\right)^{1/2}, \qquad (7.59)$$

where $n_{\rm H}$ is the average number density of hydrogen nuclei in the cloud, and an excitation temperature $T_{\rm exc} \approx 8\,{\rm K}$ is assumed. Observational attempts to find values for the X-factor usually find $X_{\rm CO} \approx 2 \times 10^{20}\,{\rm cm}^{-2}/\,{\rm K\,km\,s}^{-1}$ for nearby molecular clouds within the Milky Way Galaxy. However, within the Local Group, a larger average value of $X_{\rm CO} \approx 4 \times 10^{20}\,{\rm cm}^{-2}/\,{\rm K\,km\,s}^{-1}$ seems to hold, while in the low-metallicity Small Magellanic Cloud, an X-factor as high as $X_{\rm CO} \approx 14 \times 10^{20}\,{\rm cm}^{-2}/\,{\rm K\,km\,s}^{-1}$ appears to apply. We can only conclude that the X-factor is not universal throughout the universe, and depends on density, excitation temperature, and metallicity.

Figure 7.3 shows an all-sky map of the $^{12}$CO integrated line intensity, $W_{\rm CO}$, for the entire sky. For the molecular gas in our galaxy, the CO line strength ranges

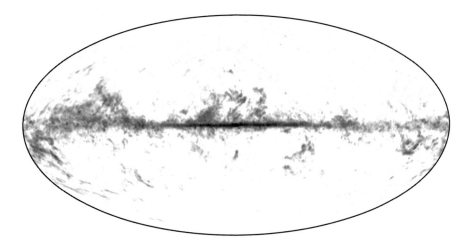

**Figure 7.3** All-sky map of the integrated line intensity $W_{\rm CO}$ of the $J = 1 \to 0$ transition of CO. The integrated line intensity is plotted on a logarithmic scale from $W_{\rm CO} = 0.1\,{\rm K\,km\,s}^{-1}$ (white) to $W_{\rm CO} = 300\,{\rm K\,km\,s}^{-1}$ (black). [Adam et al. 2016]

from $W_{CO} \sim 100\,\mathrm{K\,km\,s^{-1}}$ for lines of sight at low galactic latitude to $W_{CO} \sim 0.1\,\mathrm{K\,km\,s^{-1}}$ for lines of sight perpendicular to the galactic plane. If we assume $X_{CO} \approx 3 \times 10^{20}\,\mathrm{cm^{-2}/K\,km\,s^{-1}}$, this implies molecular column densities as high as $N(H_2) \sim 3 \times 10^{22}\,\mathrm{cm^{-2}}$, looking along the midplane of the Milky Way Galaxy. When we compare this with the column density of atomic hydrogen deduced from 21 cm emission (Figure 3.2), we find that $N(H_2) \sim N_{HI} \sim 3 \times 10^{22}\,\mathrm{cm^{-2}}$ in the midplane of our galaxy.

## 7.3 Heating and Cooling Molecular Gas

The heating of interstellar gas is usually done by fast electrons. In the lower-density portions of the ISM, fast electrons can be produced by the photoionization of atoms and ions or by the photoejection of electrons from dust grains. Molecular clouds, however, thanks to their dust content, are highly opaque to the ultraviolet photons with $h\nu > 5\,\mathrm{eV}$ that are capable of triggering photoionization and photoejection. Shock waves can produce fast electrons, but they are also effective at dissociating molecules. In practice, then, the main heating mechanism in molecular clouds is heating by **cosmic rays**.

In Section 1.3, we found that cosmic ray protons with $E < 200\,\mathrm{MeV}$ were effective at ionizing H atoms. Since the ionization energy of $H_2$ is comparable to that of H, these same cosmic ray protons can collisionally ionize $H_2$ molecules:

$$p_{cr} + H_2 \rightarrow p_{cr} + H_2^+ + e^-. \tag{7.60}$$

The kinetic energy of the ejected electron is typically $E \sim 35\,\mathrm{eV}$. This is more than enough energy to dissociate a hydrogen molecule, which has $D_0 = 4.5\,\mathrm{eV}$:

$$e^- + H_2 \rightarrow 2H + e^-. \tag{7.61}$$

The energy in excess of $D_0$ is shared as kinetic energy among the electron and the two recoiling hydrogen atoms. Alternatively, a free electron can collisionally excite a hydrogen molecule into a higher energy state. If the hydrogen molecule is then de-excited by a collision with another molecule, the excitation energy is shared as kinetic energy between the two recoiling molecules:

$$e^- + H_2 \rightarrow H_2^* + e^-, \tag{7.62}$$
$$H_2^* + H_2 \rightarrow H_2 + H_2. \tag{7.63}$$

The ejected electron typically transfers $\sim 10\,\mathrm{eV} \sim 10^{-11}\,\mathrm{erg}$ of thermal energy to the molecular gas through which it travels. The total heating gain comes to (Equation 1.32):

$$G_{cr} \sim 1 \times 10^{-27}\,\mathrm{erg\,s^{-1}} \left( \frac{\zeta_{cr}}{10^{-16}\,\mathrm{s^{-1}}} \right), \tag{7.64}$$

where $\zeta_{cr}$ is the primary cosmic ray ionization rate.

In all but the densest regions of molecular clouds, the primary cooling mechanism of molecular gas is through molecular line emission. Although $H_2$ is by far the most abundant molecule in the ISM, it's not efficient at emitting photons, thanks to its lack of an electric dipole moment. The most important molecule for cooling is CO, with its strong emission lines from rotational transitions. One obstacle to using CO for cooling is that, as shown in Section 7.1, molecular clouds are optically thick at the $J = 1 \to 0$ transition. That is, a 115 GHz photon emitted by a $J = 1 \to 0$ transition will almost immediately be absorbed in a $J = 0 \to 1$ transition, resulting in no net cooling of the gas. As we go to high values of $J$, however, the optical depth for line emission becomes smaller, primarily because of the lower column density $N_J$ at high values of $J$. The population of the $J$th rotational level is (Equation 7.32)

$$n_J \propto (2J + 1) \exp\left[ -\frac{B_0 J(J+1)}{kT_{\text{exc}}} \right], \tag{7.65}$$

where $B_0/k = 2.78\,\text{K}$ for $^{12}$CO rotational states. For $J \gg kT_{\text{exc}}/B_0$, the population of rotationally excited states drops exponentially with $J$, with

$$\frac{n_{J+1}}{n_J} = \frac{2J+3}{2J+1} \exp\left[ -\frac{B_0}{kT_{\text{exc}}} 2(J+1) \right]. \tag{7.66}$$

For the column densities and line broadening typical of molecular clouds, it's found that the $J = 5 \to 4$ line is the lowest transition at which the cloud is optically thin and hence can effectively be cooled throughout its entire volume. The $J = 6 \to 5$ line, of course, is also optically thin. However,

$$\frac{n_6}{n_5} = \frac{13}{11} \exp\left( -12 \frac{B_0}{kT_{\text{exc}}} \right) \approx 0.04, \tag{7.67}$$

assuming $T_{\text{exc}} = 10\,\text{K}$. Thus, the lower population of the $J = 6$ level (and all higher levels) prevents it from contributing much to the cooling.

The Einstein A coefficient for rotational transitions is

$$A_{J+1,J} = A_{10} \frac{(J+1)^4}{2J+1}, \tag{7.68}$$

so that

$$A_{54} = 69.4 A_{10} = 5.0 \times 10^{-6}\,\text{s}^{-1}. \tag{7.69}$$

The rotational energy of level $J$ is $E_J = B_0 J(J+1)$, so the energy of the photon emitted in a $J = 5 \to 4$ transition is

$$h\nu_{54} = 10 B_0 = 2.38\,\text{meV} = 3.82 \times 10^{-15}\,\text{erg}. \tag{7.70}$$

At an excitation temperature $T_{\text{exc}} = 10\,\text{K}$, the fraction of CO molecules in the $J = 5$ state is (from Equation 7.32)

$$f_5 \approx 6.9 \times 10^{-4}. \tag{7.71}$$

This means that about one in 1400 CO molecules is in the $J = 5$ state, ready to leap spontaneously downward to the $J = 4$ state, emitting a 575 GHz photon in the process.

The volumetric cooling rate from the $J = 5 \rightarrow 4$ transition is

$$\ell = f_5 n_{CO} A_{54} h \nu_{54} \tag{7.72}$$

$$\approx 1.3 \times 10^{-23} \text{ erg s}^{-1} n_{CO}, \tag{7.73}$$

assuming an excitation temperature $T_{exc} = 10\,\mathrm{K}$. If we make the approximation $n_{CO} \approx 0.25 n_C \approx 6.8 \times 10^{-5} n_H$, the cooling loss in ergs per second per hydrogen nucleus becomes

$$L \approx 9 \times 10^{-28} \text{ erg s}^{-1}. \tag{7.74}$$

For our assumed excitation temperature $T_{exc} = 10\,\mathrm{K}$, this loss is comparable to the cosmic ray gain, $G_{cr} \sim 10^{-27}$ erg s$^{-1}$, and the system is in temperature equilibrium. However, because of the exponential dependence of $f_5$ on temperature, the cooling loss $L$ is highly sensitive to the excitation temperature. At $T_{exc} = 9\,\mathrm{K}$, only one in 3300 CO molecules is in the $J = 5$ state, and the cooling loss is roughly halved, to $L \approx 4 \times 10^{-28}$ erg s$^{-1}$. At $T_{exc} = 11\,\mathrm{K}$, one in 740 CO molecules is in the $J = 5$ state, and the cooling loss is roughly doubled, to $L \approx 1.8 \times 10^{-27}$ erg s$^{-1}$.

## 7.4 Making and Breaking Molecules

Why does molecular hydrogen exist at all in the interstellar medium? After all, when two hydrogen atoms collide with each other, they usually bounce off instead of bonding to form a hydrogen molecule. It is, however, possible for them to bond by the process of **direct radiative association**:

$$H + H \rightarrow H_2^* \rightarrow H_2 + h\nu. \tag{7.75}$$

When two hydrogen atoms meet in this way, they create an excited hydrogen molecule ($H_2^*$) that is unbound; it must emit a photon carrying away enough energy to leave it in a bound state, or it will break apart. The lifetime of the excited hydrogen molecule until it breaks apart is roughly one vibration period ($\sim 2\pi/\omega_0 \sim 10^{-14}$ s). Since the hydrogen molecule has no electric dipole moment, the probability of its emitting a photon is small. For rotational quadrupole transitions, $A_{u\ell} \sim 10^{-11}$ s$^{-1}$ is a typical number, meaning that the excited hydrogen molecule has a probability $p \sim A_{u\ell}(2\pi/\omega_0) \sim 10^{-25}$ of emitting a photon before it falls apart. As a consequence, the rate coefficient for direct radiative association of $H_2$ is tiny ($k_{dra} < 10^{-23}$ cm$^3$ s$^{-1}$); this mechanism is too slow to produce significant amounts of $H_2$ at the low densities typical of the interstellar medium.

In a pure gas phase, with no dust present, molecular hydrogen is made primarily by a two-step process. First comes the production of negative hydrogen ions by radiative attachment:[3]

$$H^0 + e^- \rightarrow H^- + h\nu. \qquad (7.76)$$

This is a fairly slow process, with rate coefficient

$$k_{ra} \approx 1.6 \times 10^{-16}\,\text{cm}^3\,\text{s}^{-1} \left(\frac{T_{gas}}{80\,\text{K}}\right)^{0.67}. \qquad (7.77)$$

The second step is the production of molecular hydrogen by associative detachment:[4]

$$H^- + H^0 \rightarrow H_2 + e^-. \qquad (7.78)$$

This is a more rapid process, with a rate coefficient that is nearly independent of temperature at $T_{gas} \approx 80\,\text{K}$:

$$k_{ad} \approx 4.0 \times 10^{-9}\,\text{cm}^3\,\text{s}^{-1}. \qquad (7.79)$$

The production of molecular hydrogen by this two-step process is slowed by the fact that $H^-$ is a fragile ion; it takes only $I = 0.77\,\text{eV}$ to detach one of its electrons. In most of interstellar space, the main mechanism for destroying $H^-$ is photodetachment, the inverse process to radiative attachment:

$$H^- + h\nu \rightarrow H^0 + e^-. \qquad (7.80)$$

For the interstellar radiation field in the solar neighborhood, the photodetachment rate of $H^-$ is

$$\zeta_{det} \approx 2.7 \times 10^{-7}\,\text{s}^{-1} \approx 8.5\,\text{yr}^{-1}. \qquad (7.81)$$

Associative detachment, and the resulting production of hydrogen molecules, will dominate over photodetachment only when

$$n_{HI} > \frac{\zeta_{det}}{k_{ad}} \approx 70\,\text{cm}^{-3}. \qquad (7.82)$$

In most of the ISM, $H^-$ ions will undergo photodetachment before they have a chance to make $H_2$ molecules.

In reality, whenever dust grains are present the dominant mechanism for making molecular hydrogen is **grain catalysis**. The surface of a dust grain is a frenzied lab of chemical activity in comparison with low-density interstellar gas (Figure 7.4). A hydrogen atom colliding with a dust grain has some probability $p_s$ of sticking to the grain. The sticking probability $p_s$ depends on the atom's speed

---

[3] When a free electron binds to a positive ion, the process is called recombination; when a free electron binds to an atom or negative ion, it is called **attachment**.
[4] When an electron is stripped from an atom or positive ion, the process is called ionization; when an electron is stripped from a negative ion, it is called **detachment**.

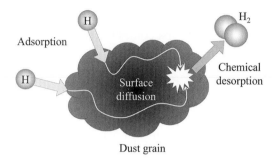

**Figure 7.4** Atoms adsorb onto the surface of a grain, diffuse until they meet another atom, and (sometimes) form a molecule. [Following Dulieu et al. 2013]

(slower atoms are more likely to stick), on the grain's temperature (hot grains are less sticky), on the grain's size (smaller grains are less sticky), and on the grain's composition. For gas temperatures and grain temperatures typical of molecular clouds, a probability $p_s \approx 0.3$ for grains with $a \sim 0.1\,\mu m$ to stick wouldn't be far wide of the mark. The process in which atoms stick to grain surfaces is called **adsorption**.[5]

A hydrogen atom that sticks to the grain is usually held by the relatively weak van der Waals force, which provides a binding energy $E_{vdw} = 0.01 \rightarrow 0.1\,eV$. Compared with the hydrogen atom's thermal energy ($\langle E \rangle \sim 0.003\,eV$ at $T = 20\,K$), this binding energy is large enough that the atom doesn't immediately evaporate from the grain surface. However, the binding energy is small enough that the atom can skitter its way across the grain surface in a thermally driven random walk. If, during its random walk, the atom encounters another H atom, they can undergo a relatively leisurely bonding process to form a hydrogen molecule. The energy released by the chemical reaction can allow the new $H_2$ molecule to escape from the grain surface; this process is called desorption. If all dust grains were identical, the rate per unit volume at which H atoms collide with grains would be

$$\Gamma_{coll} = n_{HI} n_{gr} \pi a^2 \left( \frac{8kT_{gas}}{\pi m_H} \right)^{1/2}, \tag{7.83}$$

where $\pi a^2$ is the geometric cross section of a grain. Since real dust grains have a distribution of sizes, we must average the collision rate over the distribution of grain radii:

$$\Gamma_{coll} = n_{HI} \left( \frac{8kT_{gas}}{\pi m_H} \right)^{1/2} \int_{a_{min}}^{a_{max}} \frac{dn_{gr}}{da} \pi a^2 da. \tag{7.84}$$

---

[5] An atom is **ad**sorbed if it sticks to the surface, but **ab**sorbed if it is drawn into the grain's interior.

It is customary to define

$$\sigma_{\rm gr} \equiv \frac{1}{n_{\rm H}} \int \frac{dn_{\rm gr}}{da} \pi a^2 da, \tag{7.85}$$

which is the total grain cross section per H nucleus. With this definition, the volumetric collision rate is

$$\Gamma_{\rm coll} = n_{\rm HI} n_{\rm H} \left( \frac{8kT_{\rm gas}}{\pi m_{\rm H}} \right)^{1/2} \sigma_{\rm gr}. \tag{7.86}$$

Unfortunately, the value of $\sigma_{\rm gr}$ is poorly known. For the MRN power-law distribution of grain radii, $dn_{\rm gr}/da \propto a^{-3.5}$, $\sigma_{\rm gr}$ diverges at small radii. Thus $\sigma_{\rm gr}$ depends on a poorly known value of $a_{\rm min}$, the minimum grain size. The observed UV extinction curve requires $\sigma_{\rm gr} > 10^{-21}$ cm$^2$ of dust cross section per H nucleus, but $\sigma_{\rm gr}$ could be much larger if there were enough small ($a < 0.005$ μm) grains. The dust model of Weingartner and Draine, which includes PAHs, has

$$\sigma_{\rm gr} = 6 \times 10^{-21} \ {\rm cm}^2 / {\rm H}, \tag{7.87}$$

that is, about 6 kilobarns of dust geometric cross section per H nucleus. This value of $\sigma_{\rm gr}$ gives a collision rate

$$\Gamma_{\rm coll} \approx 1.2 \times 10^{-12} \ {\rm cm}^{-3} \ {\rm s}^{-1} \left( \frac{n_{\rm H}}{40 \ {\rm cm}^{-3}} \right) \left( \frac{n_{\rm HI}}{40 \ {\rm cm}^{-3}} \right) \left( \frac{T_{\rm gas}}{80 \ {\rm K}} \right)^{1/2}, \tag{7.88}$$

scaling to properties typical of the cold neutral medium. For a neutral H atom, the typical time between collisions with a dust grain is

$$t_{\rm coll} \approx n_{\rm HI} / \Gamma_{\rm coll} \approx 1.0 \ {\rm Myr} \left( \frac{n_{\rm H}}{40 \ {\rm cm}^{-3}} \right)^{-1} \left( \frac{T_{\rm gas}}{80 \ {\rm K}} \right)^{-1/2}. \tag{7.89}$$

For a hydrogen atom, a collision with a dust grain is a once-in-a-million-years occurrence. However, for a dust grain, a collision with a hydrogen atom is a frequent occurrence. For a grain with radius $a$, the typical time between collisions with a neutral H atom is

$$t_{\rm coll} \approx \frac{1}{n_{\rm HI} \pi a^2} \left( \frac{8kT_{\rm gas}}{\pi m_{\rm H}} \right)^{-1/2} \tag{7.90}$$

$$\approx 10 \ {\rm min} \left( \frac{n_{\rm HI}}{40 \ {\rm cm}^{-3}} \right)^{-1} \left( \frac{a}{0.1 \ {\rm \mu m}} \right)^{-2} \left( \frac{T_{\rm gas}}{80 \ {\rm K}} \right)^{-1/2}. \tag{7.91}$$

Even a very small grain on the PAH borderline, with $a \approx 10^{-3}$ μm, will have a hydrogen atom bump into it every few months.

If an H atom that collides with a dust grain has a probability $p$ of ending up in a hydrogen molecule, then the volumetric rate of H$_2$ formation is

$$\frac{dn({\rm H}_2)}{dt} = \frac{1}{2} \Gamma_{\rm coll} \langle p \rangle, \tag{7.92}$$

where the factor 1/2 comes from the fact that it takes two atoms to make one molecule, and the averaging of $p$ is done over the complete distribution of grain sizes. The observed amount of molecular hydrogen in clouds argues that $\langle p \rangle$ can't be a tiny number; a plausible estimate is $\langle p \rangle \approx 0.08$. This means that dust grains are converting atomic hydrogen into molecular hydrogen on a characteristic time scale

$$t_{make} \approx \frac{n_{HI}}{\Gamma_{coll} \langle p \rangle} \approx 13\,\mathrm{Myr} \left(\frac{n_H}{40\,\mathrm{cm}^{-3}}\right)^{-1} \left(\frac{T_{gas}}{80\,\mathrm{K}}\right)^{-1/2} \left(\frac{\langle p \rangle}{0.08}\right)^{-1}. \quad (7.93)$$

Thus, the cold neutral medium should be filled with molecular hydrogen, *unless* there's a competing process that breaks the $H_2$ molecules almost as soon as grain surfaces make them.

The main process by which molecular hydrogen is destroyed is **photodissociation**. Looking at the potential energy curve for molecular hydrogen in Figure 7.1, you might think that the photodissociation of $H_2$ is a simple task; if it's in the ground state, hit it with a photon of energy $h\nu > 4.5\,\mathrm{eV}$. That will lift it to a vibrational state with quantum number $v > 14$, which is unbound; the two hydrogen atoms then fly away from each other and the dissociation is complete.[6] That is not, however, the way it usually is done. Absorbing a photon to lift the molecule to a $v > 14$ vibrational state requires an electric quadrupole transition, which has a small transition probability.

The main mechanism by which $H_2$ is actually photodissociated is a two-step process, involving the excited *electronic* states of $H_2$. Figure 7.1 shows the potential energy curve for the ground electronic state of $H_2$. Figure 7.5 shows the potential energy curve and the vibrational states for the ground electronic state and also the potential energy curves and vibrational states for the lowest excited electronic states. The energy difference between the $v = 0, J = 0$ level in the ground electronic state, and the $v = 0, J = 0$ level in the first excited electronic state is $E = 11.18\,\mathrm{eV}$, corresponding to $\lambda = 1108\,\text{Å}$. This is a higher energy than the 10.2 eV energy of the Lyman $\alpha$ line in atomic hydrogen. The transitions between the various vibrational and rotational levels in the ground electronic state and the vibrational and rotational levels in the first excited electronic states produce a series of closely spaced lines, referred to collectively as the **Lyman band**. Emission and absorption between the ground electronic state and the *second* excited electronic state is called the **Werner band**. The Lyman and Werner bands lie in the energy range $11.18 \to 13.60\,\mathrm{eV}$.

When a hydrogen molecule in its ground electronic state absorbs a photon with energy $E \sim 11.18\,\mathrm{eV}$ and leaps to the first excited electronic state, that is a permitted transition with a large transition probability. Its spontaneous decay

---

[6] Figure 7.1 implies that levels with $v < 14$ are bound; now we are saying that levels with $v > 14$ are unbound. What about $v = 14$? That's a borderline case: states with $v = 14$ and $J \leq 4$ are bound, but states with $v = 14$ and $J > 4$ are unbound.

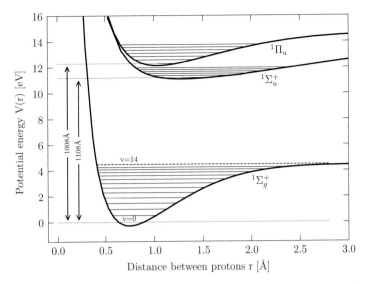

**Figure 7.5** Schematic diagram of the potential energy curves for the ground electronic states and the first and second excited electronic states of molecular hydrogen. [Data from Sharp 1971]

back down to the ground electronic state is also a permitted transition. However, a fraction $f_{dis} \approx 0.12$ of the time, the downward transition is to a vibrational state with $v > 14$, which is unbound. The two hydrogen atoms will then fly away from each other and the dissociation is complete.

The first step, radiative excitation of $H_2$ to an excited electronic state, requires a photon in the wavelength range $\lambda = 912 \to 1108$ Å. In the solar neighborhood, photons in this range are produced by hot stars and have an energy density $\varepsilon_{exc} \approx 0.0064$ eV cm$^{-3}$, representing only $\sim 1\%$ of the total energy density of starlight. Because of the numerous Lyman and Werner levels, the cross section of a hydrogen molecule for excitation by these photons is a complicated function of energy. Integrated over a plausible spectrum for the local UV background, the rate of radiative excitation for a hydrogen molecule is

$$\zeta_{exc} \approx 3.5 \times 10^{-10}\, s^{-1} \left( \frac{\varepsilon_{exc}}{0.0064\, \text{eV cm}^{-3}} \right). \qquad (7.94)$$

If radiative excitation leads to dissociation a fraction $f_{dis} \sim 0.12$ of the time (see above), the rate of photodissociation is

$$\zeta_{dis} = f_{dis}\zeta_{exc} \approx 4.2 \times 10^{-11}\, s^{-1} \left( \frac{f_{dis}}{0.12} \right) \left( \frac{\varepsilon_{exc}}{0.0064\, \text{eV cm}^{-3}} \right). \qquad (7.95)$$

Thus, the typical time scale for photodissociation of $H_2$ is $t_{dis} \sim 1/\zeta_{dis} \sim 10^3$ yr, smaller by four orders of magnitude than the $\sim 10^7$ yr time scale for the creation of $H_2$ by grain surface catalysis in the cold neutral medium.

The equilibrium abundance of $H_2$ will be determined by a balance between the slow creation of $H_2$ on grains and the swift destruction of $H_2$ by UV photons from stars. The volumetric rate of $H_2$ destruction can be written as

$$\frac{dn(H_2)}{dt} = -n(H_2)\zeta_{dis} = -\frac{1}{2}f_{mol}n_H\zeta_{dis}, \tag{7.96}$$

where $f_{mol} \equiv 2n(H_2)/n_H$ is the fraction of H atoms that are bound up in $H_2$ molecules. The volumetric rate of $H_2$ creation on dust grains can be written as

$$\frac{dn(H_2)}{dt} = k_d n_{HI} n_H = k_d(1 - f_{mol})n_H^2, \tag{7.97}$$

where the rate coefficient is (compare with Equations 7.86 and 7.92)

$$k_d = \frac{1}{2}\langle p \rangle \left(\frac{8kT_{gas}}{\pi m_H}\right)^{1/2} \sigma_{gr} \approx 3.1 \times 10^{-17}\,\mathrm{cm^3\,s^{-1}} \left(\frac{T_{gas}}{80\,\mathrm{K}}\right)^{1/2}. \tag{7.98}$$

In a region of space where $n_H$ doesn't change with time, Equations 7.96 and 7.97 can be combined in a relation that shows how the molecular fraction $f_{mol}$ changes with time:

$$\frac{df_{mol}}{dt} = -f_{mol}\zeta_{dis} + 2k_d(1 - f_{mol})n_H. \tag{7.99}$$

In equilibrium, when the creation of $H_2$ exactly balances its destruction, the molecular fraction is

$$f_{mol,eq} = \left(1 + \frac{\zeta_{dis}}{2k_d n_H}\right)^{-1}. \tag{7.100}$$

In the cold neutral medium, where $f_{mol} \ll 1$, Equation 7.100 reduces to

$$f_{mol,eq} \approx \frac{2k_d n_H}{\zeta_{dis}} \approx 5.9 \times 10^{-5} \left(\frac{n_H}{40\,\mathrm{cm^{-3}}}\right)\left(\frac{T_{gas}}{80\,\mathrm{K}}\right)^{1/2}, \tag{7.101}$$

assuming $\varepsilon_{exc} = 0.0064\,\mathrm{eV\,cm^{-3}}$ for the energy density of the exciting UV photons. By this calculation, only two out of every 34 000 hydrogen atoms in the CNM are teamed up to form an $H_2$ molecule.

At the temperature $T_{gas} \approx 15\,\mathrm{K}$ typical of molecular clouds, we can find the density $n_H$ at which $f_{mol,eq} = 0.5$ and half the hydrogen atoms have been bound into molecules. From Equation 7.100, this density is

$$n_H = \frac{\zeta_{dis}}{2k_d} \approx 1.5 \times 10^6\,\mathrm{cm^{-3}} \left(\frac{f_{dis}}{0.12}\right)\left(\frac{\varepsilon_{pe}}{0.0064\,\mathrm{eV\,cm^{-3}}}\right), \tag{7.102}$$

assuming $T_{gas} = 15\,\mathrm{K}$. If the ultraviolet radiation field were the same inside molecular clouds as outside, then molecular hydrogen wouldn't dominate over atomic hydrogen until the density reached $n_H \sim 10^6\,\mathrm{cm^{-3}}$. However, observations reveal that the molecular form dominates at densities as low as $n_H \sim 300\,\mathrm{cm^{-3}}$. This is so because the inner parts of molecular clouds are protected from UV light by **self-shielding**. That is, UV photons in the range $h\nu = 11.18 \to 13.60\,\mathrm{eV}$ are absorbed by hydrogen molecules in the outer layer of the cloud, preventing them from reaching hydrogen molecules in the cloud center.

Self-shielding becomes significant at a molecular hydrogen column density $N(H_2) \approx 10^{14}$ cm$^{-2}$; at this column density, the individual absorption lines in the Lyman and Werner bands start to become optically thick. When $\tau_0 > 1$ at line center, absorption can take place only in the Gaussian thermal wings; when $\tau_0 > \tau_{damp}$, absorption can take place only in the $v^{-2}$ wings, at frequencies where a molecule's cross section for absorbing a photon and undergoing photoexcitation is tiny. At column density $N(H_2) > 10^{14}$ cm$^{-2}$, the photodissociation rate for $H_2$ can be written as

$$\zeta_{ss} = \zeta_{dis}\beta_{ss}, \tag{7.103}$$

where $\zeta_{dis}$ is the photodissociation rate in the absence of self-shielding, as given in Equation 7.95. The self-shielding factor $\beta_{ss} \leq 1$ is difficult to calculate, in general, given how the wings of the closely spaced absorption lines start to overlap with increasing optical depth. However, a reasonable approximation is found to be

$$\beta_{ss} = \left(\frac{N(H_2)}{N_0}\right)^{-3/4} \qquad [N(H_2) > N_0], \tag{7.104}$$

where $N_0 = 10^{14}$ cm$^{-2}$ is the $H_2$ column density at which self-shielding becomes significant. In clouds thick enough for self-shielding to become important, the density at which half the mass of hydrogen is molecular becomes

$$n_H \approx 1.5 \times 10^6 \text{ cm}^{-3}\left(\frac{N(H_2)}{N_0}\right)^{-3/4}. \tag{7.105}$$

As a consequence of self-shielding, clouds with density $n_H < 1.5 \times 10^6$ cm$^{-3}$ can have $f_{mol,eq} \geq 0.5$ as long as they are protected by a minimum molecular column density

$$N(H_2) \approx 9 \times 10^4 N_0 \left(\frac{n_H}{300 \text{ cm}^{-3}}\right)^{-4/3} \tag{7.106}$$

$$\approx 9 \times 10^{18} \text{ cm}^{-2}\left(\frac{n_H}{300 \text{ cm}^{-3}}\right)^{-4/3}. \tag{7.107}$$

For illustration, consider a simplified dense cloud, with uniform density $n_H = 300$ cm$^{-3}$ and gas temperature $T_{gas} = 15$ K. The cloud is exposed to our standard ultraviolet radiation field with $\varepsilon_{pe} = 0.0064$ eV cm$^{-3}$. The outer layer of the cloud is not self-shielded; thus, it has a low molecular fraction. From Equation 7.101, we expect $f_{mol,eq} \approx 2 \times 10^{-4}$ in this outer layer, implying a number density of molecules $n(H_2) \approx 0.03$ cm$^{-3}$. Reaching the molecular column density $N_0 = 10^{14}$ cm$^{-2}$ needed for significant self-shielding requires a physical thickness

$$d = \frac{N_0}{n(H_2)} \approx \frac{10^{14} \text{ cm}^{-2}}{0.03 \text{ cm}^{-3}} \approx 3.3 \times 10^{15} \text{ cm} \sim 0.001 \text{ pc}. \tag{7.108}$$

After the UV radiation has penetrated this distance into the cloud, self-shielding becomes important and the molecular fraction starts to increase. In our uniform density cloud, the molecular fraction reaches $f_{mol,eq} = 0.5$ at a distance $d_{50} \approx 0.12$ pc from the cloud surface. (The molecular fraction rises from $f_{mol,eq} = 0.25$ at $d_{25} \approx 0.07$ pc to $f_{mol,eq} = 0.75$ at $d_{75} \approx 0.25$ pc.) Note, however, that the characteristic thickness $d_{50}$ of the self-shielding layer is strongly dependent on the density $n_H$ of the cloud. If $n_H = 1000\,\mathrm{cm}^{-3}$, the required thickness is only $d_{50} \approx 0.007$ pc; if $n_H = 100\,\mathrm{cm}^{-3}$, it rises to $d_{50} \approx 1.6$ pc.

Molecular gas can also be sheltered from UV light by dust. As we have seen, the standard relation between dust extinction and hydrogen column density is (Equation 6.9)

$$\frac{A_V}{N_H} \approx 5.3 \times 10^{-22}\,\mathrm{cm}^2\,\mathrm{mag} \qquad (7.109)$$

in the $V$ band. For the standard extinction curve in our galaxy, illustrated in Figure 6.3, the ratio of the extinction in the Lyman and Werner bands ($\lambda \sim 1000\,\text{Å}$) to the extinction in the $V$ band is $A_{1000}/A_V \approx 5$. When the hydrogen is primarily molecular, $N(H_2) = N_H/2$. The resulting dust extinction at $\lambda \sim 1000\,\text{Å}$ is

$$\frac{A_{1000}}{N(H_2)} \approx 10\frac{A_V}{N_H} \approx 5.3 \times 10^{-21}\,\mathrm{cm}^2\,\mathrm{mag}. \qquad (7.110)$$

Thus, a column density $N(H_2) \sim 2 \times 10^{20}\,\mathrm{cm}^{-2}$ will be accompanied by enough dust to produce one magnitude of extinction in the Lyman and Werner bands. However, at typical molecular cloud densities, this is larger than the column density of $H_2$ that produces effective self-shielding (Equation 7.107). Thus, in the wavelength range $\lambda = 912\,\text{Å}$ to $1108\text{Å}$, self-shielding by $H_2$ provides more protection than dust shielding.

The interiors of molecular clouds are mostly saved from photodissociation by a self-sacrificing layer of gas, $\sim 0.1$ pc thick when $n_H \sim 300\,\mathrm{cm}^{-3}$; in this layer, most $H_2$ molecules are photodissociated soon after their formation. However, the interiors of molecular clouds often host star formation. Some of the newly formed stars are of spectral types O, B, or A, and so are hot enough to pour out photons with $h\nu > 11.18\,\mathrm{eV}$. These photons are energetic enough to form **photodissociation regions** around the hot stars.

Photodissociation regions have obvious parallels to H II regions. Photons with $h\nu > 13.6\,\mathrm{eV}$ can ionize H atoms from their ground state:

$$H + h\nu \rightarrow p + e^-. \qquad (7.111)$$

Thus, extremely hot stars (O stars and the hottest B stars) are surrounded by Strömgren spheres in which photoionization is balanced by radiative recombination. Similarly, photons with $h\nu > 11.18\,\mathrm{eV}$ are able, about 12% of the time, to photodissociate molecular hydrogen from its ground state:

$$H_2 + h\nu \rightarrow H_2^* \rightarrow H + H. \qquad (7.112)$$

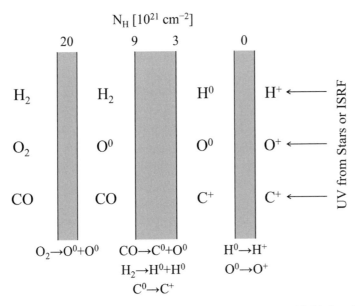

**Figure 7.6** Schematic diagram of a photodissociation region. The shielded molecular gas is on the left; the source of UV photons, either a single star or the general interstellar radiation field (ISRF), is on the right. The numbers at the top of the diagram represent the column density of hydrogen measured from the ionization front.

Thus, hot stars (O, B, and A stars) are surrounded by photodissociation regions in which photodissociation is balanced by the manufacture of molecules.

The structure of the photodissociation regions, as indicated in Figure 7.6, is made interesting by the fact that the dissociation energy of CO (11.11 eV) is nearly equal to the photodissociation energy of $H_2$ (11.18 eV), which in turn is nearly equal to the first ionization energy of carbon ($I_C = 11.26$ eV). Thus, there is a region, at a column density $N_H \approx 3 \to 9 \times 10^{21}$ cm$^{-2}$ from the ionization front, where $C^+$, $C^0$, CO, $H^0$, and $H_2$ all exist. The fractional contribution of neutral atomic carbon ($C^0$) is highest at $N_H \sim 7 \times 10^{21}$ cm$^{-2}$; closer to the ionization front, $C^+$ is the dominant form of carbon, while farther away, the plurality of carbon is in CO.

## Exercises

7.1 A molecular cloud is observed to have the following emission lines:

CO $\quad J = 1\to0$, $\nu = 115.3$ GHz, $T_A = 25$ K; $b = 6.0$ km s$^{-1}$

$^{13}$CO $\quad J = 1\to0$, $\nu = 110.2$ GHz, $T_A = 9.1$ K; $b = 3.0$ km s$^{-1}$

$C^{18}O$ $\quad J = 1\to0$, $\nu = 109.8$ GHz, $T_A = 1.6$ K; $b = 6.0$ km s$^{-1}$

For simplicity, assume the same value for the dipole moment for all three isotopologues ($\mu_0 = 0.112$ debye). From these data, deduce

(a)   the kinetic temperature of the cloud;

(b)   the $^{13}$CO and $C^{18}O$ column densities;

(c)   the total column density $N(H_2)$, and visual extinction $A_V$, through the cloud.

7.2   Consider a typical region of the cold neutral medium with density $n_H = 40\,cm^{-3}$, temperature $T = 80\,K$, and fractional ionization $x = n_e/n_H = 10^{-3}$. There is no dust present. Begin with the assumption that the fractions of hydrogen in the forms of $H^-$ and $H_2$ are tiny, with $f_{H-} \equiv n(H^-)/n_H \ll 1$ and $f_{mol} \equiv 2n(H_2)/n_H \ll 1$. (We will check this assumption later.)

(a)   Assume a state of equilibrium between the production of $H^-$ by radiative attachment and the destruction of $H^-$ by photodetachment or by associative detachment to form $H_2$. What is the equilibrium value of $f_{H-} \equiv n(H^-)/n_H$? Given the values of $k_{ra}$, $k_{ad}$, and $\zeta_{det}$ in Section 7.4, what is the numerical value of $f_{H-}$? Is the assumption that $f_{H-} \ll 1$ in a dust-free CNM justified?

(b)   Now assume a state of equilibrium between the production of $H_2$ by associative detachment and the destruction of $H_2$ by photodissociation. What is the equilibrium value of $f_{mol} \equiv 2n(H_2)/n_H$? Given the results of part (a), and the value of $\zeta_{dis}$ given in Equation 7.95, what is the numerical value of $f_{mol}$? Is the assumption that $f_{mol} \ll 1$ in a dust-free CNM justified?

7.3   In equilibrium, the molecular fraction of hydrogen $f_{mol,eq}$ in a dusty molecular cloud is given by Equation 7.100. Suppose that the molecular fraction is perturbed from its equilibrium value by some small amount, so that $f_{mol} = f_{mol,eq}(1 + \epsilon)$, with $|\epsilon| \ll 1$. Starting with Equation 7.99, show that the imposed perturbation decays as

$$\epsilon \propto \exp(-t/t_{mol}). \tag{7.113}$$

What is $t_{mol}$ as a function of $\zeta_{dis}$, $n_H$, and $k_d$? What is the numerical value of $t_{mol}$ in the CNM, where $f_{mol,eq} \ll 1$? What is the numerical value of $t_{mol}$ in the core of a molecular cloud, where self-shielding is highly effective?

# 8

# Circumgalactic and Intracluster Gas

*Falls sich dies bewahrheiten sollte, würde sich also das*
*überraschende Resultat ergeben, dass dunkle Materie in sehr*
*viel grösserer Dichte vorhanden ist als leuchtende Materie.*
*(If this should prove to be the case, the surprising result would be that*
*dark matter is present in much greater density than luminous matter.)*
Fritz Zwicky (1898–1974)
"Die Rotverschiebung von Extragalaktischen Nebeln," 1933,
Helvetica Phyica Acta 6, 110

The circumgalactic medium (CGM) is the diffuse gas, plus any stray dust, that lies outside the main body of a galaxy's stellar distribution but inside the virial radius of the galaxy's dark halo. The intracluster medium (ICM) is the diffuse hot gas that lies inside the virial radius of a cluster of galaxies but is not bound to any particular galaxy in the cluster.

The **virial radius** of a galaxy or cluster is usually defined as the radius of the largest sphere centered on the system's barycenter inside which the steady-state virial theorem holds true. If the virial theorem applies, you can safely make a mass estimate by writing

$$M \sim \frac{\sigma^2 r}{G},\tag{8.1}$$

where $\sigma$ is the velocity dispersion of the material inside radius $r$. (In Equation 7.51, for instance, the virial theorem was used to estimate the mass of a molecular cloud.)

For galaxies and clusters within our expanding universe, we adopt the commonly used approximation that the virial radius is equal to the radius $r_{200}$ within which the mean density $\overline{\rho}(r_{200})$ is equal to 200 times the critical density $\rho_{c,0}$ at which the universe is flat:

$$\rho_{c,0} \equiv \frac{3H_0^2}{8\pi G} = 9.20 \times 10^{-30}\,\mathrm{g\,cm}^{-3} h_{70}^2,\tag{8.2}$$

where $H_0$ is the Hubble constant, and $h_{70} = H_0/(70 \, \text{km s}^{-1} \, \text{Mpc}^{-1})$. The virial radius of a galaxy or cluster of galaxies is thus defined in practice as the radius inside which

$$\bar{\rho}(r_{200}) = 200\rho_{c,0} \approx 1.8 \times 10^{-27} \, \text{g cm}^{-3} \approx 2.7 \times 10^4 \, \text{M}_\odot \, \text{kpc}^{-3}. \quad (8.3)$$

For the Milky Way Galaxy, $r_{200} \approx 220 \, \text{kpc}$; the scale length of the Galaxy's stellar disk, by contrast, is only $h_R \approx 3 \, \text{kpc}$. The total mass inside $r_{200}$, called the virial mass, is $M_{200} \approx 1.1 \times 10^{12} \, \text{M}_\odot$ for the Milky Way Galaxy; the total mass of stars in the Galaxy, by contrast, is only $M_\star \approx 6 \times 10^{10} \, \text{M}_\odot$. For a rich cluster of galaxies, comparable in size to the Coma Cluster, typical values would be $r_{200} \sim 2.5 \, \text{Mpc}$ and $M_{200} \sim 2 \times 10^{15} \, \text{M}_\odot$.

## 8.1  Circumgalactic Medium: Our Galaxy

The circumgalactic medium of our galaxy is loosely defined as all the gas and dust that lies outside the bulge and disk, out as far as $r \sim r_{200} \sim 220 \, \text{kpc}$, or roughly a third the distance to our neighbor, the Andromeda Galaxy (M31). This volume embraces a large variety of gaseous components, just as the interstellar medium of our galaxy embraces a large variety of gaseous phases. The CGM is a mix of gas falling in from the intergalactic medium, gas stripped from satellite galaxies, and gas ejected from the interstellar medium of the central part of our galaxy.

Around our galaxy, the CGM was first detected in 21 cm emission. Most of the all-sky emission at 21 cm, as shown in Figure 3.2, comes from neutral hydrogen within our galaxy's disk. However, it was found that there exist discrete clouds containing H I with radial velocities higher than expected for gas clouds rotating along with the galactic disk. Thus, the circumgalactic clouds containing neutral hydrogen are called high-velocity clouds (HVCs). Figure 8.1 is an all-sky map of the column density of H I contained in HVCs; it includes neutral hydrogen with a radial velocity more than $70 \, \text{km s}^{-1}$ discrepant from the velocity range created by our galaxy's differential rotation along each line of sight.

The highest column densities in Figure 8.1 are associated with the Large Magellanic Cloud (LMC) and Small Magellanic Cloud (SMC). This gas, amounting to a neutral hydrogen mass of $M_{\text{HI}} \approx 4.4 \times 10^8 \, \text{M}_\odot$ in the LMC and $M_{\text{HI}} \approx 4.0 \times 10^8 \, \text{M}_\odot$ in the SMC, is more accurately described as the ISM of the Magellanic Clouds than as the CGM of the Milky Way Galaxy. However, there exists a long stream of high-velocity gas, called the Magellanic Stream, consisting of gas that has been removed from the Magellanic Clouds by tidal effects and ram-pressure stripping. In Figure 8.1, the Magellanic Stream (MS) is seen to stretch from its "head" at the location of the Magellanic Clouds ($\ell \approx 290°, b \approx -40°$) to the tip of its "tail" over $90°$ away ($\ell \approx 90°, b \approx -45°$). The observed radial

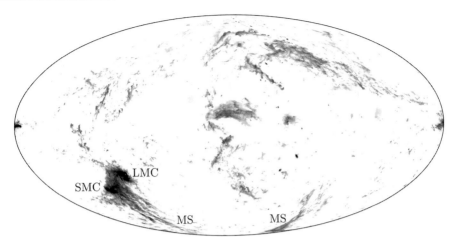

**Figure 8.1** All-sky map of H I column density for high-velocity clouds surrounding our galaxy (map is centered on the galactic anticenter). Column density is plotted on a logarithmic scale from $N_{HI} = 10^{18}\,\mathrm{cm}^{-2}$ (white) to $N_{HI} = 10^{21}\,\mathrm{cm}^{-2}$ (black). LMC = Large Magellanic Cloud, SMC = Small Magellanic Cloud, MS = Magellanic Stream. [Westmeier 2018]

velocity of the Magellanic Stream varies smoothly from $v_r \approx +250\,\mathrm{km\,s}^{-1}$ at its head to $v_r \approx -400\,\mathrm{km\,s}^{-1}$ at its tail. However, part of this observed radial velocity is due to the Sun's orbital velocity about the center of our galaxy. After subtracting the component of the radial velocity due to the Sun's motion, the remaining galactocentric velocity ranges from $v_{gal} \sim +90\,\mathrm{km\,s}^{-1}$ at the head of the Magellanic Stream to $v_{gal} \sim -240\,\mathrm{km\,s}^{-1}$ at the tail. The total H I mass of the Magellanic Stream is $M_{HI} \approx 3 \times 10^8\,M_\odot$. Adding helium and ionized hydrogen raises the total mass of the Stream to $M_{ms} \approx 1 \times 10^9\,M_\odot$, comparable to the gas mass left behind in the SMC and LMC.

Although the Magellanic Stream has a "smoking gun" association with the Magellanic Clouds, most of the HVCs shown in Figure 8.1 are not obviously associated with any of the Milky Way's satellite galaxies. The HVCs generally have galactocentric velocities in the range $-250\,\mathrm{km\,s}^{-1} < v_{gal} < +250\,\mathrm{km\,s}^{-1}$. Their 21 cm emission has a typical broadening parameter $b \sim 12\,\mathrm{km\,s}^{-1}$, indicating that the emission is from warm neutral gas with $T_{gas} \sim 9000\,\mathrm{K}$. Excluding the Magellanic Stream, about 6% of our sky is covered by HVCs at a neutral-hydrogen column density $N_{HI} > 10^{19}\,\mathrm{cm}^{-2}$, but only 0.3% is covered at a column density $N_{HI} > 10^{20}\,\mathrm{cm}^{-2}$.

Distances to HVCs can be determined approximately by seeing which halo stars of known distance are in front of the clouds and which are behind. Most HVCs are at distances $d < 15\,\mathrm{kpc}$ from the galactic center and are within $\sim 30°$ of the disk plane as seen from the galactic center. Thus, HVCs are associated with the disk of our galaxy, as opposed to being strewn widely throughout its dark halo. With distances known, we can determine the physical sizes and masses of individual HVCs. If we again exclude the massive, relatively distant Magellanic Stream, the total H I mass of the high-velocity clouds around our galaxy is $M_{\rm HI} \sim 3 \times 10^7\,M_\odot$. Adding in helium and ionized hydrogen makes the estimated total mass of HVCs $M_{\rm hvc} \sim 7 \times 10^7\,M_\odot$, less than 10% the mass of the Magellanic Stream.

Although warm neutral gas within high-velocity clouds is the most easily detected component of the circumgalactic medium, it contributes only a small fraction of the total mass of the CGM. The existence of a much hotter ionized circumgalactic medium (at $T \sim 10^6$ K) was predicted by Lyman Spitzer in a 1956 paper titled "On a possible interstellar galactic corona." Although Spitzer referred to this component of our galaxy as "interstellar," he computed that it would have a scale height $h_z \approx 8$ kpc at $T = 10^6$ K and a still greater scale height at higher temperatures. Spitzer's proposal led to a model for the CGM in which high-velocity clouds with $T \sim 10^4$ K were in pressure equilibrium with the surrounding ionized gas at $T \sim 10^6$ K.

Further observations revealed, however, that the circumgalactic medium is more complex than this simple two-phase model. Circumgalactic gas in the intermediate temperature range $10^4$ K $< T < 10^6$ K is detected through the absorption lines it produces in the spectra of background sources such as quasars. For instance, Si III has a strong absorption line at $\lambda = 1206$ Å that is a useful diagnostic for gas at $T \approx 2 \times 10^4$ K. Observations along lines of sight through our galaxy's circumgalactic medium reveal that $\sim$80% contain Si III at a detectable level. At higher temperatures, a useful diagnostic is the $\lambda = 1031.9$ Å and 1037.6 Å doublet of O VI. As we saw in Section 5.4, this doublet is a tracer for gas at $T \approx 3 \times 10^5$ K. Observations reveal that $\sim$60% of sightlines contain detectable O VI absorption from circumgalactic gas, with a mean column density $N_{\rm OVI} \approx 9 \times 10^{13}$ cm$^{-2}$.

The hottest portion of the CGM, at $T \geq 10^6$ K, can be detected by looking for O VII and O VIII absorption lines toward background X-ray sources. Helium-like O VII is the dominant form of oxygen at $T \sim 10^6$ K; it has a strong spectral line at an energy $h\nu = 0.574$ keV, corresponding to a wavelength $\lambda = 21.60$ Å. Hydrogenic O VIII is dominant at $T \sim 2.5 \times 10^6$ K; its Lyman $\alpha$ transition is at an energy $h\nu = 0.654$ keV, corresponding to a wavelength $\lambda = 18.97$Å.

Figure 8.2 shows O VII and O VIII in absorption toward the X-ray-bright quasar 1ES 1553+113, located in the constellation Serpens at a galactic latitude $b = +44°$.[1] The O VII line, shown in the left panel of Figure 8.2, is best fitted by assuming that it comes from a warm-hot component of the CGM, with $T \approx 10^6$ K.

---

[1] The data in Figure 8.3 result from 1.85 megaseconds of observation with the *XMM-Newton* Reflection Grating Spectrometer. This three-week-long exposure gives you an idea of the difficulty of detecting the hotter portions of the circumgalactic medium.

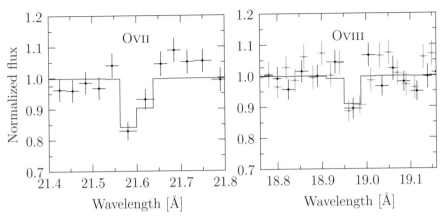

**Figure 8.2** Portions of the *XMM-Newton* X-ray spectrum of the quasar 1ES 1553+113, centered on the 21.60 Å absorption line of O VII (left) and the 18.97 Å absorption line of O VIII (right). Wavelengths are in the observer's rest frame: the lack of large redshifts shows that the absorbers are associated with our galaxy. [Data from Das et al. 2019a]

By contrast, the O VIII line in the right panel is best fitted by assuming that it comes from a hotter component of the CGM, with $T \approx 10^7$ K. The deduced column densities along this line of sight are $N_{\mathrm{OVII}} \sim 10^{16}$ cm$^{-2}$ and $N_{\mathrm{OVIII}} \sim 3 \times 10^{15}$ cm$^{-2}$. However, at the temperature of the hot CGM ($T \approx 10^7$ K), most of the oxygen will be fully ionized O IX: only $\sim$2% will be in the form of hydrogenic O VIII (see Figure 5.6). Thus, there is a large but undetected column density of O IX along this line of sight. Assuming collisional ionization equilibrium, and assuming a roughly solar oxygen-to-hydrogen ratio, the column density of our galaxy's CGM in this particular direction is $N_{\mathrm{H}} \sim 3 \times 10^{20}$ cm$^{-2}$.

Along lines of sight without a helpful X-ray source in the background, the hotter regions of the CGM can be seen in X-ray emission. Figure 8.3, for instance, shows the X-ray emission spectrum looking outward through our galaxy's CGM. (This spectrum is from an annulus surrounding the quasar 1ES 1553+113 on the sky. Thus, it represents emission from circumgalactic gas close to, but not identical with, the gas whose absorption lines are seen in Figure 8.2.) Over the plotted energy range $h\nu = 0.33 \rightarrow 7$ keV, most of the X-ray photons come from the foreground (including emission from the hot Local Bubble), the background (including unresolved distant X-ray sources), and instrumental emission (including a strong fluorescent line at 1.5 keV from the aluminum inside the telescope). However, in the narrower range $0.5 \rightarrow 0.9$ keV, at least half the emission comes from the circumgalactic medium. The fit is improved by assuming two different temperature components, corresponding to the warm-hot CGM and hot CGM deduced from the O VII and O VIII absorption lines. In Figure 8.3, the warm-hot component, at $T \approx 2 \times 10^6$ K, produces the spectrum shown by the solid

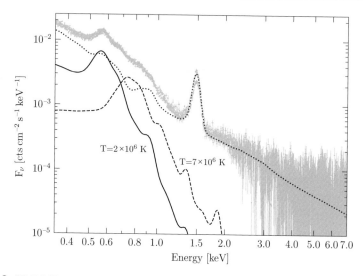

**Figure 8.3** *XMM-Newton* X-ray spectrum looking through our galaxy's CGM. Gray error bars = data. Dotted line: foreground, background, and instrumental components. Solid line: spectrum of warm-hot CGM ($T \approx 2 \times 10^6$ K). Dashed line: spectrum of hot CGM ($T \approx 7 \times 10^6$ K). [Data from Das et al. 2019]

line. The hot component, at $T \approx 7 \times 10^6$ K, produces the spectrum shown by the dashed line.

Thus, there is evidence that our galaxy's circumgalactic medium contains phases with a wide range of temperatures, from $T \sim 10^4$ K to $T \sim 10^7$ K. The total mass of circumgalactic gas associated with the Milky Way is not very well known, particularly since this mass is dominated by the imperfectly understood warm-hot and hot phases. Some estimates, making use of numerical simulations, conclude that the total mass of our galaxy's CGM is $M_{cgm} \approx 2 \times 10^{10} M_\odot$. In Chapter 1, we gave the mass of circumgalactic material in the universe as being 5% of the total baryonic mass, but it could easily be lower or higher. The distinction between circumgalactic gas and interstellar gas is a fuzzy one; so is the distinction between circumgalactic gas and intergalactic gas.

## 8.2 Circumgalactic Medium: Other Galaxies

Some of the circumgalactic gas around our galaxy, and other galaxies, is low-metallicity intergalactic gas that has fallen inward through the virial radius. Numerical simulations of structure formation in the consensus $\Lambda$CDM cosmology, to be discussed further in Chapter 10, show gas flowing inward along filaments to regions where galaxies are forming and evolving. This filamentary flow of gas tends to dilute the metallicity of the circumgalactic gas. However, the

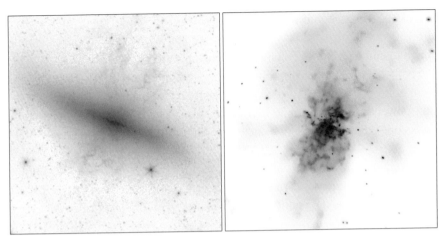

**Figure 8.4** Left: Negative near-infrared image of the starburst galaxy M82 ($d \approx 3.5\,\text{Mpc}$) taken with *Spitzer*. Image size is $8.5 \times 8.5\,\text{arcmin}$, corresponding to $\sim 9 \times 9\,\text{kpc}$ at the distance of M82. [NASA/JPL-Caltech] Right: Negative X-ray image of M82 taken with *Chandra*. Angular scale and orientation (north up) are the same in both images. [NASA/CXC/Wesleyan/R. Kilgard et al.]

CGM is also fed by a **galactic wind** that blows interstellar gas outward, away from the disk. Some galaxies, such as the starburst galaxy M82 (Figure 8.4), have extremely strong galactic winds. The left panel of Figure 8.4, showing M82 at $\lambda \sim 4\,\mu\text{m}$, emphasizes the established stellar populations in the galaxy's nearly edge-on disk. The molecular gas in M82, whose CO emission is displayed in Figure 7.2, is also largely contained within the galaxy's disk. By contrast, the right panel of Figure 8.4, showing M82 at $h\nu \sim 1\,\text{keV}$, reveals the existence of a hot ($T \sim 10^7\,\text{K}$) galactic wind, flowing in a direction perpendicular to the disk.

The galactic wind of M82 is powered primarily by supernova explosions.[2] From observations of supernova remnants in M82, it is estimated that its supernova rate is $R \sim 0.1\,\text{yr}^{-1}$. Most of the supernovae occur within a distance $d \sim 250\,\text{pc}$ of the galaxy's center; this implies that the average supernova rate per unit area in M82's central disk is $\Sigma_{SNR} \sim 0.5\,\text{pc}^{-2}\,\text{Myr}^{-1}$, about 25 000 times the feeble supernova rate in our region of the Milky Way Galaxy. Only a minority of newly formed stars will explode as supernovae. To end its existence in a core collapse supernova, a star must begin with an initial mass $M > 8\,\text{M}_\odot$. The initial mass function of stars falls fairly steeply with mass at its high-mass end, with $N(M)dM \propto M^{-2.3}dM$ for stars with $M > 1\,\text{M}_\odot$. As a result, when

---

[2] Active galactic nuclei can also drive galactic winds by injecting large amounts of energy into the interstellar and circumgalactic medium. In this section, however, we will focus on the specific case of supernova-driven winds in starburst galaxies.

you look at a mass equal to $100\,M_\odot$ of newly formed stars, you will typically find $\sim 150$ stars with $M < 8\,M_\odot$, but only one star with $M > 8\,M_\odot$ and therefore destined to end with a core collapse supernova. Thus, M82's supernova rate $\Sigma_{SNR}$ $\sim 0.5\,\mathrm{pc}^{-2}\,\mathrm{Myr}^{-1}$ implies a star formation rate $\Sigma_{sfr} \sim 50\,M_\odot\,\mathrm{pc}^{-2}\,\mathrm{Myr}^{-1}$. (Also keep in mind that a newly formed massive star harbors a time-delay fuse; it has a fusion-powered lifetime of $\sim 30\,\mathrm{Myr}$ before it explodes in a supernova. Therefore, today's core-collapse supernova rate is an echo of the star formation rate $\sim 30\,\mathrm{Myr}$ ago.)

To consider the physics behind a strong galactic wind, consider a simplified disk galaxy with a supernova rate per unit area $\Sigma_{SNR}$. The supernovae go off in the galaxy's midplane as the result of star formation in a thin layer of molecular gas. To give physically plausible numbers, a molecular gas layer with scale height $h_z \sim 50\,\mathrm{pc}$ and average midplane density $n_H \sim 1000\,\mathrm{cm}^{-3}$ will have a mass surface density $\Sigma \sim 5000\,M_\odot\,\mathrm{pc}^{-2}$, enough to fuel an M82-like starburst for up to $\sim 100\,\mathrm{Myr}$. Above and below the thin star-forming layer, the interstellar medium consists of atomic gas with number density $n_H$ and mass density $\rho \sim m_H n_H$. If we assume an energy $E \approx 10^{51}$ erg for each supernova, this implies an energy injection rate per unit area

$$\Sigma_{\dot{E}} = 1.7\,\mathrm{erg\,cm}^{-2}\,\mathrm{s}^{-1}\left(\frac{\Sigma_{SNR}}{0.5\,\mathrm{pc}^{-2}\,\mathrm{Myr}^{-1}}\right), \tag{8.4}$$

scaled to the supernova rate in M82. If the supernovae are assumed to start exploding at a time $t = 0$, and if the supernova rate remains constant after that, we can find a self-similar solution for the rate at which the resulting planar shock fronts move away from the midplane in opposite directions.[3]

If the shock fronts move through atomic gas with constant mass density $\rho$, the only combination of $\rho$, $\Sigma_{\dot{E}}$, and $t$ that gives the correct dimensionality for $r_{sh}$ (the distance of the shock fronts from the midplane) is

$$r_{sh} = \xi\left(\frac{\Sigma_{\dot{E}}}{\rho}\right)^{1/3} t, \tag{8.5}$$

where $\xi$ is the usual dimensionless factor that we expect to be of order unity. This implies a constant speed for the shock fronts,

$$u_{sh} = \xi\left(\frac{\Sigma_{\dot{E}}}{\rho}\right)^{1/3}. \tag{8.6}$$

Assuming $\xi \approx 1$, and using Equation 8.4, we find

$$r_{sh} \approx 1\,\mathrm{kpc}\left(\frac{\Sigma_{SNR}}{0.5\,\mathrm{pc}^{-2}\,\mathrm{Myr}^{-1}}\right)^{1/3}\left(\frac{n_H}{1\,\mathrm{cm}^{-3}}\right)^{-1/3}\left(\frac{t}{1\,\mathrm{Myr}}\right) \tag{8.7}$$

---

[3] This is similar in spirit to finding the Sedov–Taylor solution (Section 5.2) but with planar rather than spherical symmetry and with continuous rather than instantaneous energy injection.

and

$$u_{\rm sh} \approx 1000 \, {\rm km \, s^{-1}} \left( \frac{\Sigma_{\rm SNR}}{0.5 \, {\rm pc^{-2} \, Myr^{-1}}} \right)^{1/3} \left( \frac{n_{\rm H}}{1 \, {\rm cm^{-3}}} \right)^{-1/3} . \qquad (8.8)$$

This shock speed implies a temperature for the post-shock gas (Equation 5.19)

$$T_2 \sim 10^7 \, {\rm K} \left( \frac{\Sigma_{\rm SNR}}{0.5 \, {\rm pc^{-2} \, Myr^{-1}}} \right)^{2/3} \left( \frac{n_{\rm H}}{1 \, {\rm cm^{-3}}} \right)^{-2/3} , \qquad (8.9)$$

comparable to the temperature of the X-ray-emitting gas in M82's galactic wind (Figure 8.4). At the high temperatures of the post-shock gas, the primary cooling mechanism is free–free emission, with cooling time (Equation 5.22)

$$t_{\rm cool} \approx 30 \, {\rm Myr} \left( \frac{\Sigma_{\rm SNR}}{0.5 \, {\rm pc^{-2} \, Myr^{-1}}} \right)^{1/3} \left( \frac{n_{\rm H}}{1 \, {\rm cm^{-3}}} \right)^{-4/3} . \qquad (8.10)$$

As the planar shock front moves away from the midplane of the galaxy, it eventually runs out of gas to shock. Warm atomic gas in disk galaxies typically has a scale height $h_z \lesssim 1 \, {\rm kpc}$. Once the shock is at a distance $\sim h_z$ from the midplane, it encounters lower densities, and hence accelerates. This is usually referred to as the "breakout" of the shock and occurs after a time (Equation 8.7)

$$t_{\rm break} \approx 1 \, {\rm Myr} \left( \frac{h_z}{1 \, {\rm kpc}} \right) \left( \frac{\Sigma_{\rm SNR}}{0.5 \, {\rm pc^{-2} \, Myr^{-1}}} \right)^{-1/3} \left( \frac{n_{\rm H}}{1 \, {\rm cm^{-3}}} \right)^{1/3} . \qquad (8.11)$$

For plausible gas properties this time scale is shorter than the cooling time of Equation 8.10, so the shock reaches breakout before it has a chance to cool significantly. As the shock accelerates, the shocked gas begins to fragment as the result of instabilities. Individual knots and clumps of gas, moving faster than the galaxy's escape speed, move through the circumgalactic medium toward intergalactic space.

## 8.3 Intracluster Medium

Clusters of galaxies come in a variety of sizes. The small cluster in which we live, known with hometown pride as the **Local Group**, is dominated by a pair of spiral galaxies: the Milky Way Galaxy and the Andromeda Galaxy, also known as M31. In the $V$ band, the stellar luminosity of the Milky Way is $L_V \sim 2 \times 10^{10} \, {\rm L_{\odot, \it V}}$; although it is difficult to determine precisely the luminosity of a galaxy from a viewpoint within it, we will take $L_{\rm MW, \it V} \equiv 2 \times 10^{10} \, {\rm L_{\odot, \it V}}$ as a nominal Milky Way luminosity. The luminosity of M31 is $L_{\rm M31, \it V} \approx 1.3 L_{\rm MW, \it V}$. The total luminosity of the Local Group is $L_{\rm LG, \it V} \approx 2.7 L_{\rm MW, \it V}$, with the Milky Way and M31 together providing $\sim 85\%$ of the luminosity. Although the Local Group is known to contain more than 50 galaxies, the majority are dwarf galaxies with $L_V < 10^{-4} L_{\rm MW, \it V}$.

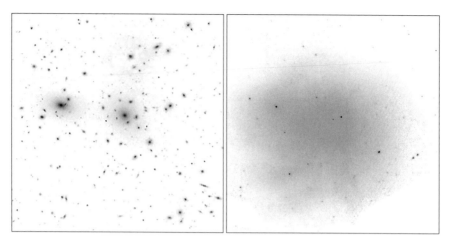

**Figure 8.5** Left: Negative image of the Coma Cluster ($d \approx 100$ Mpc) at visible wavelengths. The image size is $22 \times 22$ arcmin, corresponding to $\sim 0.5 \times 0.5$ Mpc at the distance of the Coma Cluster. [SDSS] Right: Negative image of the Coma Cluster at X-ray wavelengths, imaged with *Chandra*. The angular scale and orientation (north up) are the same in both images. [NASA/CXC/MPE/J. Sanders et al.]

The nearest extremely rich cluster of galaxies is the Coma Cluster, whose center is at a distance $d \approx 100$ Mpc from the Local Group. At this distance, the angular half-light radius of the cluster, $\theta_h \approx 52$ arcmin, corresponds to a physical radius $r_h \approx 1.5$ Mpc. At visible wavelengths, as seen in the left panel of Figure 8.5, the cluster appears like a swarm of individual, well-defined galaxies. About 1000 galaxies in the Coma Cluster have spectroscopic redshifts; when you add in the dwarf galaxies there are more than $10^4$ galaxies in the Coma Cluster. The two brightest galaxies in the Coma Cluster are a pair of elliptical galaxies, NGC 4889 and NGC 4874. These two galaxies have $V$-band luminosity $L_{4889, V} \approx 11 L_{MW, V}$ and $L_{4874, V} \approx 9 L_{MW, V}$ respectively. Although these two galaxies are extraordinarily luminous, as galaxies go, together they provide only $\sim 8\%$ of the total stellar luminosity of the Coma Cluster, $L_{CC, V} \approx 250 L_{MW, V}$.

Thus, the structure of a rich cluster like the Coma Cluster is different from that of a poor cluster like the Local Group. The Local Group is dominated by a pair of galaxies, the Milky Way and M31; each of these galaxies has its own dark halo and baryonic circumgalactic medium and possesses its own entourage of dwarf galaxies. However, the bright elliptical galaxies NGC 4889 and NGC 4874 are far from dominating the Coma Cluster. It is useful to think of a rich cluster as consisting of a single cluster-sized dark halo permeated with a hot, low-density **intracluster medium (ICM)**. Individual galaxies, even luminous ones such as NGC 4889 and NGC 4874, are small perturbations to the global structure of the cluster.

In the central regions of the Coma Cluster, the line-of-sight velocity dispersion is

$$\sigma_v \approx 880 \, \text{km s}^{-1}. \tag{8.12}$$

The characteristic size of the cluster can be taken as the half-light radius $r_h \approx 1.5 \, \text{Mpc}$. If we boldly use the virial theorem in the form of Equation 7.45, we can make a mass estimate for the Coma Cluster of

$$M_{\text{vir}} = 5 \frac{\sigma_v^2 r_h}{G} \approx 2.7 \times 10^{48} \, \text{g} \approx 1.4 \times 10^{15} \, M_\odot. \tag{8.13}$$

This estimate is not exact (in deriving Equation 7.45, we assumed uniform density, for instance), but a mass in the range $M_{vir} = 1 \rightarrow 2 \times 10^{15} \, M_\odot$ is a reasonable result for an application of the virial theorem to a rich cluster such as Coma.

The total $V$-band luminosity of all the stars in all the galaxies in the Coma Cluster is

$$L_V \approx 250 L_{\text{MW}, V} \approx 5 \times 10^{12} \, L_{\odot, V}. \tag{8.14}$$

Most of the visible light from the Coma Cluster comes from elliptical galaxies, whose stellar populations have a mass-to-light ratio $\Upsilon_\star \approx 8 \, M_\odot / L_{\odot, V}$ in the $V$ band. This yields a total stellar mass for the Coma Cluster of

$$M_\star = \Upsilon_\star L_V \approx (8 \, M_\odot / L_{\odot, V})(5 \times 10^{12} \, L_{\odot, V}) \approx 4 \times 10^{13} \, M_\odot. \tag{8.15}$$

Thus, the total mass $M_{\text{vir}}$ of the Coma Cluster, estimated from the virial theorem, is 25 to 50 times the total mass of the stars that it contains.

It was the discrepancy between the stellar mass and the virial mass of the Coma Cluster that led Fritz Zwicky, in 1933, to deduce the existence of "dunkle Materie," or "dark matter" in English translation. Although only eight redshifts of galaxies in the Coma Cluster were known at the time, it was enough for Zwicky to realize that the velocity dispersion was $\sigma_v \sim 1000 \, \text{km s}^{-1}$; this is far too large for Coma to remain bound if stars were the only matter present.

Although stars put on a pretty show, they provide a minority of the baryons in a rich cluster like the Coma Cluster. Seen in X-ray emission at $h\nu = 0.3 \rightarrow 2 \, \text{keV}$, as shown in the right panel of Figure 8.5, the Coma Cluster is revealed as containing a hot diffuse intracluster medium. In the higher energy range $h\nu = 3 \rightarrow 30 \, \text{keV}$, shown in Figure 8.6, the central regions of the Coma Cluster have a spectrum that displays a free–free (bremsstrahlung) continuum plus a broad emission line at $h\nu \sim 6.7 \, \text{keV}$. The emission line is the Fe K$\alpha$ line, associated with the electronic transition to the ground level of Fe from the first excited level. The histogram in Figure 8.6 is the best-fitting model assuming a uniform temperature; the best fit to these data yields $kT = 8.58 \, \text{keV}$, or $T \approx 1.0 \times 10^8 \, \text{K}$. The presence of the Fe K$\alpha$ line is an obvious indicator that metals are present in the intracluster medium of the Coma Cluster; more detailed studies, using

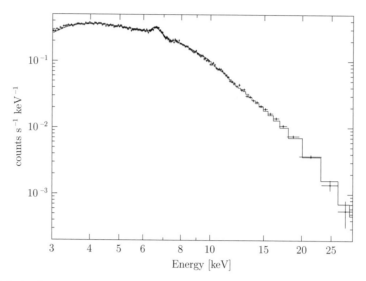

**Figure 8.6** Background-subtracted *NuSTAR* spectrum of the central $12 \times 12$ arcmin region of the Coma Cluster, corresponding to $\sim 0.35 \times 0.35$ Mpc. The best-fitting one-temperature model ($kT = 8.58$ keV) is shown as a histogram. [Data from Gastaldello et al. 2015]

emission lines of multiple elements, indicate an average metallicity $Z \approx Z_\odot/3$ for the intracluster gas.

For fully ionized hydrogen, the free–free emissivity is

$$j_{\nu,\text{ff}} = C_{\text{ff}} \left( \frac{m_e c^2}{kT} \right)^{1/2} n_e n_i \exp\left( \frac{-h\nu}{kT} \right) g_{\nu,\text{ff}}, \tag{8.16}$$

where the normalization is

$$C_{\text{ff}} \equiv \frac{8}{3} \left( \frac{2\pi}{3} \right)^{1/2} \frac{e^6}{m_e^2 c^4} = 7.070 \times 10^{-44} \text{ erg cm}^3 \text{ s}^{-1} \text{ Hz}^{-1} \text{ sr}^{-1}, \tag{8.17}$$

and $g_{\nu,\text{ff}}$ is the Gaunt factor. Aside from the exponential Boltzmann cutoff in Equation 8.16, the frequency $\nu$ enters only through the mild frequency dependence of the Gaunt factor. At temperatures $kT \gg 13.6$ eV and frequencies $h\nu < kT$, the Gaunt factor depends logarithmically on frequency, with

$$g_{\nu,\text{ff}} \approx \frac{\sqrt{3}}{\pi} \ln\left( \frac{4}{e^\gamma} \frac{kT}{h\nu} \right) \approx 0.551 \ln\left( 2.25 \frac{kT}{h\nu} \right), \tag{8.18}$$

where $\gamma \approx 0.5772$ is the Euler constant.

The power per unit volume produced by free–free emission is

$$\ell_{\text{ff}} = 4\pi \int j_{\text{ff}} d\nu = \Lambda_{\text{ff},0} \left( \frac{kT}{m_e c^2} \right)^{1/2} n_e n_i g_{\text{ff}}, \tag{8.19}$$

where the normalization for the free–free power is (compare with Equation 4.53)

$$\Lambda_{\mathrm{ff},0} = 4\pi \frac{m_e c^2}{h} C_{\mathrm{ff}} = 1.10 \times 10^{-22} \,\mathrm{erg\,cm^3\,s^{-1}}. \tag{8.20}$$

The frequency-averaged Gaunt factor $g_{\mathrm{ff}}$, for temperatures $T \sim 10^8$ K, is $g_{\mathrm{ff}} \approx 1.15$. Thus, normalized to the conditions expected in the intracluster medium, the power emitted per unit volume by free–free emission is

$$\ell_{\mathrm{ff}} \approx 1.77 \times 10^{-29} \,\mathrm{erg\,cm^{-3}\,s^{-1}} \left(\frac{kT}{10\,\mathrm{keV}}\right)^{1/2} \left(\frac{n_{\mathrm{H}}}{10^{-3}\,\mathrm{cm^{-3}}}\right)^2, \tag{8.21}$$

assuming fully ionized hydrogen gas. The mass density of the ICM is

$$\rho_{\mathrm{icm}} = m_{\mathrm{H}} n_{\mathrm{H}} \approx 1.67 \times 10^{-27} \,\mathrm{g\,cm^{-3}} \left(\frac{n_{\mathrm{H}}}{10^{-3}\,\mathrm{cm^{-3}}}\right), \tag{8.22}$$

resulting in a mass-to-light ratio for the ICM (counting only free–free emission) of

$$\Upsilon_{\mathrm{icm}} = \frac{\rho_{\mathrm{icm}}}{\ell_{\mathrm{ff}}} \approx 180 \,M_\odot/L_\odot \left(\frac{kT}{10\,\mathrm{keV}}\right)^{-1/2} \left(\frac{n_{\mathrm{H}}}{10^{-3}\,\mathrm{cm^{-3}}}\right)^{-1}, \tag{8.23}$$

where the luminosity is expressed in terms of the Sun's bolometric luminosity, $1\,L_\odot = 3.828 \times 10^{33} \,\mathrm{erg\,s^{-1}}$.

Because free–free emission is not very effective at draining energy from fast electrons in low-density gas, the cooling times for clusters are long. The thermal energy density of the intracluster medium is

$$\varepsilon_{\mathrm{icm}} = \frac{3}{2}(2n_{\mathrm{H}})kT \approx 4.8 \times 10^{-11} \,\mathrm{erg\,cm^{-3}} \left(\frac{n_{\mathrm{H}}}{10^{-3}\,\mathrm{cm^{-3}}}\right) \left(\frac{kT}{10\,\mathrm{keV}}\right). \tag{8.24}$$

The cooling time for the intracluster medium is then, combining Equations 8.21 and 8.24,

$$t_{\mathrm{cool}} = \varepsilon_{\mathrm{icm}}/\ell_{\mathrm{ff}} \approx 86\,\mathrm{Gyr} \left(\frac{kT}{10\,\mathrm{keV}}\right)^{1/2} \left(\frac{n_{\mathrm{H}}}{10^{-3}\,\mathrm{cm^{-3}}}\right)^{-1}. \tag{8.25}$$

The gas density profile of clusters is often fitted with a beta profile:

$$n_{\mathrm{H}} = n_0 \left[1 + (r/r_c)^2\right]^{-3\beta/2}. \tag{8.26}$$

For the Coma Cluster, the best-fitting beta profile has $\beta \approx 0.75$, $r_c \approx 0.3$ Mpc, and $n_0 \approx 3 \times 10^{-3} \,\mathrm{cm^{-3}}$. This implies that even in the dense central regions of the Coma Cluster, the cooling time will be longer than the Hubble time, $H_0^{-1} \sim 14$ Gyr.

For gas with a beta profile, the amount of gas within a radius $r$ is

$$M_{\mathrm{gas}}(r) \approx \frac{4\pi}{3(1-\beta)} m_{\mathrm{H}} n_0 r_c^3 \left(\frac{r}{r_c}\right)^{3(1-\beta)} \tag{8.27}$$

when $r \gg r_c$ and $\beta < 1$. For the Coma Cluster, this comes to

$$M_{\text{gas}}(r) \approx 3.4 \times 10^{13}\, M_{\odot} \left( \frac{r}{0.3\,\text{Mpc}} \right)^{0.75}. \tag{8.28}$$

To make a rough estimate of the total amount of intracluster gas in the Coma Cluster, we can assume a cutoff at the virial radius ($r_{200} \approx 3\,\text{Mpc}$). This gives a mass estimate $M_{\text{gas}} \approx 2 \times 10^{14}\, M_{\odot}$, about five times the estimated stellar mass, $M_{\star} \approx 4 \times 10^{13}\, M_{\odot}$.

Notice that when $\beta > 0.5$, the total free–free luminosity remains finite as $r \to \infty$. For $\beta \approx 0.75$, the free–free luminosity of intracluster gas is

$$L_{\text{ff}} \approx 1.9 \times 10^{11}\, L_{\odot} \left( \frac{r_c}{0.3\,\text{Mpc}} \right)^3 \left( \frac{kT}{8.6\,\text{keV}} \right)^{1/2} \left( \frac{n_0}{3 \times 10^{-3}\,\text{cm}^{-3}} \right)^2, \tag{8.29}$$

scaling to the properties of the Coma Cluster.

Seen in X-rays (as in the right panel of Figure 8.5), the central regions of the Coma Cluster look smooth; thus, we expect that Coma, at least in its central regions, is a *relaxed* cluster, close to hydrostatic equilibrium. Spherical objects in hydrostatic equilibrium, whether they are tiny stars or large clusters, obey the equation

$$\frac{dP}{dr} = -\frac{GM(r)\rho_{\text{gas}}(r)}{r^2}, \tag{8.30}$$

where $M(r)$ is the mass of everything inside a radius $r$, including gas, stars, and dark matter, while $\rho_{\text{gas}}(r)$ is the density of the gas alone, excluding the stars and dark matter. If the gas in a cluster is ionized hydrogen, the ideal gas law tells us that

$$P = nkT = \frac{2\rho_{\text{gas}}(r)kT_{\text{gas}}(r)}{m_{\text{H}}}, \tag{8.31}$$

where $\rho_{\text{gas}}$ is the mass density of the gas and $T_{\text{gas}}$ is its temperature. Combining the equation of hydrostatic equilibrium with the ideal gas law, we find a relation that tells us the total mass $M$ contained within a radius $r$:

$$M(r) = -\frac{r^2}{G\rho_{\text{gas}}} \frac{2k}{m_{\text{H}}} \frac{d}{dr}(\rho_{\text{gas}}T_{\text{gas}}) \tag{8.32}$$

$$= \frac{2rkT_{\text{gas}}}{Gm_{\text{H}}} \left( -\frac{d\ln\rho_{\text{gas}}}{d\ln r} - \frac{d\ln T_{\text{gas}}}{d\ln r} \right). \tag{8.33}$$

In reality, the temperature of the intracluster gas is not uniform, nor is the cluster perfectly spherical. If we want an order-of-magnitude estimate, however, we can pretend that the intracluster gas is isothermal, with $T_{\text{gas}} = $ constant, and assume a beta profile (Equation 8.26) for the gas density. This yields a mass estimate

$$M(r) = \frac{6\beta kT_{\text{gas}}}{Gm_{\text{H}}}r \approx \frac{4.5kT_{\text{gas}}}{Gm_{\text{H}}}r, \tag{8.34}$$

well outside the central core of a cluster with $\beta \approx 0.75$.

If a cluster is in hydrostatic equilibrium, and has a cutoff at the virial radius $r_{200}$, then its complete mass is

$$M_{\text{hyd}} \approx \frac{4.5 k T_{\text{gas}}}{G m_{\text{H}}} r_{200} \approx 2.6 \times 10^{15} \, M_\odot \left( \frac{k T_{\text{gas}}}{8.6 \, \text{keV}} \right) \left( \frac{r_{200}}{3 \, \text{Mpc}} \right), \qquad (8.35)$$

scaled to the properties of the Coma Cluster. Given Coma's virial mass estimate $M_{\text{vir}} \approx 1.4 \times 10^{15} \, M_\odot$, we can state that the total mass of the Coma Cluster is $M_{\text{tot}} \sim 2 \times 10^{15} \, M_\odot$. The mass of the stars is then $M_\star \sim 0.02 M_{\text{tot}}$ and the mass of the intracluster gas is $M_{\text{gas}} \sim 0.1 M_{\text{tot}}$.

For the Coma Cluster, with its long cooling time, the assumption of hydrostatic equilibrium is useful. However, roughly half of all clusters have a cooling time in their central regions that is shorter than the Hubble time. In these clusters the possibility exists of a **cooling flow**, as gas loses its pressure support and falls toward the cluster center. The cluster Abell 478, for example, is a rich cluster with a total mass $M_{\text{tot}} \approx 1.3 \times 10^{15} \, M_\odot$, not much smaller than the mass of the Coma Cluster. The central gas density of Abell 478, however, is $n_{\text{H}} \sim 0.1 \, \text{cm}^{-3}$, much higher than that of the Coma Cluster, while its central gas temperature is a relatively modest $kT \sim 3 \, \text{keV}$. This means that the cooling time at the center of Abell 478, from Equation 8.25, is only $t_{\text{cool}} \approx 0.5 \, \text{Gyr}$. The cooling time doesn't drop to $t_{\text{cool}} = H_0^{-1}$ until you reach a distance $r_{\text{cool}} \approx 0.2 \, \text{Mpc}$ from the cluster's center. Within this cooling radius $r_{\text{cool}}$, the mass of coolable intracluster gas is estimated to be $M_{\text{cool}} \approx 7 \times 10^{12} \, M_\odot$. These numbers imply a quasistatic inflow of cooling gas (on a time scale longer than the freefall time), with a cooling flow rate

$$\dot{M}_{\text{cool}} \sim \frac{M_{\text{cool}}}{H_0^{-1}} \sim \frac{7 \times 10^{12} \, M_\odot}{1.4 \times 10^{10} \, \text{yr}} \sim 500 \, M_\odot \, \text{yr}^{-1}. \qquad (8.36)$$

This rate is on the high end for nearby clusters with cooling flows; more typically, deduced cooling flow rates lie in the range $\dot{M}_{\text{cool}} = 30 \to 300 \, M_\odot \, \text{yr}^{-1}$.

Astronomers studying clusters with dense cool cores talk about the "cooling flow problem," which is simply the problem of what happens to the intracluster gas that cools and sinks toward the cluster center. The bright central galaxies in cooling flow clusters do not have huge reservoirs of cool gas in their ISM and CGM; nor have they been manufacturing stars out of the infalling gas. Typically the bright central galaxy in a cooling flow cluster has a star formation rate that is only $\sim 0.01 \dot{M}_{\text{cool}}$. The obvious way to prevent a buildup of cool gas at the cluster center is to reheat the gas that is falling inward. From Equation 8.29, the free–free luminosity of a rich cluster is $L_{\text{ff}} \sim 2 \times 10^{11} \, L_\odot \sim 10^{45} \, \text{erg s}^{-1}$. One possible heating source to balance the free–free cooling is supernova explosions. However, injecting energy at a rate of $10^{45} \, \text{erg s}^{-1}$ would require exploding a $\sim 10^{51} \, \text{erg}$ supernova once every 10 days; even a bright starburst galaxy like M82, with a supernova every 10 years on average, falls far short of the needed rate. A more promising heating source is an active galactic nucleus (AGN) in one or more of the cluster's bright galaxies. The black hole at the heart of the AGN accretes gas at a rate $\dot{M}_{acc}$ and converts the mass–energy of the accreted gas into photons, with an efficiency $\eta_r \sim 0.1$. This yields an AGN luminosity

$$L_{agn} = \eta_r \dot{M}_{acc}c^2 = 1.1 \times 10^{45} \text{ erg s}^{-1} \left(\frac{\eta_r}{0.1}\right)\left(\frac{\dot{M}_{acc}}{0.2 \text{ M}_\odot \text{ yr}^{-1}}\right). \qquad (8.37)$$

This luminosity, though, is not useful for reheating the ICM, since intracluster gas is highly transparent; most of the AGN luminosity escapes the cluster. However, the AGN within a brightest cluster galaxy frequently produces a pair of gas jets, flowing away from the central black hole at highly relativistic speeds. If matter is injected into a jet at a rate $\dot{M}_{jet}$ with a Lorentz factor $\Gamma$, the mechanical power of the jet is

$$Q_{jet} \approx \Gamma \dot{M}_{jet}c^2 \approx 1.1 \times 10^{45} \text{ erg s}^{-1} \left(\frac{\Gamma}{20}\right)\left(\frac{\dot{M}_{jet}}{10^{-3} \text{ M}_\odot \text{ yr}^{-1}}\right). \qquad (8.38)$$

The relativistic, supersonic jets send shock waves through the intracluster medium, reheating the gas that has been cooling by free–free emission. At high enough resolution, X-ray observations of the centers of galaxy clusters reveal the presence of heated gas side-by-side with cooling inflows, as well as structures suggestive of shock fronts, cavities, filaments, and turbulence. Just as with the ISM and CGM, when the intracluster medium is viewed carefully, it reveals a rich variety of gaseous components.

## Exercises

8.1  Suppose that the circumgalactic medium of our galaxy can be approximated as a spherical distribution, centered on the galactic center. The density is assumed to be a power law, with

$$n_H = n_0(r/r_0)^{-\alpha} \qquad (8.39)$$

when $r < r_0$, and $n_H = 0$ when $r \geq r_0$. We are located in the midplane of our galaxy's disk, at a distance $R = 8.2 \text{ kpc}$ from the galactic center.

(a)  Suppose that you look through our galaxy's CGM in a direction perpendicular to the galaxy's disk. What will be the column density $N_H$ of the CGM as a function of $n_0$, $r_0$, $\alpha$, and $R$? [You may assume $r_0 > R$ in your calculations.]

(b)  What is the total mass $M_{cgm}$ of our galaxy's CGM as a function of $n_0$, $r_0$, and $\alpha$? [State your assumptions about the metallicity of the CGM.]

(c)  If you observe a particular value of $N_H$ in the direction perpendicular to the disk, what combination of $\alpha$ and $r_0$ minimizes $M_{cgm}$ for the given $N_H$? What combination of $\alpha$ and $r_0$ maximizes $M_{cgm}$ for the given $N_H$?

8.2  At the center of a rich cluster of galaxies, the intracluster medium has density $n_H \sim 10^{-3} \text{ cm}^{-3}$ and temperature $T_{gas} \sim 10^8 \text{ K}$. Assume the intracluster medium consists of pure ionized hydrogen.

(a) At $T_{gas} \sim 10^8$ K, the mean kinetic energy per free electron is much greater than the mean photon energy $h\nu \sim 0.6$ meV of the cosmic microwave background (CMB). In this case, scattering events between free electrons and CMB photons are an example of "inverse Compton scattering," in which energy is transferred from the electron to the photon. The scattering cross section is $\sigma_e = 6.6525 \times 10^{-25}$ cm$^2$ and the mean energy transfer per scattering is

$$\Delta E = h\nu \frac{4kT_{gas}}{m_e c^2}. \tag{8.40}$$

At the center of the cluster, what is the cooling time $t_{icc}$ from inverse Compton cooling, written as a function of $T_{gas}$ and $n_H$? What is the inverse Compton cooling time expressed as a fraction of the free–free cooling time?

(b) As the universe expands, the mean photon energy and number density of cosmic background photons have the dependences $h\nu \propto 1 + z$ and $n_\gamma \propto (1+z)^3$, where $z$ is the cosmological redshift. If we assume that $n_H$ and $T_{gas}$ for the intracluster medium have remained constant with time, at what redshift $z$ was the inverse Compton cooling time equal to the free–free cooling time?

# 9

# Diffuse Intergalactic Medium

*First there was nothing.*
*Then there was something.*
*Then there was hydrogen.*
*Then there was dirty hydrogen.*

Anonymous

An epigrammatic cosmologist (whose name is lost to history) once summed up the history of the universe with the words quoted above. Some of the hydrogen and the "dirt" has accumulated in galaxies, but much remains in intergalactic space.

Observations of temperature fluctuations in the cosmic microwave background, by satellites such as *Planck*, yield a good estimate of the density of baryonic matter in the universe. The *Planck* 2018 results for a standard $\Lambda$CDM universe imply that the current mass density of baryonic matter ("dirty hydrogen") is

$$\overline{\rho}_{bary,0} = (4.21 \pm 0.03) \times 10^{-31} \, \text{g cm}^{-3}. \tag{9.1}$$

This corresponds to a mean number density of baryons (mostly protons)

$$\overline{n}_{bary,0} = 2.52 \times 10^{-7} \, \text{cm}^{-3}. \tag{9.2}$$

The amount of baryonic matter in gravitationally bound systems, including stars, ISM, CGM, and ICM, provides only $\sim 15\%$ of this mean cosmic baryon density. The remainder is provided by a tenuous **intergalactic medium (IGM)**, which today is highly ionized. The warm-hot intergalactic medium is largely collisionally ionized, while the cooler diffuse intergalactic medium (DIM) is primarily photoionized.

## 9.1 Gunn–Peterson Effect

In studying the intergalactic medium, we must finally acknowledge that the universe expands. As a consequence, the mean baryon density decreases with time, since baryon number is conserved.[1] The expansion of the universe is homogeneous and isotropic on large scales, and thus can be described by a simple scale factor, $a(t)$. The scale factor is customarily normalized so that $a(t_0) = 1$ at the present time, $t_0 \sim 14$ Gyr after the Big Bang. The expansion of the universe is frequently described in terms of the Hubble parameter, $H(t) \equiv \dot{a}/a$. The value of the Hubble parameter evaluated at the present time is the Hubble constant $H_0$. If we take $H_0 = 70$ km s$^{-1}$ Mpc$^{-1}$ $h_{70}$, we can write the Hubble constant in less baroque units as $H_0 = 2.27 \times 10^{-18}$ s$^{-1}$ $h_{70}$. This yields a Hubble time $H_0^{-1} = 13.97$ Gyr $h_{70}^{-1}$. Converting from a time to a distance, using the speed of light in a vacuum, the Hubble distance is $c/H_0 = 4280$ Mpc $h_{70}^{-1}$. (In our discussion of the intergalactic medium, we will set $h_{70} = 1$. For those who prefer $H_0 = 67$ km s$^{-1}$ Mpc$^{-1}$ or $H_0 = 73$ km s$^{-1}$ Mpc$^{-1}$, the necessary $\sim 4\%$ adjustment to the Hubble scale is left as an exercise for the reader.)

The intergalactic medium is difficult to detect. One way of searching for intergalactic gas is to look for absorption lines from the IGM along the line of sight to a distant quasar. The Lyman $\alpha$ line at $\lambda = 1216$ Å is a useful probe of intergalactic gas, as long as the gas isn't too highly ionized. Light emitted by a quasar at time $t_e$ is observed by us at $t_0 > t_e$. The scale factor today, $a(t_0) \equiv 1$, is greater than the scale factor at the time the light was emitted, $a(t_e) < 1$. The scale factor $a(t_e)$ can be found from the redshift $z_e$ of the quasar's emission lines: $a(t_e) = (1 + z_e)^{-1}$. Between the time of emission and the time of observation, the number density of baryons has been decreasing at the rate $\bar{n}_{bary}(t) = \bar{n}_{bary,0}\, a(t)^{-3} = \bar{n}_{bary,0}\,(1 + z)^3$, and the energy of the emitted photons has been decreasing at the rate $h\nu(t) = h\nu_0 a(t)^{-1} = h\nu_0(1 + z)$.

Consider the nearby quasar 3C 273, at a redshift $z_e = 0.158$; its spectrum is shown in the upper panel of Figure 9.1. If 3C 273 produces a Lyman $\alpha$ photon with an energy $h\nu_e = 10.20$ eV, by the time it reaches our telescopes, it is redshifted to the lower energy $h\nu_0 = 10.20$ eV$/1.158 = 8.81$ eV. If the continuum of 3C 273 contained a photon with an initial energy $h\nu_e = 10.20$ eV $\times 1.158 = 11.81$ eV, by the time it reaches our telescope, it is redshifted to the lower energy $h\nu_0 = h\nu_e/1.158 = 10.20$ eV and is ripe for absorption by neutral hydrogen in our own galaxy. In general, if a quasar is at a redshift $z_e$, photons with initial energy in the range $h\nu_{u\ell} < h\nu_e < (1 + z_e)h\nu_{u\ell}$ can be absorbed by a transition with energy $h\nu_{u\ell}$ (equal to 10.20 eV in the case of Lyman $\alpha$) somewhere along the line of sight from the quasar to us. A photon with energy $h\nu_e$ in this range will be absorbed at a time $t'$ with $t_e < t' < t_0$ given implicitly by the relation $a(t') = \nu_{u\ell}/\nu_e = 10.20$ eV$/h\nu_e$.

---

[1] It is possible that protons decay into non-baryonic particles, but the lower limit on the proton lifetime is $\sim 10^{24}$ times the current age of the universe.

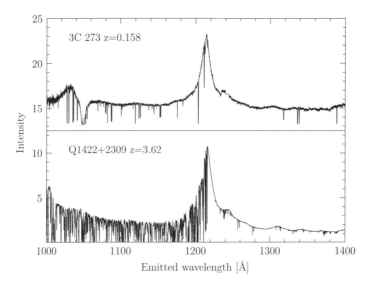

**Figure 9.1** Spectra of the quasars 3C 273 (top) at $z_e = 0.158$, and Q1422+2309 (bottom) at $z_e = 3.62$. Wavelengths are given in the emission rest frame. The Lyman alpha forest is seen to the left of the quasar's Lyman $\alpha$ emission line. [Data provided by W. Keel]

When discussing Lyman $\alpha$ absorption in our own galaxy, we wrote the optical depth as (Equation 2.43)

$$\tau_\nu = \int_0^s \kappa_\nu ds = \int_0^s n_\ell \sigma_{\ell u}(\nu) ds, \qquad (9.3)$$

where $n_\ell$ is the number density of hydrogen atoms in the ground state, and $\sigma_{\ell u}$ is the cross section for absorption of a Lyman $\alpha$ photon. In terms of the light travel time $dt = ds/c$, we can rewrite Equation 9.3 as

$$\tau_\nu = c \int_{t_e}^{t_0} n_\ell \sigma_{\ell u}(\nu) dt. \qquad (9.4)$$

As a function of frequency, the absorption cross section is

$$\sigma_{\ell u}(\nu) = \sigma_\alpha \nu_{u\ell} \Phi_\nu \qquad (9.5)$$

where

$$\sigma_\alpha \equiv \frac{g_u}{g_\ell} \frac{c^2}{8\pi \nu_{u\ell}^3} A_{u\ell} = 4.48 \times 10^{-18} \text{ cm}^2, \qquad (9.6)$$

and $\Phi_\nu$ is the Voigt profile of the Lyman $\alpha$ line, normalized as usual so that $\int \Phi_\nu d\nu = 1$. Thus, we can write

$$\tau_\nu = c\sigma_\alpha \nu_{u\ell} \int_{t_e}^{t_0} n_\ell \Phi_\nu dt, \qquad (9.7)$$

which is a useful form when $n_\ell$ and $\Phi_\nu$ can be time-variable.

In an expanding universe, we can change our variable from $t$ to $a(t)$, using the relation $dt = da/(Ha)$. The optical depth for Lyman $\alpha$ absorption at a frequency $\nu$ can then be written as

$$\tau_\nu = c\sigma_\alpha \nu_{u\ell} \int_{a(t_e)}^{1} n_\ell(a)\Phi_\nu \frac{da}{H(a)a}. \qquad (9.8)$$

If we want to use the redshift $z$ instead of the scale factor $a$ as our variable, we can use the relation $1 + z = 1/a$ to write

$$\tau_\nu = c\sigma_\alpha \nu_{u\ell} \int_{0}^{z_e} n_\ell(z)\Phi_\nu \frac{dz}{H(z)(1+z)}. \qquad (9.9)$$

This equation gives a mapping between the observed optical depth $\tau_\nu$ at a redshift $z = \nu_{u\ell}/\nu - 1$ and the underlying number density $n_\ell$ of atomic hydrogen in its ground state at a cosmological redshift $z$.

To make use of Equation 9.9, we need an appropriate functional form for the Voigt profile $\Phi_\nu$. The Lorentz function for Lyman $\alpha$ has an intrinsic width $\Delta v \sim 0.01 \text{ km s}^{-1}$. This is much smaller than the thermal broadening $b \sim 10 \text{ km s}^{-1}$ expected for atomic hydrogen at $T \sim 10^4$ K. However, even for a nearby quasar like 3C 273, at a relatively tiny redshift $z_e = 0.158$, the thermal broadening is much smaller than the "Hubble broadening" $cz_e = 47\,000 \text{ km s}^{-1}$ along the line of sight. From a physical viewpoint, this means that when we detect intergalactic absorption at a redshift $z$, we can assume that the observed redshift is due entirely to the cosmological expansion and that intrinsic and thermal broadening can be ignored. From a mathematical viewpoint, this means we can treat the Voigt profile $\Phi_\nu$ in Equation 9.9 as being very strongly peaked at $\nu = \nu_{u\ell}/(1+z)$. This assumption yields an optical depth

$$\tau_\nu = c\sigma_\alpha \frac{n_\ell(z)}{H(z)}, \qquad (9.10)$$

where $z = \nu_{u\ell}/\nu - 1$. At redshifts $z \ll 1$, Equation 9.10 implies an optical depth for Lyman $\alpha$ absorption of

$$\tau_\nu = \frac{c}{H_0}\sigma_\alpha n_{\ell,0} \approx 1.5 \times 10^4 \left( \frac{n_{\ell,0}}{n_{\text{bary},0}} \right). \qquad (9.11)$$

Thus, if the number density of neutral hydrogen atoms in their ground state is even a few parts in $10^4$ of the baryon density, the spectrum of a low-redshift quasar such as 3C 273 should be black in the observed wavelength range $\lambda_0 = 1216 \rightarrow 1216(1 + z_e)$ Å corresponding to the emitted wavelength range $\lambda_e = 1216/(1 + z_e) \rightarrow 1216$ Å. As Figure 9.1 shows, this is not the case for 3C 273.

For a quasar at redshift $z$, strong Lyman $\alpha$ absorption in the observed wavelength range $\lambda_0 = 1216 \rightarrow 1216(1 + z)$Å is known as the **Gunn–Peterson effect**. In 1965, Jim Gunn and Bruce Peterson used the absence of strong absorption in this wavelength range to show that *either* intergalactic gas has a density very much lower than the mean baryon density of the universe (with the surplus baryons somehow stuffed into galaxies) *or* intergalactic gas is very highly ionized. The absence of a **Gunn–Peterson trough** at redshifts $z < 5$ is now regarded as evidence that the IGM at low redshifts is highly ionized, not that it is nearly absent. Suppose that a fraction $f_H \approx 0.9$ of all baryons are hydrogen nuclei. We can assume that $f_H$ is constant after Big Bang Nucleosynthesis is complete, thanks to the inefficiency of stars at nucleosynthesis. The neutral fraction $f_n(z)$ is the fraction of hydrogen that consists of neutral atoms in their ground state; this can be a function of redshift. Since

$$\bar{n}_{\text{bary}} = \bar{n}_{\text{bary},0}/a^3 = \bar{n}_{\text{bary},0}(1+z)^3, \tag{9.12}$$

we can write

$$n_\ell(z) = \bar{n}_{\text{bary},0} f_H f_n(z)(1+z)^3 \approx 2.27 \times 10^{-7}\,\text{cm}^{-3} f_n(z)(1+z)^3. \tag{9.13}$$

We can then write the optical depth at an arbitrary redshift, using Equation 9.10, as

$$\tau_\nu = c\sigma_\alpha \bar{n}_{\text{bary},0} f_H \frac{f_n(z)(1+z)^3}{H(z)}. \tag{9.14}$$

In a flat $\Lambda$CDM universe, the Hubble parameter is

$$H(z) = H_0[\Omega_{m,0}(1+z)^3 + \Omega_{\Lambda,0}]^{1/2}, \tag{9.15}$$

where $\Omega_{m,0} \approx 0.31$ is the contribution of matter (both dark and baryonic) to the mass–energy density and $\Omega_{\Lambda,0} \approx 0.69$ is the contribution of dark energy, assumed to take the form of the cosmological constant $\Lambda$. It's awkward, mathematically speaking, that we live in an epoch when matter and dark energy contribute comparably to the energy density. However, at redshifts $z \gg (\Omega_{\Lambda,0}/\Omega_{m,0})^{1/3} - 1 \approx 0.3$, we can use the approximation that the universe is matter dominated, with

$$\tau_\nu \approx \frac{c}{H_0}\sigma_\alpha \bar{n}_{\text{bary},0} f_H \frac{f_n(z)(1+z)^{3/2}}{\Omega_{m,0}^{1/2}} \tag{9.16}$$

$$\approx 2.4 \times 10^4 f_n(z)(1+z)^{3/2}. \tag{9.17}$$

If we want to have a hope of seeing a Gunn–Peterson trough, we must go to high redshifts. Figure 9.2, for instance, shows the spectra of two high-redshift quasars at $z_e \sim 6$. Notice that the higher-redshift quasar, in particular, has a very pronounced Gunn–Peterson trough; absorption is nearly total at wavelengths shorter than Lyman $\alpha$ ($\lambda_e = 1216\,\text{Å}$, $\lambda_0 = 8530\,\text{Å}$). There are two reasons for the appearance of the Gunn–Peterson effect at high redshifts. First, the factor of $(1+z)^{3/2}$ in Equation 9.17 is $\sim 20$ times bigger at $z \sim 6$ than at $z \ll 1$. Second, we expect the neutral fraction $f_n$ of hydrogen to vary significantly during the history of the universe. To see why this is so, we need to study the saga of **recombination** and its sequel, **reionization**.

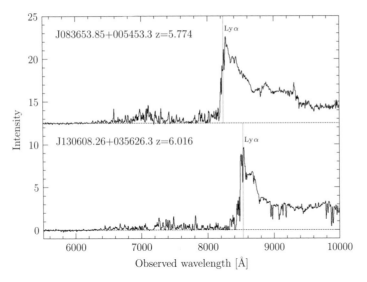

**Figure 9.2** Spectra of two quasars at $z = 5.774$ (top) and $z = 6.016$ (bottom). Wavelengths are given in the observer's rest frame. [Following Fan et al. 2001]

## 9.2 Recombination

To continue the epigrammatic history of the universe:

*First the dirty hydrogen was ionized.*
*Then it was neutral.*
*Then it was ionized again.*

The hydrogen stopped being ionized during the epoch of recombination, at $z \approx 1400$; it became ionized again during the epoch of reionization, at $z \sim 7$. The story of recombination and then of reionization is the familiar tale of the balance between photoionization and radiative recombination.

The temperature of the cosmic background radiation drops as the universe expands. Currently, the temperature is $T_0 = 2.7255 \pm 0.0006$, or $kT_0 = (2.3486 \pm 0.0005) \times 10^{-4}$ eV. At earlier times, the temperature was

$$T = \frac{T_0}{a} = T_0(1 + z).  \tag{9.18}$$

When the temperature dropped below $kT \sim 66$ keV, corresponding to an expansion factor $a \sim 3.6 \times 10^{-9}$, deuterium nuclei could form without swiftly being photodissociated. This was the starting gun for Big Bang Nucleosynthesis, which dirtied the hydrogen in the universe with significant amounts of helium and traces of lithium. During the entire radiation-dominated epoch, the hydrogen was highly ionized. (So were the helium and lithium, after they showed up to the party.)

Eventually, the temperature $T$ reached a level low enough that the fractional ionization $x$ of hydrogen dropped below $1/2$; the time when this happened is known as the epoch of recombination. For blackbody radiation, like the cosmic background radiation, the number density of photons is, from Equation 2.13,

$$n_\nu(T) = \frac{8\pi}{c^3} \frac{\nu^2}{\exp(h\nu/kT) - 1}.$$ (9.19)

The photoionization rate of hydrogen by blackbody radiation is

$$\zeta_{\text{pho}}(T) = c \int_{\nu_0}^{\infty} n_\nu(T)\sigma_{\text{pho}} d\nu,$$ (9.20)

where $\nu_0 = I_{\text{H}}/h = 3.29 \times 10^{15}\,\text{Hz}$ is the threshold ionization frequency of hydrogen. The ionization cross section of hydrogen, as we saw in Section 4.1, can be approximated as $\sigma_{\text{pho}} \approx \sigma_0(\nu/\nu_0)^{-3}$ when $\nu > \nu_0$. When $kT \ll I_{\text{H}}$, corresponding to redshifts $z \ll 6 \times 10^4$, the photoionization rate is

$$\zeta_{\text{pho}}(T) \approx \frac{8\pi}{c^2}\sigma_0\nu_0^3 \int_{\nu_0}^{\infty} \exp\left(-\frac{h\nu}{kT}\right) \frac{d\nu}{\nu}$$ (9.21)

$$\approx \frac{8\pi}{c^2}\sigma_0\nu_0^3 \left(\frac{kT}{I_{\text{H}}}\right) \exp\left(-\frac{I_{\text{H}}}{kT}\right).$$ (9.22)

Numerically, this is

$$\zeta_{\text{pho}} \approx 4.61 \times 10^8\,\text{s}^{-1} \left(\frac{kT}{1\,\text{eV}}\right) \exp\left(-\frac{13.6\,\text{eV}}{kT}\right).$$ (9.23)

Because of its exponential dependence on the temperature of the cosmic background radiation, the photoionization rate plummets from $\zeta_{\text{pho}} \sim 1.5 \times 10^7\,\text{s}^{-1}$ at $kT = 3\,\text{eV}$ (corresponding to $t \approx 4.1\,\text{kyr}$) to $\zeta_{\text{pho}} \sim 3 \times 10^{-12}\,\text{s}^{-1}$ at $kT = 0.3\,\text{eV}$ (corresponding to $t \approx 270\,\text{kyr}$).

The total number density of blackbody photons, found by integrating Equation 9.19 over all frequencies, is given by the relation

$$n_\gamma(T) = \zeta(3)16\pi \left(\frac{kT}{hc}\right)^3 \approx 3.17 \times 10^{13}\,\text{cm}^{-3} \left(\frac{kT}{1\,\text{eV}}\right)^3,$$ (9.24)

where $\zeta(3) \approx 1.202$ is the Riemann zeta function. We can combine this with Equation 9.22 to write the photoionization rate in the form[2]

$$\zeta_{\text{pho}} \approx n_\gamma \frac{\sigma_0 c}{2\zeta(3)} \left(\frac{I_{\text{H}}}{kT}\right)^2 \exp\left(-\frac{I_{\text{H}}}{kT}\right).$$ (9.25)

If we make the approximation that the hydrogen is not dirty (that is, that helium and metals don't exist), the fractional ionization $x = n_e/n_{\text{H}}$ is given by Equation 4.14:

$$1 - x - x^2 \frac{n_{\text{H}}\alpha_{\text{H}}}{\zeta_{\text{pho}}} = 0,$$ (9.26)

---

[2] We apologize profusely for the proliferation of zetas; there aren't always enough Greek letters to go around.

where $n_H$ is the number density of hydrogen nuclei and $\alpha_H$ is the appropriate radiative recombination rate. Thus, the moment when $x = 1/2$ corresponds to the moment when

$$\frac{n_H \alpha_H}{\zeta_{\text{pho}}} = 2. \tag{9.27}$$

Using our previous relation for the photoionization rate $\zeta_{\text{pho}}$, given in Equation 9.25, we find that $x = 1/2$ when

$$\left(\frac{kT}{I_H}\right)^2 \exp\left(\frac{I_H}{kT}\right) = \frac{\sigma_0 c}{\zeta(3)\alpha_H} \frac{n_\gamma}{n_H}. \tag{9.28}$$

In a pure hydrogen universe $n_H = n_{\text{bary}}$, and the baryon-to-photon ratio, $\eta = n_{\text{bary}}/n_\gamma \approx 6.1 \times 10^{-10}$, remains constant with time.[3] The appropriate radiative recombination rate is the Case B rate from Equation 4.12; we are not dealing here with an isolated gas cloud from which an ionizing photon might escape, but inescapably with the entire universe. Using the appropriate $\alpha_{B,H} \propto T^{-0.82}$ in Equation 9.28, we find that $x = 1/2$ when

$$\left(\frac{kT}{I_H}\right)^{1.18} \exp\left(\frac{I_H}{kT}\right) = 9.6 \times 10^{15}. \tag{9.29}$$

The solution to this equation is $kT = 0.0243 I_H = 0.330\,\text{eV}$, or $T \approx 3800\,\text{K}$. This corresponds to a redshift $z \approx 1400$, an age for the universe $t \approx 0.23\,\text{Myr}$, and a mean baryon density $\bar{n}_{\text{bary}} \approx 700\,\text{cm}^{-3}$. Because baryons are so badly outnumbered by photons, and because the photoionization cross section at threshold $\sigma_0$ is so large, the hydrogen in the universe remains mostly ionized until the thermal energy $kT$ drops to just 2.4% of the ionization energy of hydrogen.

## 9.3  Reionization

The recombination of hydrogen brings in what astronomers call the "dark ages." This is the era between recombination at $z \approx 1400$ ($t \approx 0.23\,\text{Myr}$) and the formation of the first stars at $z \sim 30$ ($t \sim 100\,\text{Myr}$). If you jump into your time machine and travel back to the dark ages, you won't find it completely dark, however. The cosmic background radiation is always there, at a temperature ranging from $T \approx 3800\,\text{K}$ at $z \approx 1400$ to $T \sim 80\,\text{K}$ at $z \sim 30$. At the beginning of the dark ages, you would be as well-lit as you would be in the photosphere of an M star. By the end of the dark ages, however, the photons of the cosmic background radiation are far too low in energy to ionize hydrogen atoms. The **reionization** of intergalactic hydrogen must be accomplished by ultraviolet photons that come from massive stars and from active galactic nuclei (AGN).

---

[3] Although photon number isn't a strictly conserved quantity, the epoch of recombination occurs before stars start to toss non-CMB photons into the universe.

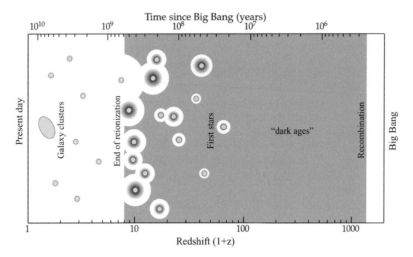

**Figure 9.3** Schematic diagram of recombination, the dark ages, and the epoch of reionization. [Following Barkana 2006]

A highly schematic picture of reionization is shown in Figure 9.3. The essential point is that reionization is a patchy process. The ionized regions around individual star-forming galaxies and galaxies with AGN gradually merge to form a single expanse of ionized gas. Thus, we cannot speak of an instant of reionization, but rather an epoch of reionization as the patches of ionized gas take over more and more of the universe.

Observing the Lyman alpha forest in the spectra of relatively low-redshift quasars ($z_e < 4$) tells us about the ionization state of intergalactic gas in the recent universe. At wavelengths shorter than a quasar's Lyman $\alpha$ emission line, we can estimate the transmission fraction $F(z)$ of the quasar's light as a function of the redshift $z$ of the intervening hydrogen gas. As seen in the lower panel of Figure 9.1, for a quasar at $z_e \sim 4$ the value of $F(z)$ ranges from $F \sim 0$ at redshifts corresponding to a Lyman $\alpha$ absorber to $F \sim 1$ at redshifts lying between absorbers. However, by combining the spectra of many quasars in the redshift range $2 < z_e < 4$ (when quasars were most abundant), we can find the mean transmission fraction $\overline{F}(z)$ averaged over many lines of sight. This can be translated into an effective optical depth for Lyman $\alpha$ absorption, $\tau_{\text{eff}} \equiv -\ln \overline{F}$.

It is customary to fit $\tau_{\text{eff}}$ as a power law in $1 + z$. One study using $\sim$40 000 quasars found

$$\tau_{\text{eff}} = 0.93 \left(\frac{1+z}{5}\right)^{3.18}, \tag{9.30}$$

shown as the solid line in Figure 9.4. Our theoretical expectation, from Equation 9.17, is that Lyman $\alpha$ will have an optical depth

$$\tau_\nu \approx 2.7 \times 10^5 f_n(z) \left(\frac{1+z}{5}\right)^{1.5}, \tag{9.31}$$

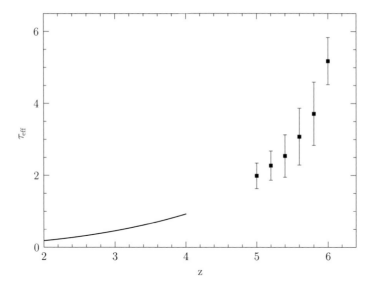

**Figure 9.4** Solid line: Effective optical depth for Lyman $\alpha$ in the redshift interval $2 < z < 4$. [Data from Kamble et al. 2020] Squares with error bars: Mean and standard deviation in Lyman $\alpha$ optical depth along lines of sight to high-redshift quasars. [Data from Bosman et al. 2018]

where $f_n$ is the fraction of hydrogen in its neutral state. By combining Equations 9.30 and 9.31, we find that the neutral fraction must be tiny for $z < 4$:

$$f_n(z) \approx 3.4 \times 10^{-6} \left( \frac{1+z}{5} \right)^{1.68}. \tag{9.32}$$

The relatively small number of quasars at $z_e > 5$ typically have pronounced Gunn–Peterson troughs, with $\tau > 2$ and thus a transmission fraction $F < 0.14$. (See, for instance, the high-redshift quasars in Figure 9.2.) By looking at the spectra of high-redshift quasars, we can determine how the Lyman $\alpha$ optical depth $\tau$ evolves with redshift in the range $5 < z < 6$, at the tail end of reionization. At each value of $z$ in this range, the distribution of $\tau$ can be fairly well fitted by a Gaussian (implying a log-normal distribution for the transmission fraction $F$). Figure 9.4 shows how the mean value $\overline{\tau}$ of $\tau$ increases with redshift, from $\overline{\tau} \approx 2.0$ at $z = 5$ to $\overline{\tau} \approx 5.2$ at $z = 6$. The standard deviation of $\tau$ also tends to increase with redshift, illustrating the patchy nature of reionization.

It is hard to use quasar spectra to determine the ionization state of gas at $z > 6$. First, quasars at $z_e > 6$ are rare. Second, we expect an average optical depth $\overline{\tau} > 5.2$ at $z > 6$; this corresponds to a transmission fraction $F = \exp(-\overline{\tau}) < 0.005$, which is difficult to distinguish from zero in a noisy spectrum. Thus, quasar spectra are useful only for looking at the tail end of reionization; at the practical limit of $z \sim 6$, where $\tau \sim 5$, the neutral fraction of hydrogen has already dropped to $f_n \sim 10^{-5}$ (this is found using Equation 9.31). To observe the earlier phases of reionization, when the neutral fraction was still close to unity, we have to take a different approach.

The cosmic microwave background (CMB) contains information about the epoch of reionization. We think most frequently about the CMB as telling us about the epoch of recombination ($z \sim 1400$). However, it also tells us about the low-redshift reionized universe. The reionized gas of the IGM at low redshift provides the population of free electrons between us and the CMB. These free electrons scatter the photons of the CMB via Thomson scattering, with cross section $\sigma_e = 6.6525 \times 10^{-25}$ cm$^2$. If the optical depth from Thomson scattering were $\tau_e \gg 1$ then the temperature fluctuations of the CMB would be thoroughly smeared out, since CMB photons would random-walk their way to us, losing all information about their original direction of motion. From the observed temperature fluctuations in the CMB, we deduce that the free electrons in the reionized gas provide an optical depth $\tau_e \ll 1$. The fact that a small fraction of CMB photons have undergone scattering has imposed a slight blurring of the hot and cold spots in the cosmic microwave background. Figure 9.5 shows the amplitude of CMB temperature fluctuations as a function of the angular multipole moment $\ell$, approximately related to angle by the relation $\theta \sim 180°/\ell$. The heavy solid line represents the temperature fluctuation spectrum in the absence of Thomson scattering by free electrons. The other curves show how the amplitude of the fluctuations is suppressed on small angular scales, for values of $\tau_e$ ranging from 0.05 to 0.5.

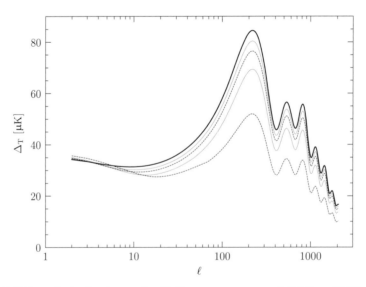

**Figure 9.5** Effect of reionization on the CMB power spectrum. Solid line: CMB spectrum without Thomson scattering by reionized gas. The other lines, in order of decreasing height of main spectral peak, represent $\tau_e = 0.05, 0.1, 0.2$, and 0.5. [Calculations made using CAMB]

The actual CMB spectrum shows only a modest suppression of the small-scale temperature fluctuations due to scattering from free electrons in the reionized IGM. The *Planck* 2018 results give $\tau_e \approx 0.056$. From the measured value of $\tau_e$, we can estimate the time when reionization of the IGM occurred. The optical depth from Thomson scattering can be written as (compare with Equation 9.4 for the optical depth from Lyman $\alpha$ absorption):

$$\tau_e = c \int_{t_*}^{t_0} n_e(t)\sigma_e dt = c\sigma_e \int_{a(t_*)}^{1} n_e(a)\frac{da}{aH}, \tag{9.33}$$

where $t_*$ is the time when reionization begins, and $H \equiv \dot{a}/a$ as usual. For simplicity, let's assume a universe made of pure hydrogen ($n_H = n_{bary}$) that undergoes complete reionization at the instant $t_*$.[4] Then, at $t > t_*$, the number density of free electrons is $n_e = \bar{n}_{bary,0}a^{-3}$ and

$$\tau_e = \frac{c}{\lambda_{mfp,0}} \int_{a(t_*)}^{1} \frac{da}{a^4 H}, \tag{9.34}$$

where

$$\lambda_{mfp,0} \equiv \frac{1}{\bar{n}_{bary,0}\sigma_e} = 1.93 \times 10^6\,\mathrm{Mpc} \tag{9.35}$$

is a photon's mean free path for scattering from a free electron, assuming ionized hydrogen at the current mean baryon density. Converting the variable of integration from the scale factor $a$ to the redshift $1 + z = 1/a$,

$$\tau_e = \frac{c}{\lambda_{mpf,0}} \int_0^{z_*} \frac{(1+z)^2 dz}{H(z)} \tag{9.36}$$

$$= 5.03 \times 10^{-21}\,\mathrm{s}^{-1} \int_0^{z_*} \frac{(1+z)^2 dz}{H(z)}. \tag{9.37}$$

Using the Hubble parameter for a universe with both matter and a cosmological constant (Equation 9.15), we find the analytic solution

$$\tau_e = \frac{\tau_0}{\Omega_{m,0}} \left( \left[ \Omega_{m,0}(1+z_*)^3 + \Omega_{\Lambda,0} \right]^{1/2} - 1 \right), \tag{9.38}$$

where

$$\tau_0 \equiv \frac{2}{3}\frac{c/H_0}{\lambda_{mpf,0}} \approx 1.48 \times 10^{-3}. \tag{9.39}$$

Solving for the redshift $z_*$ of reionization, we find

$$1 + z_* = \frac{1}{\Omega_{m,0}^{1/3}} \left[ \left( \Omega_{m,0}\frac{\tau_e}{\tau_0} + 1 \right)^2 - \Omega_{\Lambda,0} \right]^{1/3}. \tag{9.40}$$

---

[4] Although complete instantaneous reionization is not a particularly good approximation to the patchy reionization of the real universe, it's an assumption that lets us calculate a characteristic time $t_*$ for reionization.

Using $\Omega_{m,0} = 0.31$, $\Omega_{\Lambda,0} = 0.69$, and the observed optical depth $\tau_e = 0.056$, this yields $z_* \approx 7.0$ for the redshift of reionization. Since reionization occurs when the universe is still matter-dominated, we can use the relation between cosmic time and redshift in a matter-dominated universe:

$$1 + z = \left( \frac{3}{2} \Omega_{m,0}^{1/2} H_0 t \right)^{-2/3}.$$ (9.41)

This translates to

$$t = \frac{2}{3} H_0^{-1} \Omega_{m,0}^{-1/2} (1+z)^{-3/2} = 0.74 \, \text{Gyr} \left( \frac{1+z}{8} \right)^{-3/2},$$ (9.42)

assuming $\Omega_{m,0} = 0.31$ and $h_{70} = 1$. Of course, we have adopted highly simplifying assumptions for calculating the redshift of reionization, but the estimate $z_* \sim 7$ and $t_* \sim 0.7 \, \text{Gyr}$ is still useful in giving a time for the transition of the baryonic universe from neutral to ionized.

To explain in more detail how reionization occurs, we must identify the objects that produced large quantities of ionizing photons ($h\nu > 13.6 \, \text{eV}$) at redshift $z \geq 7$. We know that galaxies existed at these high redshifts (at the time of writing, the record high galaxy redshift was $z \approx 11$). If a galaxy contains an active galactic nucleus then the AGN is a source of ionizing photons. An AGN luminous enough to be labeled as a "quasar" typically emits ionizing photons at a rate $Q_0 > 10^{56} \, \text{s}^{-1}$, millions of times the rate at which a single hot star emits ionizing photons. However, quasars are extremely rare at high redshifts; thus, in searching for the photons that reionized the universe, it is more fruitful to look at the much more common star-forming galaxies at $z \geq 7$.

Suppose that a galaxy in the early universe has a star formation rate $R_{sf}$. Of the newly formed stars in the galaxy, only the most massive stars, with initial mass $M_\star > 25 \, M_\odot$, will produce significant amounts of ionizing radiation. Stars with $M_\star > 25 \, M_\odot$ are hot O stars while on the main sequence, and later spend time as hot blue supergiants. Integrated over its entire lifetime, a star with $M_\star > 25 \, M_\odot$ produces a number of ionizing photons $N_\gamma$ given by the relation

$$\frac{N_\gamma}{M_\star} \approx 5 \times 10^{61} \, M_\odot^{-1}.$$ (9.43)

Stars with $M_\star > 25 \, M_\odot$ represent a fraction $f_m \approx 0.13$ of the mass of newly formed stars. Thus, an ongoing star formation rate $R_{sf}$ will lead to an ionizing-photon production rate[5]

$$Q_0 = \frac{N_\gamma}{M_\star}(f_m R_{sf}) \approx 2 \times 10^{53} \, \text{s}^{-1} \left( \frac{R_{sf}}{1 \, M_\odot \, \text{yr}^{-1}} \right).$$ (9.44)

---

[5] A more detailed derivation of this relation is given by Topping & Shull 2015, who note the fairly sensitive dependence of ionizing photon production on stellar metallicity and rotation as well as on the initial mass function of stars.

This rate implies that for every $1\,M_\odot$ of stars that are formed, enough ultraviolet photons are emitted to photoionize $\sim 5000\,M_\odot$ of hydrogen.

If the gas surrounding the star-forming galaxy is atomic hydrogen with density $n_H$, our expectation is that a supersized Strömgren sphere will be created on a time scale comparable to the recombination time,

$$t_{rec} = \frac{1}{\alpha_{B,H} n_H}, \tag{9.45}$$

where $\alpha_{B,H}$ is the Case B recombination rate. However, let's look more closely at what happens when the ionized region extends into the pristine intergalactic gas surrounding the star-forming galaxy. The density of the IGM decreases as the universe expands, with $n_H \approx \bar{n}_{bary} = \bar{n}_{bary,0}(1+z)^3$. Since the intergalactic medium is nearly metal-free at early times, we can assume that the Strömgren sphere has a high temperature, $T \sim 3 \times 10^4$ K, given the absence of line cooling. This implies a relatively low recombination rate, $\alpha_{B,H} \approx 1.0 \times 10^{-13}\,\mathrm{cm^3\,s^{-1}}$, and a recombination time

$$t_{rec} \approx \frac{1}{\alpha_{B,H} \bar{n}_{bary,0}} (1+z)^{-3} \approx 2.5\,\mathrm{Gyr} \left(\frac{1+z}{8}\right)^{-3}. \tag{9.46}$$

For redshifts $z < 17$, the recombination time $t_{rec}$ is longer than the age of the universe (Equation 9.42). Thus, at $z < 17$, corresponding to $t > 0.22$ Gyr, an ionization region doesn't have enough time to reach ionization equilibrium and become a classic Strömgren sphere.

As a first approximation, let's assume that intergalactic hydrogen atoms that are photoionized at $z > 17$ (when $t_{rec} < t$) all recombine, while intergalactic hydrogen atoms that are photoionized at $z < 17$ (when $t_{rec} > t$) never recombine. During the epoch between $z = 17$ and $z_* = 7$, corresponding to a time interval $\Delta t \approx 0.52$ Gyr, we can compute what the average star formation rate must be in order to completely reionize the intergalactic medium by $z_* = 7$. Consider a comoving volume of intergalactic gas (that is, one expanding along with the Hubble expansion); its volume at the present day is $V_0 = 1\,\mathrm{Mpc^3}$. The total number of hydrogen atoms in the comoving volume, assuming a pure hydrogen baryonic universe, is

$$N_H = n_{bary,0} V_0 = 7.4 \times 10^{66}, \tag{9.47}$$

representing $\sim 6 \times 10^9\,M_\odot$ of hydrogen. To produce this many ionizing photons during a time interval $\Delta t \approx 0.52\,\mathrm{Gyr} \approx 1.6 \times 10^{16}$ s, the (comoving) volumetric rate of ionizing photon production must be

$$\frac{Q_0}{V_0} = \frac{N_H}{\Delta t V_0} \approx 4.5 \times 10^{50}\,\mathrm{s^{-1}\,Mpc^{-3}}. \tag{9.48}$$

If the ionizing photons are produced by young stars within galaxies, there is a final complicating factor. Galaxies contain relatively dense interstellar gas, as well as dust, which absorb many of the ultraviolet photons before they can escape into

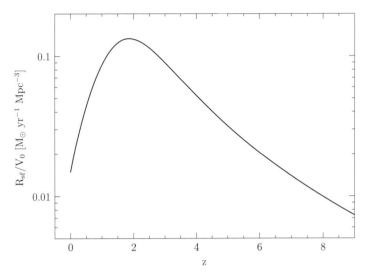

**Figure 9.6** Cosmic star formation rate in units of solar masses per year per comoving cubic megaparsec. [Data from Madau & Dickinson 2014]

intergalactic space. The average fraction $f_{esc}$ of ultraviolet photons that escape the galaxy of their birth is not well known, particularly at $z > 7$. Let's take $f_{esc} = 0.2$ as a first guess. Using Equation 9.44, the required rate of ionizing-photon production translates into a star formation rate per comoving volume

$$\frac{R_{sf}}{V_0} \approx 0.011\, M_\odot \, yr^{-1} \, Mpc^{-3} \left( \frac{0.2}{f_{esc}} \right). \tag{9.49}$$

Note that early reionization ($z_* > 7$) and small escape fractions for ionizing photons ($f_{esc} \ll 0.2$) would both require high star formation rates in the early universe (or perhaps alternative sources of ionizing photons).

The cosmic star formation rate deduced from observations is shown in Figure 9.6. The maximum rate in solar masses per year per comoving cubic megaparsec occurred at a redshift $z = 1.9$, corresponding to an age for the universe $t \approx 3.4\,Gyr$. However, the peak about this maximum is fairly broad; the comoving star formation rate was within a factor two of its maximum value during the redshift range $z = 3.5 \to 0.8$, corresponding to $t \approx 2 \to 7\,Gyr$.[6] The comoving star formation rate at $z \sim 7$ was comparable to its value today, with $R_{sf}/V_0 \sim 0.014\, M_\odot \, yr^{-1} \, Mpc^{-3}$. However, the exact star formation rate at high redshifts is still somewhat conjectural, owing to lack of data. The picture is further muddied by the fact that the earliest generation of stars (Population III stars) had few or no metals, and their properties are poorly understood.

---

[6] The era of prodigious star formation at $z \sim 2$ is known as "cosmic noon," in contrast with the "cosmic dawn" at the end of the dark ages, and the "cosmic cocktail hour" during which we now live.

Quasars also emit ionizing photons. The comoving number density of bright quasars reached a maximum at a redshift $z = 2.4$, corresponding to an age for the universe $t \approx 2.7\,\mathrm{Gyr}$. The peak about this maximum is narrower than the peak in the cosmic star formation rate. The comoving quasar density was within a factor two of its maximum value in the redshift range $z = 3.4 \to 1.6$, corresponding to $t \approx 2 \to 4\,\mathrm{Gyr}$. Because of the relatively steep rise in the number of quasars at early times, the number density of bright quasars at $z \sim 7$ ($t \sim 0.7\,\mathrm{Gyr}$) wasn't nearly enough to reionize the IGM. The lower-luminosity AGN may have been numerous enough to contribute significantly to the photoionizing background, but the luminosity function of quasars and AGN at high redshift is poorly constrained.

## 9.4 Lyman Alpha Forest

A uniformly black Gunn–Peterson trough is seen only at redshifts $z > 6$, as shown in Figure 9.2. However, at lower redshifts, there exists a "Lyman alpha forest" of absorption lines, as shown in Figure 9.1. The idea of the Lyman alpha forest was planted in 1970, when Roger Lynds observed the spectrum of 4C 05.34, whose redshift $z_e = 2.877$ was the largest then known for any quasar. Lynds found many absorption lines, most of which were at observed wavelengths shorter than the Lyman $\alpha$ emission line of the quasar itself. Of the 27 strongest absorption lines he detected in the range $\lambda = 3500\,\text{Å}$ to $6000\,\text{Å}$, only two were at a wavelength longer than the observed wavelength $\lambda = 4713\,\text{Å}$ of the redshifted Lyman $\alpha$ emission from the quasar. Lynds concluded that most of the absorption lines that he saw were Lyman $\alpha$ lines from hydrogen along the line of sight to the quasar; the other absorption lines were from relatively common metals (such as O, C, N, and Si) at the same redshifts as the absorbing hydrogen. As similar distributions of short-wavelength absorption lines began to be seen in the spectra of additional quasars, astronomers began using the metaphor of a Lyman alpha "forest" of absorption lines.

Consider a gas cloud along a line of sight to a background quasar, producing a single Lyman $\alpha$ absorption line. The absorbing cloud has a column density $N_{\mathrm{HI}}$ of neutral hydrogen. Although observed Lyman $\alpha$ absorbers have a range of column densities, certain values of $N_{\mathrm{HI}}$ are the most physically significant. For example, consider the effect of ionizing photons penetrating the cloud from outside. At cosmic noon ($z \sim 2$), when massive stars, AGN, and quasars were pouring out ionizing photons, the photoionization rate for an unshielded hydrogen atom was $\zeta_{\mathrm{pho}} \sim 10^{-12}\,\mathrm{s}^{-1} \sim 30\,\mathrm{Myr}^{-1}$. The minimum photon energy for ionizing hydrogen, $h\nu = 13.60\,\mathrm{eV}$, corresponds to a wavelength $\lambda = 912\,\text{Å}$, known as the "Lyman limit." At this wavelength, the cross section for the photoionization of hydrogen is $\sigma_0 = 6.30 \times 10^{-18}\,\mathrm{cm}^2$ (from Equation 4.3). Thus, a cloud with column density

$$N_{\text{HI}} > \frac{1}{\sigma_0} = 1.6 \times 10^{17}\,\text{cm}^{-2} \tag{9.50}$$

is optically thick to Lyman limit photons. Above this column density self-shielding starts to protect the hydrogen in the cloud's interior from the ionizing photon background.

In Section 7.4 we found that a molecular cloud starts to experience self-shielding from dissociating photons at a column density $N(\text{H}_2) \approx 10^{14}\,\text{cm}^{-2}$. However, the molecular fraction at the cloud's center doesn't rise to $f_{\text{mol}} = 0.5$ until the column density reaches a much higher value: $N(\text{H}_2) \approx 2 \times 10^{18}\,\text{cm}^{-2}$ at $n_{\text{H}} = 1000\,\text{cm}^{-3}$. Similarly, although a neutral atomic cloud starts to experience self-shielding from ionizing photons at a column density $N_{\text{HI}} \approx 1.6 \times 10^{17}\,\text{cm}^{-2}$, the neutral fraction at the cloud's center doesn't rise to $f_n = 0.5$ until the column density is significantly higher. For the typical densities of Lyman $\alpha$ absorbers, the neutral fraction rises above $f_n = 0.5$ at a column density $N_{\text{HI}} \sim 2 \times 10^{20}\,\text{cm}^{-2}$.

Absorbing systems with a column density in the range $1.6 \times 10^{17}\,\text{cm}^{-2} < N_{\text{HI}} < 2 \times 10^{20}\,\text{cm}^{-2}$ are referred to as **Lyman limit systems**.[7] Observed Lyman limit systems typically have a temperature $T \sim 3 \times 10^4$ K, producing a thermal broadening parameter $b \sim 20\,\text{km}\,\text{s}^{-1}$. The Lyman $\alpha$ absorption line produced by a Lyman limit system will thus have an optical depth at line center (Equation 2.49)

$$\tau_0 \approx 6000 \left(\frac{b}{20\,\text{km}\,\text{s}^{-1}}\right)^{-1} \left(\frac{N_{\text{HI}}}{1.6 \times 10^{17}\,\text{cm}^2}\right). \tag{9.51}$$

At a temperature $T \sim 3 \times 10^4$ K, the damping optical depth for Lyman $\alpha$ is $\tau_{\text{damp}} \sim 15\,000$. As a result, the Lyman $\alpha$ absorption lines from Lyman limit systems with $N_{\text{HI}} < 10^{18}\,\text{cm}^{-2}$ are on the flat part of the curve of growth and have equivalent width $W_\lambda \sim 0.5$ Å in the absorber's rest frame.

By the time the column density of a hydrogen absorber reaches $N_{\text{HI}} = 2 \times 10^{20}\,\text{cm}^{-2}$, the Lyman $\alpha$ absorption line is strongly damped and the equivalent width of the line in the absorber's frame is (using Equation 2.73)

$$W_\lambda \approx 10\,\text{Å} \left(\frac{N_{\text{HI}}}{2 \times 10^{20}\,\text{cm}^{-2}}\right)^{1/2}, \tag{9.52}$$

appropriate for a line in the square-root part of the curve of growth. Absorbing systems with a neutral hydrogen column density $N_{\text{HI}} > 2 \times 10^{20}\,\text{cm}^{-2}$ are called **damped Lyman alpha systems**. Figure 9.7 shows the absorption line arising from a damped Lyman alpha system along the line of sight to the quasar QSO J0121741.8-370100, which lies at a redshift $z_e = 2.910$. The redshift of the damped Lyman alpha system is $z_{\text{abs}} = 2.429$; the curved line in Figure 9.7

---

[7] In practice, the minimum column density needed for a system to qualify as a Lyman limit system is often rounded down to $10^{17}\,\text{cm}^{-2}$.

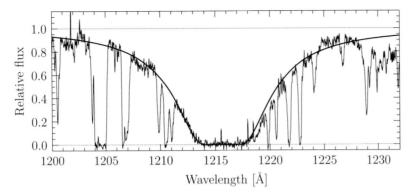

**Figure 9.7** A strongly damped Lyman $\alpha$ line in the spectrum of the quasar QSO J021741.8-370100. Wavelengths are given in the absorber's rest frame. To convert to observed wavelength, multiply by $1 + z_{\mathrm{abs}}$, where $z_{\mathrm{abs}} = 2.429$ is the redshift of the damped Lyman alpha system. [Data from ESO VLT UVES, processed by S. Frank]

is the best-fitting Lyman $\alpha$ line, which corresponds to a column density $N_{\mathrm{HI}} \approx 4.2 \times 10^{20}\,\mathrm{cm}^{-2}$.

Damped Lyman alpha systems are interesting to astronomers in part because of their relation to galaxies. By definition, a damped Lyman alpha system has $N_{\mathrm{HI}} > 2 \times 10^{20}\,\mathrm{cm}^{-2}$. For comparison, H I 21 cm maps of our galaxy (such as Figure 3.2) imply column densities $N_{\mathrm{HI}} \sim 10^{20}\,\mathrm{cm}^{-2}$ perpendicular to the disk at our location, and $N_{\mathrm{HI}} \sim 10^{23}\,\mathrm{cm}^{-2}$ in the disk plane. Thus, damped Lyman alpha systems have column densities of neutral hydrogen comparable with a large galaxy like our own. The metallicity of damped Lyman alpha systems is typically in the range $Z \approx 0.01 \rightarrow 0.3\,Z_{\odot}$, larger than the values $Z \approx 0.001 \rightarrow 0.01\,Z_{\odot}$ characteristic of the lower-optical-depth systems of the Lyman alpha forest. Damped Lyman alpha systems can be thought of as gravitationally bound protogalaxies containing gas (and dark matter), which haven't yet been highly effective at converting gas into stars.

The lower-column-density absorption lines in the Lyman alpha forest, which are vastly more numerous than the damped Lyman alpha systems, cannot be associated with individual gravitationally bound gas clouds. The most common lines in the Lyman alpha forest, those with $N_{\mathrm{HI}} < 10^{17}\,\mathrm{cm}^{-2}$, are not dense enough to represent gravitationally collapsed, virialized systems with an enhanced neutral fraction of hydrogen. Instead, the absorption lines of lower column density are produced from highly ionized regions of gas that are broadened primarily by the Hubble expansion. The Lyman alpha forest, aside from the rare damped Lyman alpha and Lyman limit systems, shouldn't be thought of as resulting from discrete clouds along the line of sight to a quasar. Instead, it is more useful to think of it as a mapping from a smoothly fluctuating density field $n(z)$ along the line of sight to an observed optical depth $\tau(z)$ for Lyman $\alpha$ absorption.

To see how density maps to optical depth for Lyman $\alpha$ absorption, consider a simplified universe in which the gas is pure hydrogen. In this universe, photoionization by quasars, AGN, and massive stars is balanced by radiative recombination. The fractional ionization $x = n_e/n_H$ is then given by the quadratic equation (see Equation 9.26)

$$1 - x - x^2 \frac{n_H \alpha_H(T)}{\zeta_{\text{pho}}} = 0, \tag{9.53}$$

where $\alpha_H(T)$ is the radiative recombination rate and $\zeta_{\text{pho}}$ is the photoionization rate. After the epoch of reionization at $z_* \sim 7$, the neutral fraction $f_n = 1 - x$ will be much smaller than one everywhere outside galaxies and damped Lyman alpha systems. In the limit $f_n \ll 1$, the solution of the quadratic in Equation 9.53 is

$$f_n \approx \frac{n_H \alpha_H(T)}{\zeta_{\text{pho}}}, \tag{9.54}$$

yielding a number density of neutral hydrogen atoms

$$n_{\text{HI}} = f_n n_H \approx \frac{n_H^2 \alpha_H(T)}{\zeta_{\text{pho}}}. \tag{9.55}$$

Outside the damped Lyman alpha and Lyman limit systems, the intergalactic medium is largely transparent to Lyman $\alpha$ photons, so the appropriate radiative recombination rate is the Case A rate (Equation 4.10)

$$\alpha_{A,H} \approx 4.9 \times 10^{-13} \text{ cm}^3 \text{ s}^{-1} \left(\frac{T}{8000 \text{ K}}\right)^{-0.71}. \tag{9.56}$$

Thus, the number density of neutral hydrogen atoms is $n_{\text{HI}} \propto n_H^2 T^{-0.71} \zeta_{\text{pho}}^{-1}$.

In simulations of the evolution of intergalactic gas, it is found that there is a correlation between $n$ and $T$ at $T < 10^5$ K, with higher densities corresponding to higher temperatures. The origin of this correlation lies in the balance between heating and cooling of the gas. An analytic approximation for this correlation can be derived in the case of our hydrogen-only universe. With only hydrogen present, heating is done by the electrons ejected during the photoionization of hydrogen. The resulting volumetric heating rate, assuming ionization equilibrium, is

$$g_{\text{pho}} = n_{\text{HI}} \zeta_{\text{pho}} \langle E \rangle \tag{9.57}$$

$$= n_e n_{\text{HII}} \alpha_{A,H} \langle E \rangle, \tag{9.58}$$

where $\langle E \rangle$ is the average kinetic energy of an ejected electron. In a highly ionized hydrogen gas, with $n_e = n_{\text{HII}} \approx n_H$, the heating rate can be written as

$$g_{\text{pho}} \approx n_H^2 \alpha_{A,H} \langle E \rangle. \tag{9.59}$$

The regions that give rise to low-column-density Lyman $\alpha$ lines are never much denser than the average density of the universe. Thus, they cool mainly through adiabatic cooling as the universe expands. During adiabatic expansion, the thermal energy density $\varepsilon$ has the dependence $\varepsilon \propto V^{-\gamma}$, where $V$ is the volume of a gas element and the adiabatic index is $\gamma = 5/3$ for a monatomic gas. The dependence of volume on the scale factor $a(t)$ of the universe is $V \propto a(t)^3$, which means that the volumetric cooling rate is

$$\ell_{\text{adi}} = -\frac{d\varepsilon}{dt} = 3\gamma \varepsilon(t) H(t), \qquad (9.60)$$

where $H(t) \equiv \dot{a}/a$ is the Hubble parameter for the expanding universe.

Equating the heating rate from Equation 9.59 with the cooling rate from Equation 9.60, we find temperature equilibrium for the gas when

$$n_{\text{H}}^2 \alpha_{\text{A,H}} \langle E \rangle = 3\gamma \, (3 n_{\text{H}} kT) \, H(t), \qquad (9.61)$$

or

$$n_{\text{H}} = 9\gamma \frac{kT}{\alpha_{\text{A,H}} \langle E \rangle} H(t). \qquad (9.62)$$

This implies $n_{\text{H}} \propto T^{1.71}$ for temperatures near $T \sim 8000\,\text{K}$, where $\alpha_{\text{A,H}} \propto T^{-0.71}$. We can thus write

$$T = T_0 \left( \frac{n_{\text{H}}}{\overline{n}_{\text{bary}}} \right)^{0.58}, \qquad (9.63)$$

where $T_0$ is the equilibrium temperature expected when the number density of atoms is equal to the mean number density of baryons in the universe. More sophisticated calculations, such as the numerical simulations we'll look at in Section 10.1, indicate that $T_0 \sim 5000\,\text{K}$ is an appropriate temperature level for diffuse intergalactic gas in the present-day universe.

The optical depth $\tau$ for Lyman $\alpha$ absorption at a given redshift is proportional to the number density of neutral hydrogen at that redshift. Thus, given both photoionization equilibrium and temperature equilibrium,

$$\tau \propto n_{\text{HI}} \propto n_{\text{H}}^2 T^{-0.71} \zeta_{\text{pho}}^{-1} \propto n_{\text{H}}^{1.58} \zeta_{\text{pho}}^{-1}, \qquad (9.64)$$

implying that the production of Lyman $\alpha$ absorption lines will be strongly weighted to denser regions of the intergalactic medium with no nearby ionizing sources such as quasars. Properly normalized, the mapping between number density $n(z)$ and Lyman $\alpha$ optical depth $\tau(z)$ is

$$\tau = A \left( \frac{n}{\overline{n}_{\text{bary}}} \right)^{1.58}, \qquad (9.65)$$

where the constant $A$ depends on the assumed cosmology as well as on the amount of ionizing radiation present. In the matter-dominated epoch of expansion $(z > 0.31)$, the amplitude $A$ can be approximated as

$$A(z) \approx 1.6 \left( \frac{T_0}{5000\,\text{K}} \right)^{-0.71} \left( \frac{\zeta_{\text{pho}}}{10^{-12}\,\text{s}^{-1}} \right)^{-1} \left( \frac{1+z}{4} \right)^{4.5}. \qquad (9.66)$$

Since Equation 9.65 describes the equivalent of Gunn–Peterson absorption for a non-uniform photoionized medium, it is referred to as the **fluctuating Gunn–Peterson approximation**. This approximation breaks down in the densest regions of intergalactic gas ($n \gg \bar{n}_{bary}$), where shock heating occurs; however, it provides a good description of Lyman $\alpha$ absorption at lower densities.

## Exercises

9.1    In discussing the Gunn–Peterson effect (Section 9.1), we state that the Voigt profile $\Phi_\nu$ in an expanding universe can be treated as being "very strongly peaked" at a frequency $\nu = \nu_{u\ell}/(1+z)$. The ultimate in strong peaks is the Dirac delta distribution $\delta(x)$, which is infinitesimally narrow, with $\delta(x) = 0$ for $x \neq 0$, but which has the property that $\int f(x)\delta(x)dx = f(0)$ for any function $f(x)$ integrated over any range of $x$ that includes $x = 0$.

Start with Equation 9.9, which gives the general relation for the optical depth $\tau_\nu$ from Lyman $\alpha$ absorption in an expanding universe. Demonstrate that using a Dirac delta distribution for the Voigt profile $\Phi_\nu$ results in the optical depth usually quoted for the Gunn–Peterson effect (Equation 9.10):

$$\tau_\nu = c\sigma_\alpha \frac{n_\ell}{H(z)}. \tag{9.67}$$

[Hint: in this problem, what is the appropriate choice for $x$ in the Dirac delta distribution $\delta(x)$?]

9.2    A quasar at a redshift $z_{em} = 2.60$ is observed to have an absorption line in its spectrum at a central wavelength $\lambda_0 = 3675\,\text{Å}$ (in the observer's rest frame).

   (a)   If this absorption feature is a Lyman $\alpha$ absorption line from hydrogen in a galaxy along the line of sight to the quasar, what is the redshift $z_{abs}$ of the intervening galaxy?

   (b)   The absorption feature has an observed equivalent width $W_{\lambda, obs} = 6.5\,\text{Å}$. What is the equivalent width in the rest frame of the intervening galaxy? Estimate the optical depth $\tau_0$ required to produce this equivalent width in a Lyman $\alpha$ line. (State the thermal broadening $b$ you assume for the absorbing hydrogen gas.)

   (c)   Estimate the column density $N_{HI}$ of neutral hydrogen in the intervening galaxy. How does this compare with the column density of neutral hydrogen you would see looking through the Milky Way Galaxy from outside?

9.3    A quasar spectrum has two absorption lines at observed wavelengths $\lambda_0 = 5048.65\,\text{Å}$ and $5057.06\,\text{Å}$, with observed equivalent widths of $W_\lambda = 0.04\,\text{Å}$ and $0.02\,\text{Å}$, respectively. The two lines are identified as arising from a C IV doublet with wavelengths $\lambda \approx 1548\,\text{Å}$ and $1550\,\text{Å}$ in the absorber's frame.

Atomic data for the C IV doublet are as follows:

| Line | $\lambda_0$ | $g_u$ | $g_\ell$ | $A_{u\ell}$ |
|---|---|---|---|---|
| C IV $\lambda$1548 | 1548.19 Å | 4 | 2 | $2.65 \times 10^8\ \mathrm{s}^{-1}$ |
| C IV $\lambda$1550 | 1550.77 Å | 2 | 2 | $2.64 \times 10^8\ \mathrm{s}^{-1}$ |

(a) Using the atomic data above and assuming these lines are in the linear part of the curve of growth, derive an expression like Equation 2.60 for the column density $N_\ell$ as a function of $W_\lambda$ in units of $W_\lambda = 0.01$ Å for C IV.

(b) What is the redshift of the absorbing cloud?

(c) Using your scaling relation from part (a), compute the column density of C IV through the absorbing cloud.

(d) Further observations reveal a spiral galaxy at the same redshift as the C IV absorber along the line of sight to the quasar. Assuming that this galaxy has the same C/H ratio as in the ISM of the Milky Way near the Sun ($n_C/n_H \approx 2.7 \times 10^{-4}$), estimate the column density of H I. Given your answer, is this galaxy a damped Lyman alpha system, a Lyman limit system, or a lower column density system?

(e) Is the H I column density estimate from part (d) an upper or lower limit? Explain your reasoning.

# 10

# Warm-Hot Intergalactic Medium

*There appears to be an absorption of light in intergalactic space*
*which is such that the coefficient of transmission for the light of*
*the most distant nebulae whose images are impressed upon our*
*photographic plates is perhaps of the order of about 8 per cent.*

Frank W. Very (1852–1927)
"Are the White Nebulae Galaxies?"
1911, Astronomische Nachrichten 189, 441

The properties of intergalactic gas have been debated ever since some "nebulae" in the night sky were suspected to be external galaxies comparable in size with the Milky Way Galaxy. Unfortunately for early twentieth century astronomers such as Frank Very, the intergalactic medium is extremely low in density, and attempts to determine its absorption properties using images on photographic plates gave erroneous answers. Throughout the twentieth century, the intergalactic medium remained difficult to detect. By the 1990s, astronomers began talking about a "missing baryon problem." The current mean baryon density, $\bar{n}_{\mathrm{bary},0}$, was fairly well known at the time from the results of Big Bang Nucleosynthesis. However, the density of easily detected baryons – in galaxies, clusters, and Lyman alpha systems – accounted for roughly half the mean baryon density. Notice that people referred to the missing baryon problem, not the missing baryon *crisis*. People had a good notion that the unobserved baryons were in a hot, relatively low-density gas spread through intergalactic space. It was observing this warm-hot intergalactic medium (WHIM) that was the problem.

Much of our knowledge of the warm-hot intergalactic medium is deduced from numerical cosmological simulations. These simulations predict the existence of a WHIM that is distributed in a filamentary structure (the "cosmic web"). The temperature of the WHIM lies in the range $10^5$ K $< T < 10^7$ K. Its density generally lies in the range $\bar{n}_{\mathrm{bary},0} < n_{\mathrm{H}} < 400\bar{n}_{\mathrm{bary},0}$, or from $n_{\mathrm{H}} \sim 2 \times 10^{-7}$ cm$^{-3}$ to $n_{\mathrm{H}} \sim 10^{-4}$ cm$^{-3}$. At the temperatures predicted by numerical simulations, the hydrogen in the WHIM is almost completely ionized and thus difficult to detect.

## 10.1 Simulations

Numerical cosmological simulations give useful insight into the properties of the warm-hot intergalactic medium, as long as they contain all the relevant physics. Consider, for instance, the numerical simulation whose results are shown in Figure 10.1. The "Illustris TNG100" simulation, as it is called, uses a cubic box with sides of comoving length $\ell \sim 110\,\mathrm{Mpc}$; it is part of a $\Lambda$CDM universe, with cosmological parameters taken from the *Planck* 2016 results. The comoving box contains $\sim 6 \times 10^9$ dark matter particles, each of mass $m_{\mathrm{dm}} \sim 7 \times 10^6\,\mathrm{M_\odot}$, which interact only through gravity. It also contains an equal number of baryonic particles, each of mass $m_{\mathrm{bar}} \sim 1.4 \times 10^6\,\mathrm{M_\odot}$, which include prescriptions for star formation, stellar evolution, chemical enrichment, gas cooling, stellar feedback, magnetohydrodynamics, and other complications of baryonic physics.

The two panels in Figure 10.1 show a slice through the box at the present day (redshift $z = 0$). The left panel shows the distribution of the diffuse intergalactic medium (DIM), defined as gas with $T < 10^5\,\mathrm{K}$ and $n_{\mathrm{H}} < 10^{-4}\,\mathrm{cm^{-3}}$. The right panel shows the distribution of the warm-hot intergalactic medium (WHIM), defined as gas with $10^5\,\mathrm{K} < T < 10^7\,\mathrm{K}$ and $n_{\mathrm{H}} < 10^{-4}\,\mathrm{cm^{-3}}$.

The warm-hot intergalactic medium is found primarily in long filaments, in contrast with the more widely distributed diffuse intergalactic medium. The WHIM, as it flows along the filaments to the clusters where filaments meet, is

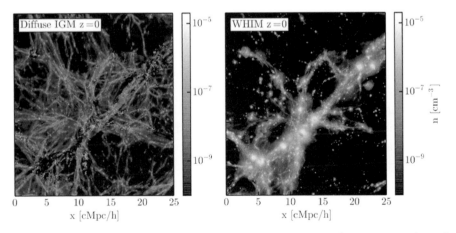

**Figure 10.1** Left: Distribution of diffuse intergalactic gas ($T < 10^5\,\mathrm{K}, n_{\mathrm{H}} < 10^{-4}\,\mathrm{cm^{-3}}$) at $z = 0$. Right: Distribution of warm-hot intergalactic gas ($10^5\,\mathrm{K} < T < 10^7\,\mathrm{K}, n_{\mathrm{H}} < 10^{-4}\,\mathrm{cm^{-3}}$) at $z = 0$. In the figure labels, "cMpc" is "comoving megaparsec," and $h$ is the Hubble constant in units of $100\,\mathrm{km\,s^{-1}\,Mpc^{-1}}$. [Adapted from Martizzi et al. 2019]

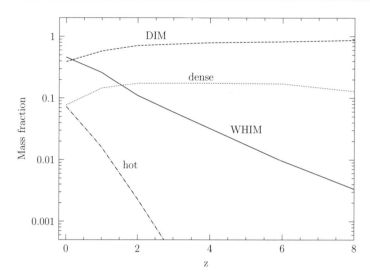

**Figure 10.2** Redshift evolution of the mass fraction of baryonic matter in each of four components. Dashed line: Diffuse intergalactic medium. Solid line: Warm-hot intergalactic medium. Dotted line: Dense gas ($n_H > 10^{-4}(1 + z)\,\mathrm{cm}^{-3}$, $T < 10^7$ K). Dot-dash line: Hot gas ($T > 10^7$ K). [Data from Martizzi et al. 2019, with additional data provided by D. Martizzi]

shocked and heated to higher temperatures than the photoionized DIM. As the large-scale structure of the universe develops with time, more and more of the baryonic gas is converted into shock-heated warm-hot gas. Figure 10.2 shows how different components of the universe evolve with time (or equivalently, redshift) in the Illustris simulation.

The dashed line in Figure 10.2 shows how the mass fraction of the diffuse intergalactic medium evolves with time, where the DIM is defined as having $T < 10^5$ K and $n_H < 10^{-4}(1 + z)\,\mathrm{cm}^{-3}$. This combination of temperature and redshift-dependent density selects the phase in which the fractional overdensity that occurs in some regions evolves in the linear regime, and helium and heavier elements are not fully ionized. At $z = 7$, around the time of reionization, about 84% of the baryonic gas in the universe was in the diffuse intergalactic medium. However, as the structure went nonlinear and collapsed, more and more of the baryonic gas became shock-heated to temperatures $T > 10^5$ K. The solid line in Figure 10.2 shows the growth in the warm-hot intergalactic medium, defined as gas with $10^5$ K $< T < 10^7$ K and $n_H < 10^{-4}(1 + z)\,\mathrm{cm}^{-3}$. At $z = 7$, only 0.6% of the baryonic mass was in the WHIM, defined in this manner, but, at $z = 0$, roughly 45% of the baryonic mass is in the shock-heated WHIM.[1] The dotted

---

[1] The diffuse intergalactic medium still dominates by *volume* fraction today; ~90% of the volume of the universe is filled by the DIM at $z = 0$ and ~10% is filled by the WHIM.

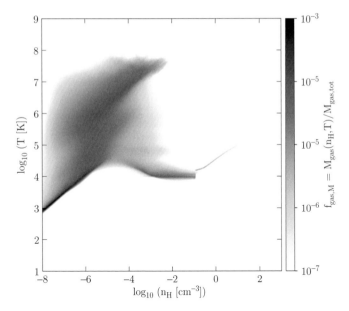

**Figure 10.3** Phase diagram of gas in the universe at redshift $z = 0$, from the Illustris TNG100 simulation. [Adapted from Martizzi et al. 2019]

line in Figure 10.2 represents the dense baryonic matter with $n_H > 10^{-4}(1 + z)\,\mathrm{cm}^{-4}$ and $T < 10^7$ K; this component represents the stars and interstellar gas within galaxies, as well as the circumgalactic medium. Finally, the dot-dash line represents all the gas with $T > 10^7$ K; this represents the hot intracluster medium.

Figure 10.3 shows the phase diagram of gas in the Illustris simulation at a redshift $z = 0$. The plot is of the mass fraction of gas as a function of $n_H$ and $T$. In the density and temperature range characteristic of the diffuse intergalactic medium ($n_H < 10^{-4}\,\mathrm{cm}^{-3}$, $T < 10^5$ K), there is a pronounced ridge line in the phase diagram corresponding to the relationship for which the photoionized DIM is in temperature equilibrium (Equation 9.63):

$$T_{\mathrm{dim}} \approx 5000\,\mathrm{K} \left( \frac{n_H}{2.5 \times 10^{-7}\,\mathrm{cm}^{-3}} \right)^{0.58}. \tag{10.1}$$

In the WHIM temperature range, $10^5$ K to $10^7$ K, the temperature also tends to be positively correlated with density. However, at a given density there is a fairly broad range of temperatures within the WHIM. The temperature range of the warm-hot intergalactic medium embraces the temperature $T \sim 10^6$ K that is typical of the hot interstellar medium in our galaxy and of the corona of the Sun. However, the most probable density $n_{\mathrm{whim}} \sim 3 \times 10^{-6}\,\mathrm{cm}^{-3}$ of the warm-hot intergalactic medium at $T = 10^6$ K is smaller by three orders of magnitude than the density $n_{\mathrm{him}} \sim 4 \times 10^{-3}\,\mathrm{cm}^{-3}$ of the hot interstellar medium, which in turn is 12 orders of magnitude smaller than the density at the base of the Sun's corona.

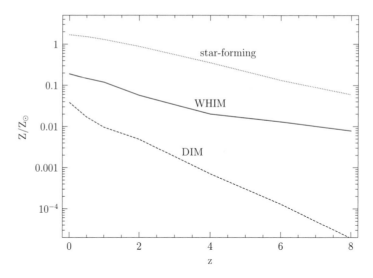

**Figure 10.4** Redshift evolution of the mean metallicity $Z$ of baryonic matter. Dashed line: Diffuse intergalactic medium. Solid line: Warm-hot intergalactic medium. Dotted line: Star-forming gas ($n_H > 0.13\,\mathrm{cm}^{-3}$, $T < 10^7\,\mathrm{K}$, SFR $> 0$). [Data provided by D. Martizzi]

Another striking difference between the intergalactic medium and the interstellar medium of our galaxy is the low metallicity of the intergalactic medium. Figure 10.4 shows the evolution of metallicity in the Illustris TNG100 simulation. In the diffuse intergalactic medium, the mean metallicity was only $Z \sim 5 \times 10^{-5}\,Z_\odot$ at a redshift $z = 7$, increasing to $Z \sim 0.04\,Z_\odot$ at the present day. The warm-hot intergalactic medium, concentrated in the denser filaments, has a higher metallicity at any given redshift, ranging from $Z \sim 0.01\,Z_\odot$ at $z = 7$ to $Z \sim 0.2\,Z_\odot$ at the present. In the simulated universe, the metallicity in the dense star-forming regions of the universe ($n_H > 0.13\,\mathrm{cm}^{-3}$, star formation rate $> 0$) was already as high as $Z \sim 0.1\,Z_\odot$ at a redshift $z = 7$, and is (on average) supersolar today, with $Z \sim 1.7\,Z_\odot$.

As shown in Figure 10.5, the cooling function in the temperature range $10^5\,\mathrm{K} < T < 10^7\,\mathrm{K}$ is highly sensitive to metallicity. In the hot ionized medium of our own galaxy, the metallicity is near the solar value. We saw in Section 5.3 that cooling in the HIM is dominated by emission lines of carbon, oxygen, neon, and iron until a temperature $T \sim 2 \times 10^7\,\mathrm{K}$ is reached, and free–free emission (bremsstrahlung) takes over. In the lowest-metallicity regions of the warm-hot intergalactic medium, free–free emission dominates the cooling down to a temperature as low as $T \sim 10^6\,\mathrm{K}$.

The free–free luminosity density, normalized to the properties of the WHIM, is

$$\ell_{\mathrm{ff}} \approx 5.4 \times 10^{-35}\,\mathrm{erg\,cm}^{-3}\,\mathrm{s}^{-1} \left( \frac{kT}{0.1\,\mathrm{keV}} \right)^{1/2} \left( \frac{n_H}{5 \times 10^{-6}\,\mathrm{cm}^{-3}} \right)^2. \quad (10.2)$$

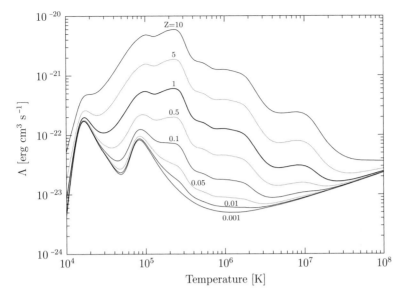

**Figure 10.5** Cooling function for atomic gas in collisional-ionization equilibrium at a range of metallicities (the numbers next to each line are the metallicity relative to solar). Compare with Figure 5.7 for a solar-metallicity gas. [Calculated using CHIANTI v9]

(Compare this with Equation 8.21 for the much higher luminosity density in the hotter, denser intracluster medium.) The energy density of the WHIM is

$$\varepsilon \approx 3n_\mathrm{H}kT \approx 2.4 \times 10^{-15}\,\mathrm{erg\,cm}^{-3}\left(\frac{kT}{0.1\,\mathrm{keV}}\right)\left(\frac{n_\mathrm{H}}{5 \times 10^{-6}\,\mathrm{cm}^{-3}}\right). \quad (10.3)$$

This yields a very long cooling time for the warm-hot intergalactic medium:

$$t_\mathrm{cool} = \varepsilon/\ell_\mathrm{ff} \approx 1400\,\mathrm{Gyr}\left(\frac{kT}{0.1\,\mathrm{keV}}\right)^{1/2}\left(\frac{n_\mathrm{H}}{5 \times 10^{-6}\,\mathrm{cm}^{-3}}\right)^{-1}. \quad (10.4)$$

The recombination time for low-density hydrogen at $T \sim 10^6\,$K, or $kT \sim 0.1\,$keV, is also a time much longer than the Hubble time (Equation 5.76):

$$t_\mathrm{rec} = \frac{1}{n_e\alpha} \approx 500\,\mathrm{Gyr}\left(\frac{kT}{0.1\,\mathrm{keV}}\right)^{3/2}\left(\frac{n_\mathrm{H}}{5 \times 10^{-6}\,\mathrm{cm}^{-3}}\right)^{-1}. \quad (10.5)$$

## 10.2 Observations

Observing free–free emission from the warm-hot intergalactic medium is "challenging." The reason is that neutral hydrogen in our own galaxy prevents free–free emission from all but the hottest portions of the WHIM from reaching us. The optical depth of neutral hydrogen to ionizing photons is $\tau_\nu = N_\mathrm{HI}\sigma_\mathrm{pho}$, where $\sigma_{pho} \propto \nu^{-3}$ is the photoionization cross section of a hydrogen atom. Using the cross section as normalized in Equation 5.80, we find that a line of sight is optically thick to ionizing photons of energy $h\nu$ when the foreground column density is

$$N_{HI} > 6.3 \times 10^{19} \, \text{cm}^{-2} \left( \frac{h\nu}{0.1 \, \text{keV}} \right)^3 . \tag{10.6}$$

Thus, the neutral hydrogen in our own galaxy (whose column density is plotted in Figure 3.2) is optically thick to ionizing photons with

$$h\nu < 0.12 \, \text{keV} \left( \frac{N_{HI}}{10^{20} \, \text{cm}^{-2}} \right)^{1/3} . \tag{10.7}$$

The Lockman Hole, with $N_{HI} \approx 6 \times 10^{19} \, \text{cm}^{-2}$, is optically thin to ionizing photons with energies as low as $h\nu \approx 0.1 \, \text{keV}$. However, along a more typical line of sight, with $N_{HI} \approx 2 \times 10^{21} \, \text{cm}^{-2}$, the hydrogen becomes optically thick at a photon energy $h\nu \approx 0.3 \, \text{keV}$. Along such a line of sight, extragalactic free–free emission at a temperature $kT < 0.3 \, \text{keV}$, or $T < 3.5 \times 10^6 \, \text{K}$, is mostly absorbed. Only the scarce high-energy photons from its exponentially drooping Boltzmann tail (Equation 8.16) can safely traverse our galaxy's neutral hydrogen.

An alternative way of observing the WHIM is by looking for absorption and emission lines from highly ionized metals. Consider oxygen, for instance, the most abundant element heavier than helium. The O VI doublet at $h\nu = 11.95 \, \text{eV}$ and $12.01 \, \text{eV}$, as discussed in Section 5.4, can be seen in absorption toward hot white dwarfs within the Local Bubble of hot ionized gas. Similarly, as mentioned in Section 8.1, the O VI doublet can be seen in absorption toward some hot stars in the halo of our galaxy, revealing the presence of ionized circumgalactic gas at $T \approx 3 \times 10^5 \, \text{K}$. If a region of the warm-hot ionized medium has $T \sim 3 \times 10^5 \, \text{K}$, it should produce O VI absorption lines in the spectra of background ultraviolet sources.

Detecting O VI in the warm-hot intergalactic medium is made difficult by the low number density of that ion. The WHIM at $T \sim 3 \times 10^5 \, \text{K}$ has a fairly low number density, typically $n_H \sim 10^{-6} \, \text{cm}^{-3}$. If the metallicity is $Z \approx 0.15 \, Z_\odot$, characteristic of the low-redshift WHIM, then the oxygen-to-hydrogen ratio is $n_O/n_H \approx 9 \times 10^{-5}$; this leads to an oxygen number density $n_O \sim 9 \times 10^{-11} \, \text{cm}^{-3}$. Even at its maximum relative abundance, at $T \approx 3 \times 10^5 \, \text{K}$, O VI accounts for only 25% of all the oxygen, leading to $n_{OVI} \sim 2 \times 10^{-11} \, \text{cm}^{-3}$. (This density is equivalent to a pair of OVI ions within the Royal Albert Hall.)

Since the WHIM is concentrated along filaments of the cosmic web (see Figure 10.1), a line of sight passing through a single filament whose thickness is $\ell \sim 1 \, \text{Mpc}$ will contribute a column density

$$N_{OVI} \sim n_{OVI}\ell \sim 7 \times 10^{13} \, \text{cm}^{-2} \left( \frac{\ell}{1 \, \text{Mpc}} \right) , \tag{10.8}$$

assuming a temperature $T \approx 3 \times 10^5$ K. As we saw when considering O VI absorption lines toward local white dwarfs, measuring column densities of this magnitude is difficult, but not impossible. Danforth and Shull, in a 2008 study of UV absorption lines toward 31 bright AGN, found O VI absorption systems with column densities ranging from $N_{\text{OVI}} \sim 8 \times 10^{12}$ cm$^{-2}$ to $N_{\text{OVI}} \sim 5 \times 10^{14}$ cm$^{-2}$. The conclusion of Danforth and Shull was that warm-hot intergalactic gas in the temperature range $10^5$ K $< T < 10^6$ K, where O VI absorption is strongest, provides $\sim$10% of the baryonic material in the universe. This still leaves a large amount of "missing" baryons in the intergalactic medium.

Detecting the hotter WHIM, in the range $10^6$ K $< T < 10^7$ K, is more difficult. Helium-like O VII is the dominant form of oxygen at $T \approx 10^6$ K. As discussed in Section 8.1, O VII has an X-ray line at $\lambda = 21.60$ Å ($h\nu = 0.574$ keV). In principle, this line can be detected in absorption toward distant X-ray sources. In practice, given the relatively low density of the warm-hot intergalactic medium, detecting O VII lines requires a long exposure. In Section 8.1 we discussed how a three-week observation with the *XMM-Newton* satellite revealed O VII and O VIII absorption lines at $z \approx 0$, associated with the circumgalactic medium of our galaxy. That same observation revealed higher-redshift O VII absorption lines, as shown in Figure 10.6.

The two absorption lines in Figure 10.6 are associated with two separate absorbers along the sightline to the background source, the X-ray-bright quasar 1ES 1553+113. The closer absorbing system is at a redshift $z = 0.355$, corresponding to a distance $d \sim 1500$ Mpc; its absorption line is best described by

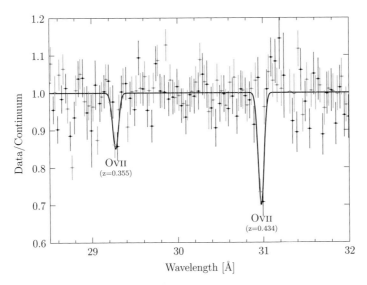

**Figure 10.6** Portion of the *XMM-Newton* X-ray spectrum of the quasar 1ES 1553+113. Intervening absorption lines from O VII are seen at redshifts $z = 0.355$ and $z = 0.434$. [Data from Nicastro et al. 2018]

a system with $T \sim 0.8 \times 10^6$ K and column density $N_H \sim 0.6 \times 10^{20}$ cm$^{-2}$, assuming a metallicity $Z = 0.15\,Z_\odot$. The more distant absorbing system is at a redshift $z = 0.434$, corresponding to a distance $z \sim 1800$ Mpc; its absorption line is best described by a system with $T \sim 1.1 \times 10^6$ K and a column density $N_H \sim 1 \times 10^{20}$ cm$^{-2}$, again assuming $Z = 0.15\,Z_\odot$.

Although the line of sight to the quasar 1ES 1553+113 represents just one stab through the intergalactic medium, it does reveal a component of the WHIM that has $T \sim 10^6$ K. The hottest component of the WHIM, with $T > 3 \times 10^6$ K, remains elusive. At a temperature $T = 10^7$ K, or $kT = 0.8$ keV, photons produced by free–free emission are energetic enough to traverse the neutral hydrogen within our own galaxy. However, the portion of the WHIM at $T = 10^7$ K has a typical density $n_H \approx 10^{-4}$ cm$^{-3}$. Although we can detect free–free emission from the intracluster medium of the Coma Cluster (see Figure 8.6), the gas in the hottest WHIM is lower in temperature by a factor $\sim 0.1$, and lower in density by a factor $\sim 0.03$. This means that the WHIM free–free power per unit volume is smaller than that of Coma by a factor (Equation 8.21)

$$\frac{\ell_{ff}(\text{whim})}{\ell_{ff}(\text{Coma})} = \left(\frac{T_{\text{whim}}}{T_{\text{Coma}}}\right)^{1/2}\left(\frac{n_{\text{whim}}}{n_{\text{Coma}}}\right)^2 \sim 3 \times 10^{-4}. \tag{10.9}$$

This low power density makes the detection of free–free emission from the hottest WHIM a difficult task. At $T = 10^7$ K, absorption lines from highly ionized iron are an alternative diagnostic tool. In particular, Fe xx has an X-ray line at $h\nu = 0.967$ keV that is potentially useful. Once again, however, the low density of intergalactic space poses a challenge for detection. In collisional ionization equilibrium at $T \approx 10^7$ K, the Fe xx ion represents at most 20% of the iron atoms. Assuming a metallicity $Z = 0.15\,Z_\odot$, the typical density $n_H \sim 10^{-4}$ cm$^{-3}$ at $T = 10^7$ K leads to a number density of Fe xx ions that is only $n_{\text{Fexx}} \sim 10^{-10}$ cm$^{-3}$.

Although X-ray observations have been tremendously useful in tracking down the "missing" baryons, the highest-temperature portion of the warm-hot intergalactic medium is still elusive (at the time of writing). These baryons are not truly missing, however; they are patiently waiting for us to develop higher-throughput X-ray spectrographs with high energy resolution. Then, at last, they will be able to get their message through to us.

## Exercises

10.1    Figure 10.6 shows two absorption lines. at $\lambda = 29.27$Å and $30.97$Å, and asserts that they represent the O vii 21.60Å absorption line at redshifts $z = 0.355$ and $z = 0.434$. An alternative hypothesis would be that they represent the O viii 18.97Å absorption line at redshifts $z = 0.543$ and $z = 0.633$. Give at least two observations you could make, at any wavelength, that would help to distinguish between these two hypotheses (O vii versus O VIII). Now look at the original paper in which the data were presented and interpreted (Nicastro et al. 2018, Nature, 558, 406). Do the reasons given by Nicastro et al. for the O VII hypothesis seem convincing to you? Explain why or why not.

10.2 Numerical cosmological simulations usually don't go as far back in time as the "deep" dark ages, at a redshift range $100 < 1+z < 1000$. The reason is that the deep dark ages (DDA) were physically simple. The gas filling the universe was nearly homogeneous, and the radiation filling the universe consisted solely of the cosmic background radiation, well described as blackbody radiation at a temperature

$$T_{\text{rad}} = 2725.5 \, \text{K} \left( \frac{1+z}{1000} \right). \tag{10.10}$$

There was no significant photoionization during the DDA and no significant radiative recombination. Thus, the fractional ionization $x = n_e/n_H$ was constant with time at a very low level, $x \sim 10^{-3}$.

(a) As a function of redshift, what was the number density of photons, $n_\gamma(z)$, during the DDA? Assuming a pure hydrogen baryonic gas, what were the number density of hydrogen atoms, $n_{H1}(z)$, and the number density of free electrons, $n_e(z)$? [Scale your results to a fractional ionization $x = 10^{-3}$.]

(b) If the hydrogen gas didn't interact with the cosmic background radiation, its temperature would be dictated purely by adiabatic cooling as the universe expands. If $T_{\text{gas}} = T_{\text{rad}}$ at the start of the DDA, what would be $T_{\text{gas}}(z)$ in the absence of heating?

(c) The main mechanism by which the hydrogen gas interacted with the cosmic background radiation was scattering by free electrons. For a free electron during the DDA, what was the mean time between scattering events, $t_{\text{scat}}$, as a function of redshift?

(d) The scattering of photons by non-relativistic electrons is a very inefficient way of transferring energy between photons and electrons. If a partially ionized gas has temperature $T_{\text{gas}} \ll T_{\text{rad}}$, then the typical energy transferred to a free electron by a single scattering event will be

$$\Delta E \approx \frac{(kT_{\text{rad}})^2}{m_e c^2}. \tag{10.11}$$

Assuming that no cooling mechanisms are at work, what is the typical time $t_{\text{heat}}$ required to heat the gas from $T_{\text{gas}} \ll T_{\text{rad}}$ to $T_{\text{gas}} = T_{\text{rad}}$? Give a relation for $t_{\text{heat}}$ as a function of redshift during the DDA.

(e) As long as $t_{\text{heat}} < t$, where $t$ is the age of the universe, the hydrogen gas is thermally coupled to the cosmic background radiation, and remains at a temperature $T_{\text{gas}} = T_{\text{rad}}$. At what redshift $z_{\text{dec}}$ and time

$t_{dec}$ does the gas thermally decouple from the cosmic background radiation? What is the temperature $T_{rad}$ at this redshift?

(f) Bonus round: using Equation 1.13, verify that a hydrogen atom's mean time between collisions with a free electron, $t_{coll}$, is short compared with the age of the universe during the deep dark ages, and thus that the hydrogen atoms are thermally coupled to the free electrons.

# Bibliography, References, and Figure Credits

*Explicit hoc totum; Pro Christi da mihi potum*
*(The job is done, I think; For Christ's sake, give me a drink)*

Copyist's colophon found at the end
of a number of medieval manuscripts

## Bibliography

There are a number of classic advanced textbooks and monographs on the physics of the interstellar and intergalactic medium that you will find on the bookshelves of astrophysicists working in this field. We referred to the following works in the preparation of this book.

*The Astrophysics of Gaseous Nebulae and Active Galactic Nuclei* by Donald E. Osterbrock and Gary J. Ferland, 2005, second edition, University Science Books. Classic text on ionized nebulae from H II regions to active galactic nuclei with a strong emphasis on the physical processes in ionized gas.

*Physics of the Interstellar and Intergalactic Medium* by Bruce T. Draine, 2011, Princeton Series in Astrophysics, Princeton University Press. Encyclopaedic, particular strong on interstellar dust (Draine's research speciality). Presents empirical fitting formulas more complex and accurate than the simple power laws used in this text. Online ancillaries include an extensive collection of diverse and challenging problems.

*The Physics and Chemistry of the Interstellar Medium* by Alexander G. G. M. Tielens, 2005, Cambridge University Press. The main emphasis is on chemistry and dust properties plus molecules and photodissociation regions. A very lucid overview of areas we do not cover in depth in this book.

*Astrophysics of the Diffuse Universe* by Michael A. Dopita and Ralph S. Sutherland, 2003, Astronomy and Astrophysics Library, Springer Verlag. A more advanced monograph, with a thorough discussion of atomic spectroscopy and notation not covered in this text and a detailed exploration of shocks in interstellar gas.

*Physical Processes in the Interstellar Medium* by Lyman G. Spitzer, 1978, John Wiley & Sons. An out-of-print classic, available as a 1998 reprint from Wiley-VCH. Compact yet deep discussion of physical processes; often employs idiosyncratic notation compared with modern texts.

We also drew upon course notes taken by RWP when he attended a graduate course given by Donald Osterbrock, and a set of notes given to RWP by Harriet Dinerstein at UT Austin in 1989.

# References

In our reference list, we use compact abbreviations for the most frequently cited journals in astronomy. These abbreviations, as used by the SAO/NASA Astrophysics Data System, are:

- A&A = Astronomy and Astrophysics
- AJ = The Astronomical Journal
- ApJ = The Astrophysical Journal
- ApJS = The Astrophysical Journal Supplement
- ARA&A = Annual Review of Astronomy and Astrophysics
- MNRAS = Monthly Notices of the Royal Astronomical Society
- PASP = Publications of the Astronomical Society of the Pacific

Adam, R., et al. 2016, "*Planck* 2015 results. X. Diffuse component separation: Foreground maps," A&A, 594, A10

Baker, J. G., & Menzel, D. H. 1938, "Physical processes in gaseous nebulae. III. The Balmer decrement," ApJ, 88, 52

Barkana, Rennan, 2006, "The first stars in the universe and cosmic reionization," Science, 313, 931

Barnard, E. E. 1899, "Photograph of the Milky Way near the star Theta Ophiuchi," ApJ, 9, 157

Barnard, E. E., & Ranyard, A. C. 1894, "Structure of the Milky Way," Knowledge: An Illustrated Magazine of Science, 17, 253

Ben Bekhti, N., et al. [HI4PI collaboration] 2016, "HI4PI: A full-sky HI survey based on EBHIS and GASS," A&A, 594, A116

Bosman, S. E. I., Fan, X., Jiang, L., Reed, S., Matsuoka, Y., Becker, G., & Haehnelt, M. 2018, "New constraints on Lyman-$\alpha$ opacity with a sample of 62 quasars at $z > 5.7$," MNRAS, 479, 1055

Bowen, I. S. 1927, "The origin of the nebulium spectrum," Nature, 120, 473

Boyle, Robert, 1669, *A Continuation of New Experiments Physico-Mechanical, Touching the Spring and Weight of the Air, and their Effects,* Oxford: Henry Hall, printer

Cardelli, J. A., Clayton, G. C., & Mathis, J. S. 1989, "The relationship between infrared, optical, and ultraviolet extinction," ApJ, 345, 245

Carruthers, George R. 1970, "Rocket observation of interstellar molecular hydrogen," ApJ, 161, L81

Cox, D. P., & Smith, B. W. 1974, "Large-scale effects of supernova remnants on the galaxy: Generation and maintenance of a hot network of tunnels," ApJ, 189, L105

Croxall, K. V., Pogge, R. W., Berg, D. A., Skillman, E. D., & Moustakas, J. 2015, "CHAOS. II. Gas-phase abundances in NGC 5194," ApJ, 808, 42

Danforth, C. W., & Shull, J. M. 2008, "The low-$z$ intergalactic medium. III. HI and metal absorbers at $z < 0.4$," ApJ, 679, 194

Dartois, E., et al. 1998, "Dectection of the '44 μm' band of water ice in absorption in combined ISO SWS-LWS spectra," A&A, 338, L21

Das, S., Mathur, S., Nicastro, F., & Krongold, Y. 2019a, "Discovery of a very hot phase of the Milky Way circumgalactic medium with non-solar abundance ratios," ApJ, 882, L23

Das, S., Mathur, S., Gupta, A., Nicastro, F., & Krongold, Y. 2019b, "Multiple temperature components of the circumgalactic medium of the Milky Way," ApJ, 887, 257

Dickey, J. M., Kulkarni, S. R., van Gorkom, J. H., & Heiles, C. E. 1983, "A survey of HI absorption at low latitudes," ApJS, 53, 591

D'Onghia, E., & Fox, A. J. 2016, "The Magellanic Stream: Circumnavigating the Galaxy," ARA&A, 54, 363

Dulieu, F., Congiu, E., Noble, J., Baouche, S., Chaabouni, H., Moudens, A., Minissale, M., & Cazaux, S. 2013, "How micron-sized dust particles determine the chemistry of our universe," Scientific Reports, 3, 1338

Dunham, Theodore, Jr. 1937, "Interstellar neutral potassium and neutral calcium," PASP, 49, 26

Ewen, H. I., & Purcell, E. M. 1951, "Observation of a line in the galactic radio spectrum: Radiation from galactic hydrogen at 1,420 Mc/sec," Nature, 168, 356

Fan, X., et al. 2001, "A survey of $z > 5.8$ quasars in the Sloan Digital Sky Survey. I. Discovery of three new quasars and the spatial density of luminous quasars at $z \sim 6$," AJ, 122, 2833

Ferland, G. 2003, "Quantitative spectroscopy of photoionized clouds," ARA&A, 41, 517

Ferrière, Katia M. 2001, "The interstellar environment of our galaxy," Rev. Mod. Phys., 73, 1031

Field, G. B., Goldsmith, D. W., & Habing, H. J. 1969, "Cosmic-ray heating of the interstellar gas," ApJ, 155, L149

Finkbeiner, Douglas P. 2003, "A full-sky H$\alpha$ template for microwave foreground prediction," ApJS, 146, 407

Gastaldello, F., et al. 2015, "A NuSTAR observation of the center of the Coma Cluster," ApJ, 800, 139

Gnedin, N. Y., & Hollon, N. 2012, "Cooling and heating functions of photoionized gas," ApJS, 202, 13

Green, Martin A. 2008, "Self-consistent optical parameters of intrinsic silicon at 300 K including temperature coefficients," Solar Energy Materials and Solar Cells, 92, 1305

Groenewegen, M. A. T., Whitelock, P. A., Smith, C. H., & Kerschbaum, F. 1998, "Dust shells around carbon Mira variables," MNRAS, 293, 18

Gunn, J. E., & Peterson, B. A. 1965, "On the density of neutral hydrogen in intergalactic space," ApJ, 142, 1633

Hartmann, J. 1904, "Investigations on the spectrum and orbit of $\delta$ Orionis," ApJ, 19, 268

Heger, Mary Lea, 1919, "Stationary sodium lines in spectroscopic binaries," PASP, 31, 304

Heyer, M. H., & Brunt, C. M. 2004, "The universality of turbulence in galactic molecular clouds," ApJ, 615, L48

Hollenbach, D. J., & Tielens, A. G. G. M., 1997, "Dense photodissociation regions (PDRs)," ARA&A, 35, 179

Huggins, William, 1864, "On the spectra of some of the nebulae," Phil. Trans. R. Soc., 154, 437

Kahn, F. D. 1954, "The acceleration of interstellar clouds," Bull. Astron. Inst. Netherlands, 12, 187

Kamble, K., Dawson, K., du Mas des Bourboux, H., Bautista, J., & Schneider, D. P. 2020, "Measurements of effective optical depth in the Ly $\alpha$ forest from the BOSS DR12 quasar sample," ApJ, 892, 70

Kamm, J. R. 2000, "Evaluation of the Sedov–von Neumann–Taylor blast wave solution," Los Alamos National Laboratory Report LA-UR-00-6055

Landau, L. D., & Lifshitz, E. M. 1987, *Fluid Mechanics*, second edition, Butterworth-Heinemann

Li, A., & Draine, B. T. 2001, "Infrared emission from interstellar dust. II. The diffuse interstellar medium," ApJ, 554, 778

Linsky, J. L., & Wood, B. E. 1996, "The $\alpha$ Centauri line of sight: D/H ratio, physical properties of local interstellar gas, and measurement of heated hydrogen (the 'hydrogen wall') near the heliopause," ApJ, 463, 254

Lodders, Katharina, 2010, "Solar system abundances of the elements," in *Principles and Perspectives in Cosmochemistry. Astrophysics & Space Science Proceedings,* Goswami, A., & Reddy, B., eds., Springer-Verlag

Lynds, Roger, 1971, "The absorption-line spectrum of 4C 05.34," ApJ, 164, L73

Lynds, R., & Wills, D. 1970, "The unusually large redshift of 4C 05.34," Nature, 226, 532

Madau, P., & Dickinson, M. 2014, "Cosmic star-formation history," ARA&A, 52, 415

Martins, F., Schaerer, D., & Hillier, D. J. 2005, "A new calibration of stellar parameters of Galactic O stars," A&A, 436, 1049

Martizzi, D., et al. 2019, "Baryons in the Cosmic Web of IllustrisTNG – 1: Gas in knots, filaments, sheets, and voids," MNRAS, 486, 3766

Mathis, J. S., Mezger, P. G., & Panagia, N. 1983, "Interstellar radiation field and dust temperatures in the diffuse interstellar matter and in giant molecular clouds," A&A, 128, 212

Mathis, J. S., Rumpl, W., & Nordsieck, K. H. 1977, "The size distribution of interstellar grains," ApJ, 217, 425

McKee, C. F., & Ostriker, J. P. 1977, "A theory of the interstellar medium: Three components regulated by supernova explosions in an inhomogeneous substrate," ApJ, 218, 148

Molaro, P., Vladilo, G., Monai, S., D'Odorico, S., Ferlet, R., Vidal-Madjar, A., & Dennefeld, M. 1993, "Interstellar Ca II and Na I in the SN 1987A field. I. Foreground and intermediate velocity gas," A&A, 274, 505

Nicastro, F., et al. 2018, "Observations of the missing baryons in the warm-hot intergalactic medium," Nature, 558, 406

O'Donnell, James E. 1994, "$R_V$-dependent optical and near-ultraviolet extinction," ApJ, 422, 158

Paglione, T. A. D., et al. 2001, "A mapping survey of the $^{13}CO$ and $^{12}CO$ emission in galaxies," ApJS, 135, 183

Paton, C., Hellstrom, J., Paul, B., Woodhead, J., & Hergt, J. 2011, "A PCA study to determine how features in meteorite reflectance spectra vary with the samples' physical properties," J. Quant. Spectrosc. Radiat. Transfer, 112, 1803

Porter, T. A., Jóhannesson, G., & Moskalenko, I. V. 2017, "High-energy gamma rays from the Milky Way: Three-dimensional spatial models for the cosmic-ray and radiation field densities in the interstellar medium," ApJ, 846, 67 [GALPROP spectral energy distribution]

Richter, P., et al. 2017, "An HST/COS legacy survey of high-velocity ultraviolet absorption of the Milky Way's circumgalactic medium and the Local Group," A&A 607, A48 [Si III in CGM]

Savage, B. D., & Lehner, N. 2006, "Properties of O VI absorption in the local interstellar medium," ApJS, 162, 134

Sedov, L. I. 1946, "Rasprostraneniya sil'nykh vzryvnykh voln [Propagation of strong blast waves]," Prikladnaya matematika i mekhanika, 10, 241

Sembach, K. R., et al. 2003, "Highly ionized high-velocity gas in the vicinity of the Galaxy," ApJS, 146, 165 [O VI in CGM]

Sharp, T. E. 1971, "Potential-energy curves for molecular hydrogen and its ions," Atomic Data, 2, 119

Shu, Frank H. 1992, *The Physics of Astrophysics. Volume II. Gas Dynamics*, University Science Books

Slipher, V. M. 1913, "On the spectrum of the nebula in the Pleiades," Popular Astronomy, 21, 186

Speck, A. K., Barlow, M. J., Sylvester, R. J., & Hofmeister, A. M. 2000, "Dust features in the 10 μm infrared spectra of oxygen-rich evolved stars," A&AS, 146, 437

Spitzer, Lyman, 1956, "On a possible interstellar galactic corona," ApJ, 124, 20

Spitzer, L., & Fitzpatrick, E. L. 1993, "Composition of interstellar clouds in the disk and halo. I. HD 93521," ApJ, 409, 299

Strömgren, Bengt, 1939, "The physical state of interstellar hydrogen," ApJ, 89, 526

Swings, P., & Rosenfeld, L. 1937, "Considerations regarding interstellar molecules," ApJ, 86, 483

Taylor, Sir Geoffrey, 1950, "The formation of a blast wave by a very intense explosion. I. Theoretical discussion," Proc. R. Soc. London, Series A, 201, 159 [Taylor wrote this paper in 1941; it was declassified in 1949.]

Topping, M. W., & Shull, J. M. 2015, "The efficiency of stellar reionization: Effects of rotation, metallicity, and initial mass function," ApJ, 800, 97

Trumpler, Robert J. 1930, "Preliminary results on the distance, dimensions and space distribution of open star clusters," Lick Observatory Bulletin no. 420, 14, 154

van de Hulst, H. C. 1945, "Herkomst der radiogolven uit het wereldruim [Origin of the radio waves from space]," Nederlandsch Tijdschrift voor Natuurkunde, 11, 210 [English translation in *Classics in Radio Astronomy*, edited by W. T. Sullivan III, 1982, D. Reidel Publishing, p. 302]

Very, Frank W. 1911, "Are the white nebulae galaxies?" Astron. Nachr., 189, 441

Weingartner, J. C., & Draine, B. T. 2001, "Dust grain-size distributions and extinction in the Milky Way, Large Magellanic Cloud, and Small Magellanic Cloud," ApJ, 548, 296

Westheimer, Tobias, 2018, "A new all-sky map of Galactic high-velocity clouds from the 21 cm PI4PI survey," MNRAS, 474, 289

Zwicky, F. 1933, "Die Rotverschiebung von extragalaktischen Nebeln [The redshift of extragalactic nebulae]," Helvetica Physica Acta, 6, 110

## Figure Credits

Astronomical images were obtained from a variety of sources with a preference for public-domain images or those licensed under Creative Commons. When we used figures from peer-reviewed scientific papers, we followed the publishers' guidelines for securing the necessary permissions. A number of the all-sky maps were created from data obtained from the Legacy Archive for Microwave Background Data Analysis (LAMBDA) at the NASA Goddard Space Flight Center (lambda.gsfc.nasa.gov), replotting them to ensure a common presentation throughout, using a Mollweide projection of galactic coordinates centered on the galactic center (except where noted).

Plots created by the authors were composed in Jupyter notebooks with code written in Python using the open-source Anaconda Scientific Python 3 Distribution (anaconda.com) and the matplotlib plotting package (Hunter 2007, CiSE, 9, 90). All-sky data in FITS HEALPix format were processed using the Python healpy package (healpy.readthedocs.io). Additional illustrations created by the authors were composed in Microsoft PowerPoint and post-processed using the GNU Image Manipulation Program (GIMP v2.8.18).

Collisional ionization equilibrium calculations were performed using ChiantiPy v0.9.5 (github.com/chianti-atomic/ChiantiPy) and the CHIANTI Atomic Database v9.0 (www.chiantidatabase.org). CHIANTI is a collaborative project involving George Mason University, the University of Michigan, the University of Cambridge, and NASA Goddard Space Flight Center; Dere et al. 1997, A&AS, 125, 149 and Landi et al. 2013, ApJ, 763, 86. Photoionization equilibrium calculations were performed using Cloudy C17.02; Ferland et al. 2017, RevMexAA, 53, 385 (www.nublado.org). Analysis of nebular emission lines was performed using PyNeb v1.1.12; Luridiana et al. 2015, A&A, 573, 42 (research.iac.es/proyecto/PyNeb).

**Dedication:** Detail of the presentation scene from the Talbot Shrewsbury Book, AD 1444/45. Miniature attributed to the Talbot Master. Royal MS 15 E VI, fol. 2v, held and digitized by the British Library.

**Figure 1.1:** Drawn by the authors using data from Nicastro et al. 2018, Nature, 558, 406, and Martizzi et al. 2019, MNRAS, 486, 3766.

**Figure 1.2:** Reproduced from Plate X, Figure 5 of Huggins 1864, Phil. Trans. R. Soc., 154, 437. [Huggins' notes on this object are: "A planetary nebula; very bright; pretty small; suddenly brighter in the middle, very small nucleus."]

**Figure 1.3:** Image of the Pleiades from the Palomar Observatory – STScI Digitized Sky Survey, based on scans of the Palomar Observatory Sky Survey II (POSS-II), ©1993–95 California Institute of Technology. Image center is $\alpha_{J2000}$=03:46.6, $\delta_{J2000}$=+24:06.2 in the digitized red (Kodak IIIa-F) plate image. Processed by the authors using data obtained from the ESO Online Digital Sky Survey (archive.eso.org/dss/dss).

**Figure 1.4:** Image of Barnard 68 obtained with the FORS1 multimode instrument at the 8.2m VLT ANTU telescope. FORS team, VLT, ESO. ESO image 0102a, ©ESO (www.eso.org/public/news/eso0102/). Converted to grayscale by the authors.

**Figure 1.5:** Image of the Orion Nebula taken with the Advanced Camera for Surveys on the *Hubble Space Telescope* (spacetelescope.org image heic0601a). NASA, ESA, M. Robberto, and the HST Orion Treasury Project Team. Converted to grayscale by the authors.

**Figure 1.6:** Composite image of the Ring Nebula taken with the Wide Field Camera 3 on the *Hubble Space Telescope* (hubblesite.org image 3170). NASA, ESA, and the Hubble Heritage (STScI/AURA) – ESA/Hubble Collaboration. Converted to grayscale by the authors.

**Figure 1.7:** Image of the Crab Nebula taken with the Wide Field Planetary Camera 2 on the *Hubble Space Telescope* (hubblesite.org image 3885). NASA, ESA, and the Hubble Heritage Team (STScI/AURA). Converted to grayscale by the authors.

**Figure 1.8:** Image of the Cygnus Loop taken with the MOSAIC camera on the WIYN 0.9-meter telescope on Kitt Peak. T. A. Rector (University of Alaska, Anchorage) and WIYN/NOAO/AURA/NSF. Converted to inverse grayscale by the authors.

**Figure 1.9:** Composite all-sky H$\alpha$ map, using data from WHAM, VTSS, and SHASSA, as combined by Finkbeiner 2003, ApJS, 146, 407. Processed by the authors using FITS healpix data obtained from the NASA GSFC LAMBDA database.

**Figures 1.10–1.12:** Calculated and plotted by the authors using the Lodders 2010 C and O abundances.

**Figure 1.13:** Redrawn by the authors; the figure is based on Figures 1 and 2 of McKee & Ostriker 1977, ApJ, 218, 148.

**Figure 2.1:** Reproduced from the bottom three panels of Figure 7 in Molaro et al. 1993, A&A, 274, 505. ©ESO, reproduced with permission.

**Figure 2.2:** Reproduced from Figure 2 in Spitzer & Fitzpatrick 1993, ApJ, 409, 299. ©AAS, reproduced with permission.

**Figure 2.3:** Plotted by the authors using spectra taken with the Large Binocular Telescope Multi-Object Double Spectrograph 1 (MODS1) by K. Croxall & R. Pogge, Ohio State University.

**Figures 2.4–2.7:** Calculated and plotted by the authors.

**Figure 2.8:** Plotted by the authors using spectra taken with the High-Resolution Spectrograph on the *Hubble Space Telescope*, following Linsky & Wood 1996, ApJ, 463, 254. Data were retrieved from the Mikulski Archive for Space Telescopes (MAST; archive.stsci.edu).

**Figure 3.1:** Drawn by the authors.

**Figure 3.2:** Plotted by the authors using data from the HI4PI survey, following Figure 2 of Ben Bekhti et al. 2016, A&A, 594, A116. The HI4PI survey is constructed from the Effelsberg–Bonn HI Survey, made with the 100 m radio telescope at Effelberg, Germany, and the Galactic All-Sky Survey, made with the 64 m radio telescope near Parkes, Australia. Data for the plot were obtained from the NASA GSFC LAMBDA database and processed by the authors using the healpy package.

**Figure 3.3:** Reproduced from the upper right panel of Figure 12 in Dickey et al. 1983, ApJS, 53, 591, with axis labels added by the authors. ©AAS, reproduced with permission.

**Figure 3.4:** Plotted by the authors using the interstellar radiation field of the GALPROP v56 code for cosmic-ray transport and diffuse emission production (galprop.stanford.edu). Data from Porter et al. 2017, ApJ, 846, 67, using the model file for x,y,z=(8.5,0,0) kpc, plus the radiative transfer dust and stellar model of Robataille et al. 2012, A&A, 545, 39.

**Figures 4.1–4.3:** Calculated and plotted by the authors.

**Figure 4.4:** Plotted by the authors using a Cloudy C17.02 photoionization equilibrium calculation for a $T = 40\,000$ K blackbody, $n_e = 100$ cm$^{-3}$, and a solar mix of elements. (This is meant to be merely representative; see Ferland 2003, ARA&A, 41, 517 or Gnedin & Hollon 2012, ApJS, 202, 13 for more sophisticated treatments.)

**Figure 4.5:** Plotted by the authors using spectra obtained with the MODS1 spectrograph on the Large Binocular Telescope (from Croxall et al. 2015, ApJ, 808, 42.)

**Figures 4.6–4.7:** Drawn by the authors.

**Figure 4.8:** Calculated and plotted by the authors using the PyNeb v1.1.12 python package.

**Figure 4.9:** Drawn by the authors.

**Figure 4.10:** Calculated and plotted by the authors.

**Figure 4.11:** Drawn by the authors.

**Figures 5.1–5.3:** Drawn by the authors.

**Figure 5.4:** Calculated and plotted by the authors using the numerical evaluation of Kamm 2000 for the Sedov–Taylor solution. A spherical geometry was assumed ($j = 3$), with an initially uniform distribution ($\omega = 0$) of monatomic gas ($\gamma = 5/3$).

**Figure 5.5:** Drawn by the authors, following Shu 1992, *Physics of Astrophysics, Volume II*, Figure 17.4.

**Figure 5.6:** Calculated and plotted by the authors using ChiantiPy and the CHIANTI atomic database v9, assuming collisional ionization equilibrium.

**Figure 5.7:** Calculated and plotted by the authors using ChiantiPy and the CHIANTI atomic database v9, assuming a solar abundance of elements and collisional ionization equilibrium.

**Figure 5.8:** Simulated spectrum calculated and plotted by the authors using ChiantiPy and the CHIANTI atomic database v9, assuming $T = 10^6$ K, $n = 0.005$ cm$^{-3}$, and solar abundance.

**Figure 5.9:** Plotted by the authors, following Figure 1 of Savage & Lehner 2006, ApJS, 162, 134, and using data in their Tables 1 and 3. Data were replotted in a Mollweide projection.

**Figure 6.1:** Reproduced from Plate II of Barnard 1899, ApJ, 9, 157, rotated 90° counterclockwise, and cropped by the authors.

**Figure 6.2:** Plotted by the authors using data from Trumpler 1930, Lick Observatory Bulletin no. 420, 14, 154, Table 3.

**Figure 6.3:** Calculated and plotted by the authors using the parameterization of Cardelli, Clayton, & Mathis 1989, ApJ, 345, 245 and the updated optical/IR coefficients of O'Donnell 1994, ApJ, 422, 158.

**Figure 6.4:** Plotted by the authors using data from Green 2008, Solar Energy Materials and Solar Cells, 92, 1305, Table 1.

**Figure 6.5:** Calculated and plotted by the authors.

**Figure 6.6:** Plotted by the authors using data from the USGS Spectral Library Version 7 (Kokaly et al. 2017, US Geological Survey Data Series 1035, doi.org/10.3133/ds1035), following Figure 1 of Paton et al. 2011, J. Quant. Spectrosc. Radiat, Transfer, 112, 1803.

**Figure 6.7:** Plotted by the authors from ISO archival SWS/LWS spectral data provided by Adwin Boogert, University of Hawaii, originally published by Dartois et al. 1998, A&A, 338, L21.

**Figure 6.8:** Plotted by the authors following Figure 8 of Li & Draine 2001, ApJ, 554, 778, using data provided by Aigen Li, University of Missouri.

**Figure 6.9:** Wire-frame molecular structures from the Chemical Entities of Biological Interest (ChEBI) Database (www.ebi.ac.uk/chebi).

**Figure 6.10:** Calculated and plotted by the authors, following the parameterization of Weingartner & Draine 2001, ApJ, 548, 296, using a Python version of the Fortran subroutine written by J. Weingartner (physics.gmu.edu/~joe/sizedists .html).

**Figure 6.11:** Reproduced from the top two rows of Figure 4 in Speck et al. 2000, A&AS, 146, 437 with axis labels added by the authors. ©ESO, reproduced with permission.

**Figure 6.12:** Reproduced from the bottom panel of Figure 3 in Groenewegen et al. 1998, MNRAS, 293, 18. ©Oxford University Press, reproduced with permission.

**Figure 7.1:** Plotted by the authors using data from Sharp 1971, Atomic Data, 2, 119.

**Figure 7.2:** Reproduced from Figure 1f in Paglione et al. 2001, ApJS, 135, 183. ©AAS, reproduced with permission.

**Figure 7.3:** Plotted by the authors, following Figure 13 of Adam et al. 2016, A&A, 594, A10, and using the CO line emission map from the ESA Planck Legacy Archive (wiki.cosmos.esa.int/planck-legacy-archive).

**Figure 7.4:** Drawn by the authors, following Figure 1 of Dulieu et al. 2013, Scientific Reports, 3, 1338.

**Figure 7.5:** Plotted by the authors using data from Sharp 1971, Atomic Data, 2, 119.

**Figure 7.6:** Drawn by the authors, following the analysis of Hollenbach & Tielens 1997, ARA&A, 35, 179.

**Figure 8.1:** Plotted by the authors, following Figure A1 of Westmeier 2018, MNRAS, 474, 299 (www.atnf.csiro.au/people/Tobias.Westmeier). Unlike the other all-sky maps in this book, this Mollweide projection is centered on the galactic anticenter.

**Figure 8.2:** Plotted by the authors using data from Figure 2 of Das et al. 2019a, ApJ, 882, L23, with the assistance of S. Das and S. Mathur, Ohio State University.

**Figure 8.3:** Replotting of Figure 2a of Das et al. 2019b, ApJ, 887, 257, using data provided to the authors by S. Das and S. Mathur, Ohio State University.

**Figure 8.4:** Near-IR image was adapted from *Spitzer* sig14-004a (a composite of 3.6 μm and 4.5 μm data). The X-ray image was adapted from *Chandra* ACIS image (a composite of data from the 0.3 → 1.1 keV, 0.7 → 2.2 keV, and 2.2 → 6 keV bands). The images were converted to inverse grayscale and cropped to the same field of view by the authors.

**Figure 8.5:** Visible image is from the Sloan Digital Sky Survey. The X-ray image is from *XMM/Newton*. NASA/CXC/MPE/J. Sanders et al. The images were converted to inverse grayscale and cropped to the same field of view by the authors.

**Figure 8.6:** Replotting of the upper panel of Figure 3 of Gastaldello et al. 2015, ApJ, 800, 139, using data provided by F. Gastaldello, INAF-IASF Milan.

**Figure 9.1:** Plotted by the authors using data provided by W. Keel, University of Alabama (www.astr.ua.edu/keel/agn/forest.html).

**Figure 9.2:** Plotted by the authors using data provided by F. Wang, University of Arizona, following Figure 5 of Fan et al. 2001, AJ, 122, 2833.

**Figure 9.3:** Drawn by the authors, inspired by Figure 1 of Barkana 2006, Science, 313, 931.

**Figure 9.4:** Plotted by the authors using data from Kamble et al. 2020, ApJ, 892, 70 ($z < 4$) and Bosman et al. 2018, MNRAS, 479, 1055 ($z > 5$).

**Figure 9.5:** Plotted by the authors using calculations made using Code for Anisotropies in the Microwave Background, obtained from the NASA GSFC LAMBDA website (lambda.gsfc.nasa.gov/toolbox).

**Figure 9.6:** Calculated and plotted by the authors using Equation 15 from Madau & Dickinson 2014, ARA&A, 52, 415.

**Figure 9.7:** Plotted by the authors using archival data from the ESO VLT UVES spectrograph obtained from the ESO Science Archive Facility (archive.eso.org) and processed by S. Frank, Ohio State University.

**Figure 10.1:** Reproduced, with modification, from Figure 8 of Martizzi et al. 2019, MNRAS, 486, 3766. ©Oxford University Press, reproduced with permission.

**Figure 10.2:** Plotted by the authors using data from Martizzi et al. 2019, MNRAS, 486, 3766, with additional data provided by D. Martizzi, University of Copenhagen.

**Figure 10.3:** Reproduced, with modification, from Figure 10 of Martizzi et al. 2019, MNRAS, 486, 3766. ©Oxford University Press, reproduced with permission.

**Figure 10.4:** Plotted by the authors using data from the Illustris TNG100 simulation provided by D. Martizzi.

**Figure 10.5:** Calculated and plotted by the authors using ChiantiPy and the CHIANTI atomic database v9, assuming collisional ionization equilibrium.

**Figure 10.6:** Plotted by the authors using data from Figures 1 and 2 of Nicastro et al. 2018, Nature, 556, 406, with the assistance of S. Das and S. Mathur, Ohio State University.

**Bibliography header:** Copyist and illustrator from a fourteenth century manuscript of the *Roman de la Rose*, perhaps a portrait of Richard de Montbaston (copyist) and a self-portrait of Jeanne de Montbaston (illustrator). Bibliothèque nationale de France (BnF) département des manuscrits, MS français 25526, fol. 77v.

# Constants and Units

Fundamental Physical Constants

| Name | Symbol | Value | Units |
|---|---|---|---|
| speed of light in vacuum | $c$ | $2.99792458 \times 10^{10}$ | $\text{cm s}^{-1}$ |
| gravitation constant | $G$ | $6.67430 \times 10^{-8}$ | $\text{cm}^3\,\text{g}^{-1}\,\text{s}^{-2}$ |
| Planck constant | $h$ | $6.62607015 \times 10^{-27}$ | erg s |
| | | $4.135667696 \times 10^{-15}$ | eV s |
| reduced Planck constant | $\hbar$ | $1.054571817 \times 10^{-27}$ | erg s |
| | | $6.582119569 \times 10^{-16}$ | eV s |
| Boltzmann constant | $k$ | $1.380649 \times 10^{-16}$ | $\text{erg K}^{-1}$ |
| | | $8.61733326 \times 10^{-5}$ | $\text{eV K}^{-1}$ |
| elementary charge | $e$ | $1.602176635 \times 10^{-19}$ | C |
| Stefan–Boltzmann constant | $\sigma_{\text{sb}}$ | $5.670375519 \times 10^{-5}$ | $\text{erg cm}^{-2}\,\text{s}^{-1}\,\text{K}^{-4}$ |
| Thomson cross section | $\sigma_e$ | $6.6524587321 \times 10^{-25}$ | $\text{cm}^2$ |
| proton mass | $m_p$ | $1.67262193269 \times 10^{-24}$ | g |
| electron mass | $m_e$ | $9.1093837015 \times 10^{-28}$ | g |
| proton magnetic moment | $\mu_p$ | $1.40160679736 \times 10^{-23}$ | $\text{erg G}^{-1}$ |
| electron magnetic moment | $\mu_e$ | $-9.2847647043 \times 10^{-21}$ | $\text{erg G}^{-1}$ |
| Bohr radius | $a_0$ | $5.29177210903 \times 10^{-9}$ | cm |

*Source: 2018 CODATA recommended values*

Astronomical Constants

| Name | Symbol | Value | Units |
|---|---|---|---|
| astronomical unit | au | $1.495978707 \times 10^{13}$ | cm |
| solar radius | $R_\odot$ | $6.957 \times 10^{10}$ | cm |
| solar luminosity | $L_\odot$ | $3.828 \times 10^{33}$ | $\text{erg s}^{-1}$ |
| solar mass parameter | $GM_\odot$ | $1.3271244 \times 10^{26}$ | $\text{cm}^3\,\text{s}^{-2}$ |
| solar effective temperature | $T_\odot$ | 5772 | K |

*Source: IAU 2012 resolution B2 on the redefinition of the astronomical unit of length; IAU 2015 resolution B3 on recommended nominal conversion constants for solar and planetary properties*

Astronomical and Physical Units

| Name | Symbol | Value | Units |
|------|--------|-------|-------|
| parsec | pc | 206 264.8063 | au |
| | | $3.08567758 \times 10^{18}$ | cm |
| day | d | 86 400 | s |
| Julian year | yr | 365.25 | d |
| | | $3.15576 \times 10^{7}$ | s |
| angstrom | Å | $10^{-8}$ | cm |
| electron volt | eV | $1.602176634 \times 10^{-12}$ | erg |
| standard atmosphere | atm | $1.01325 \times 10^{6}$ | $\mathrm{dyn\,cm^{-2}}$ |
| rayleigh | R | $10^{6}$ | $\mathrm{photons\,cm^{-2}\,s^{-1}}$ |

*Source: IAU 2012 resolution B2; IAU 2015 resolution B3; 2018 CODATA recommended values*

# Index